Forest
Mensuration

THIRD EDITION

Forest
Mensuration

BERTRAM HUSCH
Food and Agriculture Organization
of the United Nations

CHARLES I. MILLER
Purdue University

THOMAS W. BEERS
Purdue University

JOHN WILEY & SONS

New York • *Chichester* • *Brisbane* • *Toronto* • *Singapore*

Library of Congress Cataloging in Publication Data:

Husch, Bertram, 1923–
    Forest mensuration.

    Bibliography: p. 337
    Includes index.
    1. Forests and forestry—Mensuration. 2. Forest
surveys.   I. Miller, Charles I.   II. Beers, Thomas W.
III. Title.

SD555.H8   1982   634.9'285   82–4811
ISBN 0-471-04423-7   AACR2

Printed in the United States of America

10 9 8 7 6 5 4 3 2 1

Preface

During the decade that has elapsed since the second edition of this book was published, our experience in teaching forest mensuration, in applying the techniques of forest mensuration, and in planning forest inventories has led us to make changes in this book that we think will substantially increase its value as a textbook for undergraduate students and as a reference guide for graduate students and professional resource managers.

Although we have followed the approach of the widely used first and second editions, we have put more emphasis on explaining the fundamental principles of forest mensuration and on the application of forest mensuration to measurement problems in all aspects of multiple-use forestry. Our approach rejects the idea that a forest mensuration book should include only the bare facts, and that motivation and explanation should be left to the instructor. Our feeling is that students and professionals read textbooks outside the classroom without an instructor on hand to answer questions; therefore, both theory and practice are covered.

Because the importance of *inventory using sampling with varying probabilities* has increased in recent years, Chapter 14 has been enlarged. The need for practical information on *forest surveying, other mensurative considerations*, and *aerial photographs in forest inventory* has led us to add new chapters on the first two subjects and to completely rewrite the chapter on aerial photographs.

Although the sections on growth and yield functions have been made to conform to recent developments, these sections have not been expanded because we feel the time for a more extensive discussion of this subject is after developments have crystallized; the place is in an advanced book.

Throughout the book a large number of changes have been made to correct errors, clarify explanations, and update information. Moreover, we have included information on both the metric system and the U.S. system of weights and measures that will enable the student or the professional resource manager to work confidently with either system and to convert readily from one system to the other.

Bertram Husch
Charles I. Miller
Thomas W. Beers

Contents

1 Introduction *1*

2 Linear Measurement *17*

3 Forest Surveying *40*

4 Time Measurement *54*

5 Weight Measurement *63*

6 Area Measurement *80*

7 Volume Measurement *90*

8 Cubic Volume and Measures of Form *97*

9 Log Rules and Volume Tables *114*

10 Grading Forest Products *142*

11 Forest Inventory *150*

12 Sampling in Forest Inventory *156*

13 Aerial Photographs in Forest Inventory *193*

14 Inventory Using Sampling with Varying Probabilities *213*

15 Growth of the Tree *276*

16 Stand Growth *291*

17 Stand Structure, Density, Site Quality, and Yield *320*

18 Other Mensurative Considerations *352*

Appendix *365*

References Cited *377*

Index *397*

Forest
Mensuration

1

Introduction

In 1906, Henry S. Graves wrote: "Forest mensuration deals with the determination of the volume of logs, trees, and stands, and with the study of increment and yield." Since that definition was written, however, the scope of forestry has widened. Although some foresters feel that Grave's definition is still adequate, others feel that mensuration should embrace the new measurement problems that have been created as the horizons of forestry have expanded.

If we accept the challenge of a broader field of study, we must ask: To what degree should mensuration be concerned with the measurement problems of wildlife management, recreation, watershed management, and the other aspects of multiple-use forestry? Furthermore, one might argue that it is unrealistic to imagine that forest mensuration can take as its domain such a diverse group of subjects.

The question becomes irrelevant and the objection allayed if we recognize forest mensuration as a subject that provides principles applicable to all measurement problems. Consequently, this book will include principles that provide a foundation for solving measurement problems in all aspects of forestry. However, the traditional measurement problems of forestry will not be neglected.

Through the years a number of textbooks on forest mensuration have been prepared in the United States and in foreign lands. In the United States the most recent are Bruce and Schumacher (1950); Spurr (1952); Meyer (1953); Husch (1963); Husch, Miller, and Beers (1972); and Avery (1975). In other countries the most recent are Prodan (1965), Germany; Seip (1964), Norway; Giurgiu (1968), Romania; and Carron (1968), Australia. Also, Loetsch, Zöhrer, and Haller (1973) give a comprehensive and imaginative treatment of many aspects of forest mensuration.

Since the end of World War II the application of statistical theory and the use of computers and programmable calculators in forest mensuration have wrought a revolution in the solution of forest measurement problems. Consequently, the mensurationist must be competent in these areas as well as in basic mathematics. A knowledge of calculus is also desirable. In addition, a knowledge of systems analysis and operations research, approaches to problem solving that depend on model building and techniques that include simulation and mathematical programming, is also helpful.

1-1 IMPORTANCE OF FOREST MENSURATION

Forest mensuration is one of the keystones in the foundation of forestry. Forestry, in the broadest sense, is a management activity involving forest land, the plants and animals on the land, and humans as they use the land. Thus, the forester is faced with many decisions in the management of a forest. The following questions convey a general idea of the problems that must be solved for a particular forest.

1. What silvicultural treatment will result in the best regeneration and growth?
2. What species is most suitable for reforestation?
3. Is there sufficient timber for an economical harvesting operation?
4. What is the value of the timber and land?
5. What is the recreational potential?
6. What is the wildlife potential?

A forester needs information to make intelligent decisions for these and countless other questions. Whenever possible, this information should be in quantifiable terms. As it has aptly been said, "You can't efficiently make, manage, or study anything you don't locate and measure." In this sense, forest mensuration is the application of measurement principles to obtain quantifiable information for decision making.

1-2 PRINCIPLES OF MEASUREMENT

Knowledge is to a large extent the result of the acquisition and systematic accumulation of observations, or measurements, of concrete objects and natural phenomena. Thus, measurement is a basic requirement for the extension of knowledge.

In its broadest sense, measurement consists of the assignment of numbers to measurable properties. Ellis (1966) gives this definition: "Measurement is the assignment of numerals to things according to any determinative, non-degenerate rule." ("Determinative" means that the same numerals, or range of numerals, are always assigned to the same things under the same conditions. "Non-degenerate" allows for the possibility of assigning different numerals to different things, or to the same thing under different conditions.) This definition implies that we have a scale that allows us to use a rule, and that each scale inherently has a different rule that must be adhered to in representing a property by a numerical quantity. Stevens (1946) summarized the problem and formulated a classification for different kinds of scales. In spite of some shortcomings, Stevens classification is useful in understanding the measurement process (Table 1-1).

1-2.1 Scales of Measurement

The four scales of measurement are nominal, ordinal, interval, and ratio (Table 1-1).

The *nominal scale* is used for numbering objects for identification (e.g., numbering of forest types in a stand map), and for numbering a class when each member of the class is assigned the same numeral (e.g., the assignment of code numbers to species).

Table 1-1

Classification of Scales of Measurements[1]

Scale	Basic Operation	Mathematical Group Structure	Permissible Statistics	Examples
Nominal	Determination of equality (numbering and counting)	Permutation group $X' = f(X)$ where $f(X)$ means any one-to-one substitution	Number of cases Mode Contingency correlation	Numbering of forest types on a stand map Assignment of code numbers to tree species in studying stand composition
Ordinal	Determination of greater or less (ranking)	Isotonic group $X' = f(X)$ where $f(X)$ means any increasing monotonic function	Median Percentiles Order correlation	Lumber grading Tree and log grading Site class estimation
Interval	Determination of the equality of intervals or of differences (numerical magnitude of quantity, arbitrary origin)	Linear group $X' = \alpha X + b$ $\alpha > 0$	Mean Standard deviation Correlation coefficient	Fahrenheit temperature Calendar time Available soil moisture Relative humidity
Ratio	Determination of the equality of ratios (numerical magnitude of quantity, absolute origin)	Similarity group $X' = cX$ $c > 0$	Geometric mean Harmonic mean Coefficient of variation	Length of objects Frequency of items Time intervals Volumes Weights Absolute temperature Absolute humidity

[1] Columns 2, 3, 4, and 5 are cumulative in that all characteristics listed opposite a particular scale are additive to those above it. In the column which records the group structure of each scale are listed the mathematical transformations which leave the scale invariant. Thus any numeral X on a scale can be replaced by another numeral X', where X' is the function of X listed in this column. The criterion for the appropriateness of a statistic is invariance under the transformations in column 3. Thus the case that stands at the median of a distribution maintains its position under all transformations which preserve order (isotonic group), but an item located at the mean remains at the mean only under transformations as restricted as those of the linear group. The ratio expressed by the coefficient of variation remains invariant only under the similarity transformation (multiplication by a constant). The rank-order correlation coefficient is usually considered appropriate to the ordinal scale, although the lack of a requirement for equal intervals between successive ranks really invalidates this statistic.

SOURCE: Adapted from S. S. Stevens, "On Theory of Scales of Measurement," *Science* 103(2684): 677–680.

The *ordinal scale* is used to express degree, quality, or position in a series, such as first, second, and third. In a scale of this type, the successive intervals on the scale are not necessarily equal. This scale is used for lumber grades, log grades, tree grades, and site classifications.

The *interval scale* includes a series of graduations marked off at uniform intervals from a reference point of fixed magnitude. There is no absolute reference point or true origin for the scale. The origin is arbitrarily chosen. The Celsius temperature scale is a good example of an interval scale. Equal intervals of temperature are scaled off by noting equal volumes of expansion referenced to an arbitrary zero.

The *ratio scale* is similar to the interval scale in that there is equality of intervals between successive points on the scale. However, an absolute zero of origin is always present or implied. Ratio scales are the most commonly employed and the most versatile in that all types of statistical measures are applicable. It is convenient to consider ratio scales as fundamental and derived.

- *Fundamental* scales are represented by such things as frequency, length, weight, and time intervals.
- *Derived* scales are represented by such things as stand volume per acre, stand density, and stand growth per unit of time. (These are derived scales in that the values on the scale are functions of two or more fundamental values.)

1-3   UNITS OF MEASUREMENT

To describe a physical quantity, one must establish a unit of measure and determine the number of times the unit occurs in the quantity. Thus, if an object has a length of 3 meters, the meter has been taken as the unit of length, and the length dimension of the object contains three of these standard units.

The *fundamental units* in mechanics are measures of length, mass, and time. These are regarded as independent and fundamental variables of nature, although they have been chosen arbitrarily by scientists. Other fundamental units have been established for thermal, electrical, and illumination quantity measurement.

Derived units are expressed in terms of fundamental units or in units derived from fundamental units. Derived units include ones for the measurement of volume (cubic feet or meters), area (acres or hectares), velocity (miles per hour, meters per second), force (kilogram-force), etc. Derived units are often expressed in formula form. For example, the area of a rectangle is defined by the equation

$$\text{Area} = WL$$

where W and L are fundamental units of length.

Physical quantities such as length, mass, and time are called scalar quantities or scalars. Physical quantities that require an additional specification of direction for their complete definition are called *vector* quantities or vectors.

1-4 SYSTEM OF UNITS

There are two methods of establishing measurement units. We may select an arbitrary unit for each type of quantity to be measured, or we may select fundamental units and formulate from them a consistent system of derived units. The first method was employed extensively in our early history. For example, units for measuring the length of cloth, the height of a horse, or land distances were all different. Reference units were objects such as the width of a barleycorn, the length of a man's foot, the length of a man's forearm (a cubit), and so on. However, these primitive units lacked uniformity. Vestiges of this system still exist, particularly in English-speaking countries (foot, yard, pound, etc.), although the units are now uniform.

The second method of establishing a system of units is illustrated by the metric system. In this system, an arbitrary set of units has been chosen that is uniformly applicable to the measurement of any object. Moreover, there is a logical, consistent, and uniform relationship between the basic units and their subdivisions.

1-4.1 Metric System

This system of weights and measures was formulated by the French Academy of Sciences in 1790. The system was adopted in France in 1799 and made compulsory in 1840. In 1875, the International Metric Convention, which was established by treaty, furnished physical standards of length and mass to the 17 member nations. The General Conference on Weights and Measures (referred to as CGPM from the French "Conférence Générale des Poids et Mesures") is an international organization established under the Convention. This organization meets periodically. The CGPM controls the International Bureau of Weights and Measures (BIPM), which is headquartered at Sèvres, near Paris, and maintains the physical standards of units. The United States Bureau of Standards represents the United States on the CGPM and maintains our standards of measure.

The metric system has been adopted by most of the technologically developed countries of the world. Although conversion in Great Britain and in the United States has met with some resistance, a gradual changeover is taking place. In 1975, the President signed the Metric Conversion Act in which the United States adopted a policy of actively encouraging a voluntary changeover to the metric system of weights and measures.

Discussions of the problems affecting forestry in converting to the metric system in the United States and Great Britain are presented by White (1971), Bruce (1974, 1976), and Hamilton (1974). Furthermore, an excellent summary of the system as now internationally accepted is given in the National Bureau of Standards Publication 330 (1977). The following information is abstracted from this publication.

The 11th meeting of the CGPM in 1960 adopted the name "International System of Units" with the international abbreviation SI (from the French "Le Système International d'Unités"). This is now the accepted form of the metric system. Other adaptations of metric units such as in the CGS, MTS, and MKS systems are discouraged.

The SI considers three classes of units: (1) base units, (2) derived units, and (3) supplementary units. There are seven *base units*, which by convention are considered dimensionally independent. These are the meter, kilogram, second, ampere, kelvin, mole, and candela. The *derived units* are formed by combining base units according to algebraic statements that relate the corresponding quantities. The *supplementary units* are those that the CGPM established without stating whether they are base or derived units.

Here is a list of the dimensions that are measured by these base units, along with the definitions of the units. The conventional symbol for each unit is shown in parentheses.

1. Length—meter (m). The meter is equal to 1,650,763.73 wavelengths in a vacuum of the orange-red light given off by krypton-86.

2. Mass—kilogram (kg).[1] The kilogram is equal to the mass of the international prototype standard, a cylinder of platinum-iridium alloy kept at the BIPM.

3. Time—second (s). The second was originally defined as 1/86,400 part of a mean solar day. In 1956, it was redefined as 1/31,556,925.9747 of the tropical year for 1900. More recently, it has been calculated by atomic standards to be 9,192,631,770 periods of vibration of the radiation emitted at a specific wavelength by an atom of cesium-133.

4. Electric current—ampere (A). The ampere is the current in a pair of equally long, parallel, straight wires (in vacuum and 1 meter apart) that produces a force of 2×10^{-7} newtons between the wires for each meter of their length.

5. Temperature—kelvin (K). The kelvin is 1/273.15 of the thermodynamic temperature of the triple point of water. The temperature 0 K is called absolute zero. The kelvin degree is the same size as the Celsius degree (also called centigrade). The freezing point of water (0°C) and the boiling point of water (100°C) correspond to 273.15 K and 373.15 K, respectively. On the Fahrenheit scale, 1.8 degrees are equal to 1.0°C or 1.0 K. The freezing point of water on the Fahrenheit scale is 32°F.

6. Amount of substance—mole (mol). The mole is a base unit used to specify the quantity of chemical elements or compounds. It is the amount of substance of a system that contains as many elementary entities as there are atoms in 0.012 kilogram of carbon-12. When the mole is used, the elementary entities must be qualified. They may be atoms, molecules, ions, electrons, or other particles, or specified groups of such particles.

---

[1] The term "weight" is commonly used for mass although this is, strictly speaking, incorrect. The weight of a body means the force caused by gravity, acting on a mass, which varies in time and space and which differs according to the location on earth. Since it is important to know whether mass or force is being measured, the SI has established two units: the kilogram for mass and the newton for force.

7. Luminous intensity—candela (cd). The candela is 1/600,000 of the intensity, in the perpendicular direction, of one square meter of a black body radiator at the temperature at which platinum solidifies (2045 K) under a pressure of 101,325 newtons per square meter.

Derived units are expressed algebraically in terms of the base units by means of mathematical symbols of multiplication and division. Some examples of derived units are given here.

Quantity	SI Unit for the Quantity	Symbol
Area	square meter	m^2
Volume	cubic meter (the liter, 0.001 cubic meter, is not an SI unit although commonly used to measure fluid volume)	m^3
Specific volume	cubic meter per kilogram	m^3/kg
Force	newton (1 N = 1 kg·m/s²)	N
Pressure	pascal (1 Pa = 1 N/m²)	Pa
Work	joule (1 J = 1 N·m)	J
Power	watt (1 W = 1 J/s)	W
Speed	meter per second	m/s
Acceleration	(meter per second) per second	m/s^2
Voltage	volt (1 V = 1 W/A)	V
Electric resistance	ohm (1 Ω = 1 V/A)	Ω
Concentration (amount of substance)	mole per cubic meter	mol/m^3

At present, there are only two *supplementary units*, the radian and the steradian. They are defined as follows.

- The radian (rad) is the plane angle between two radii of a circle that cuts off on the circumference an arc equal to the radius. It is 57.29578 degrees for every circle.
- The steradian (sr) is the solid angle at the center of a sphere subtending a section on the surface equal in area to the square of the radius of the sphere.

There is a number of widely used units that are not part of SI. These units, which the International Committee on Weights and Measures (CIPM) recognized in 1969, are shown below.

Unit	Symbol	Equivalence in SI Units
minute	min	1 min = 60 s
hour	h	1 h = 60 min = 3600 s

Unit	Symbol	Equivalence in SI Units
day	d	$1\text{ d} = 24\text{ h} = 86{,}400\text{ s}$
degree (angular)	°	$1° = (\pi/180)\text{ rad}$
minute (angular)	'	$1' = (1/60)° = (\pi/10{,}800)\text{ rad}$
second (angular)	"	$1'' = (1/60)' = (\pi/648{,}000)\text{ rad}$
liter	L	$1\text{ L} = 1\text{ dm}^3 = 10^{-3}\text{ m}^3$
metric ton	t	$1\text{ t} = 10^3\text{ kg}$

To form decimal multiples or fractions of SI units, the following prefixes or abbreviations are used.

Prefix	Factor	Abbreviation
tera	10^{12}	T
giga	10^{9}	G
mega	10^{6}	M
kilo	10^{3}	k
hecto	10^{2}	h
deka	10^{1}	da
deci	10^{-1}	d
centi	10^{-2}	c
milli	10^{-3}	m
micro	10^{-6}	μ
nano	10^{-9}	n
pico	10^{-12}	p

1-4.2 English System

The English system of weights and measures is still widely used in the United States but it should be supplanted gradually by the metric system (SI). This may be a long process since custom and tradition are strong, and the conversion is currently voluntary. Great Britain has taken a more vigorous stance and conversion to the metric system is proceeding rapidly (Hamilton, 1974).

The units of the English system still commonly used in the United States are practically the same as those employed in the American colonies prior to 1776. The names of these units are generally the same as those of the British Imperial System. However, their values differ slightly.

In the English system, the fundamental length is the yard. The British yard is the distance between two lines on a bronze bar kept in the Standards Office, Westminster, London. Originally, the United States yard was based on a prototype bar, but in 1893 the United States yard was redefined as 3600/3937 meter = 0.9144018 meter. The

British yard, on the other hand, was 3600/3937.0113 meter = 0.914399 meter. The foot is defined as the third part of a yard, and the inch as a twelfth part of a foot. In 1959, it was agreed by Canada, United States, New Zealand, United Kingdom, South Africa, and Australia that they would adopt the value of 1 yard = 0.9144 meter. In the United Kingdom, these new values were used only in scientific work, the older, slightly different values being used for other measurements. And in the United States, the results of geodetic surveys are still expressed in feet based on the former definition of the yard (3600/3937 meter).

In the British system, the unit of mass, the avoirdupois pound, is the mass of a certain cylinder of platinum in the possession of the British government (1 pound = 0.45359243 kilogram). In the United States, the unit of mass is also the pound, but in 1893 it was defined in terms of the mass unit of the metric system (1 pound = 0.4535924277 kilogram). Thus, the United States pound and the British pound were not exactly equal. The 1959 agreement of English-speaking countries established a new value of 1 pound = 0.45359237 kilogram.

In the British system, the second is the base unit of time, as in the metric system.

Secondary units in the English system also have different values in the United States and the United Kingdom. For example, the U.S. gallon is defined as 231 cubic inches. The British gallon is defined as 277.42 cubic inches. There are other secondary units used in English-speaking countries that are as arbitrary as the gallon, but some are derived from the fundamental units. For example, the unit of work in English and American engineering practice has been the foot-pound.

Very commonly, conversions from one system of measurement to another are required. Appendix Table A-1 shows unit conversions for length, area, volume, and mass in the English and the metric systems.

1-5   NUMBERING SYSTEMS

The numbering system in general use throughout the world is the *decimal system*. This can probably be attributed to the fact that human beings have ten fingers. But the decimal system is merely one of many possible numbering systems that could be utilized. In fact, there are examples of other numbering systems used by earlier civilizations (e.g., the vigesimal system, based on twenty, utilized by the Mayas, and the sexagesimal system of the Babylonians, based on sixty). Our own system of measuring time and angles in minutes and seconds comes from the sexagesimal system. Systems to other bases, such as the duodecimal system, based on twelve (which seems to have lingered on in the use of dozen and gross), may also have been used. For a discussion of the history of number theory, the student is referred to Ore (1948).

With the development of the electronic computer, interest has been revived in numbering systems using bases other than ten. Of primary interest is the binary system, because electronic digital computers that use two basic states have been found most practical.

1-6 CONTINUOUS AND DISCRETE VARIABLES

A *variable* is a characteristic that may assume any given value or set of values. Some variables are continuous in that they are capable of exhibiting every possible value within a given range. For example, height, weight, and volume are continuous variables. Other variables are discontinuous or discrete in that they only have values which jump from one number or position to the next. The number of employees in a company, of trees in a stand, or of deer per unit area are examples of discrete variables.

Data pertaining to continuous variables are obtained by measuring. Data pertaining to discrete variables are obtained by counting. The problems of measuring continuous variables are dealt with in Chapters 2, 3, 4, 5, 6, and 7 and will not be discussed here. But we will consider the measurement of discrete variables at this time.

The process of measuring according to nominal and ordinal scales, as shown in Table 1-1, consists of counting the frequencies of occurrences of specified events. Discrete variables describe these events. The general term "event" can refer to a discrete physical standard, such as a tree, which exists as a tangible object occupying space, or to an occurrence that cannot be thought of as spatial, such as a timber sale. In either case, the measurement consists of defining the variable and then counting the number of its occurrences. There is no choice for the unit of measurement—the frequency is the only permissible numerical value.

It is important not to confuse a class established for convenience in continuous-type measurement with a discrete variable. Classes for continuous variables are often established to facilitate the handling of data in computations. Frequencies may then be assigned to these classes. These frequencies represent the occurrence or recurrence of certain measurements of a continuous variable that have been placed in a group or class of defined limits for convenience.

A discrete variable can thus be characterized as a class or series of classes of defined characteristics with no possible intermediate classes or values. A few examples of discrete variables used in forestry are species, lumber grades, and forest-fire danger classes.

At times, it may not be clear as to whether a discrete or continuous variable is being measured. For example, in counting the number of trees per acre or per hectare, the interval, one tree, is so small and the number of trees so large that analyses are made of the frequencies as though they described a continuous variable. This has become customary and may be considered permissible if the true nature of the variable is understood.

1-6.1 Forest Measurements on a Nominal Scale

Species names or forest types are examples of the use of discrete variables on nominal scales. Tree species, for example, can be the classes for the variable. The order in which the classes are recorded does not affect the discrete variable, the species. Indeed, the order could be changed without changing the meaning.

It is frequently convenient to assign a code number or letter to each class of a discrete variable. Code numbers are especially useful if data are to be entered on punch cards for machine computation. It is important to be aware that the code numbers have no intrinsic meaning but are merely identifying labels. No meaningful mathematical operations can be performed on such code numbers.

1-6.2 Forest Measurement on an Ordinal Scale

Ordinal scales for discrete variables abound in forestry. Examples are lumber grades, log grades, piece products grades, Christmas tree grades, nursery stock grades, and site quality classes.

The order in which the classes of a discrete variable are arranged on an ordinal scale has an intrinsic meaning. The classes are arranged in order of increasing or decreasing qualitative rank, so the position on the scale affords an idea of comparative rank. The continuum of the variable consists of the range between the limits of the established ranks or grades. As many ranks or grades can be established as are deemed suitable. An attempt may be made to have each grade or rank occupy an equal interval of the continuum. However, this will rarely be achieved, since the ranks are subjectively defined with no assurance of equal increments between ranks.

Chapter 10 discusses in more detail the principles of quality measurement and its applications in forestry.

1-7 PRECISION, ACCURACY, AND BIAS

The terms "precision" and "accuracy" are frequently used interchangeably in nontechnical parlance and often with varying meaning in technical usage. In this text they will have two distinct meanings. *Precision* as used here (and generally accepted in forest mensuration) means the degree of agreement in a series of measurements.[2] *Accuracy*, on the other hand, is the closeness of a measurement to the true value. Of course, the ultimate objective is to obtain accurate measurements.

Bias refers to systematic errors that may result from faulty measurement procedures, instrumental errors, flaws in the sampling procedure, errors in the computations, mistakes in recording, and so forth.

In sampling, accuracy refers to the size of the deviation of a sample estimate from the true population value. Precision, expressed as a standard deviation, refers to the deviation of sample values about their own mean, which, if biased, does not correspond to the true population value. It is possible to have a very precise estimate in that the deviations from the sample mean are very small; yet, at the same time, the estimate may not be accurate if it differs from the true value due to bias. For example, one

---

[2] The term is also used to describe the resolving power of a measuring instrument or the smallest unit in observing a measurement. In this sense, the more decimal places used in the measurement the more precise the measurement.

might carefully measure a tree diameter repeatedly to the nearest millimeter with a caliper that reads about 5 mm low (Section 2-1.3). The results of this series of measurements are precise because there is little variation between readings, but they are biased and inaccurate because of improper adjustment of the instrument.

Bruce (1975) has shown that bias B and precision P can be equated with accuracy A as follows.

$$A^2 = B^2 + P^2$$

This indicates that, if we reduce B^2 to zero, accuracy equals precision.

1-8 SIGNIFICANT DIGITS

A significant digit is any digit denoting the true size of the unit at its specific location in the overall number. The term significant as used here should not be confused with its use in reference to statistical significance. The significant figures in a number are the digits reading from left to right beginning with the first nonzero digit and ending with the last digit written, which may be a zero. The numbers 24, 2.5, 0.25, and 0.025 all have two significant figures, the 2 and the 5. The numbers 25.0, 0.250, and 0.0250 all have three significant figures, the 2, 5, and 0. When one or more zeros occur immediately to the left of the decimal position and there is no digit to the right of the point, the number of significant digits may be in doubt. Thus, the number 2500 may have two, three, or four significant digits, depending on whether one or both zeros denote an actual measurement or have been used to round off a number and indicate the position of the decimal point. Thus, zero can be a significant figure if used to show the quantity in the position it occupies and not merely to denote a decimal place. A convention sometimes used to indicate the last significant digit is to place a dot above it. Thus, 5,12$\dot{1}$,000 indicates four significant digits and 5,121,$\dot{0}$00 indicates five significant digits. Another method is to first divide a number into two factors, one of them being a power of ten. A number such as 150,000,000 could be written as 1.5×10^8 or in some other form such as 15×10^7. A convention frequently used is to show the significant figures in the first factor and to use one nonzero digit to the left of the decimal point. Thus, the numbers 156,000,000 (with three significant figures), 31.53, and 0.005301 would be written as 1.56×10^8, 3.153×10, and 5.301×10^{-3}.

If a number has a significant zero to the right of the decimal place following a nonzero number, it should not be omitted. For example, 1.05010 indicates six significant digits including the last zero to the right. To drop it would reduce the precision of the number. Zeros when used to locate a decimal place are not significant. In the number 0.00530, only the last three digits 5, 3, and 0 are significant; the first two to the right of the decimal place are not.

When the units used for a measurement are changed, it may change the number of decimal places but not the number of significant digits. Thus, a weight of 355.62

grams has five significant figures, as does the same weight expressed as 0.35562 kilograms, although the number of decimal places has increased. This emphasizes the importance of specifying the number of significant digits in a measurement rather than simply the number of decimal places.

1-9   SIGNIFICANT DIGITS IN MEASUREMENTS

The numbers used in mensuration can be considered as arising from pure numbers, from direct measurements, and from computations involving pure numbers and values from direct measurements.

Pure numbers can be the result of a count in which a number is exact, or they can be the result of some definition. Examples of pure numbers are the number of sides on a square, the value of π, or the number of meters in a kilometer.

Values of direct measurements are obtained by reading a measuring instrument (e.g., measuring a length with a ruler). The numerical values obtained in this way are approximations in contrast to pure numbers. The precision of the approximation is indicated by the number of significant digits used. For example, measurement of a length could be taken to the nearest one, tenth, or hundredth of a foot, and recorded as 8, 7.6, or 7.60. Each of these measurements implies an increasing standard of precision. A length of 8 feet means a length closer to 8 than to 7 or 9 feet. The value of 8 can be considered to lie between 7.5 and 8.5. Similarly, a length of 7.6 means a measurement whose value is closer to 7.6 feet than to 7.5 or 7.7. The value of 7.6 lies between 7.5500 . . . 01 and 7.6499 . . . 99, or, conventionally, 7.55 and 7.65. In the measurement 7.60, the last digit is significant and the measurement implies a greater precision. The value 7.60 means the actual value lies anywhere between 7.59500 . . . 01 and 7.60499 . . . 99, or, conventionally, 7.595 and 7.605.

It is incorrect to record more significant digits than were observed. Thus, a length measurement of 8 feet taken to the nearest foot should not be recorded as 8.0 feet since this may mislead the reader into thinking the measurement is more precise than it actually is. On the other hand, one should not omit significant zeros in decimals. For example, one should write 112.0 instead of 112 if the zero is significant.

Since the precision of the final results is limited by the precision of the original data, it is necessary to consider the numbers of significant figures to take and record in original measurements. It is well to keep in mind that using greater precision than needed is a waste of time and money. A few suggestions follow.

1. Do not try to make measurements to a greater precision (more significant digits) than can be reliably indicated by the measuring process or instrument. For example, it would be illogical to try to measure the height of a standing tree to the nearest tenth of a foot with an Abney level.

2. The precision needed in original data may be influenced by how large a difference is important in comparing results. Thus, if the results of a series of silvicultural treatments are to be compared in terms of volume growth

response to the nearest tenth of a cubic meter, then there would be no need to estimate volumes more exactly than the nearest tenth of a cubic meter.

3. The variation in a population sampled and the size of the sample influence the precision chosen for original measurement. If the population varies greatly, or if there are few observations in the sample, then high measurement precision is not worthwhile.

1-10 ROUNDING OFF

When dealing with the numerical value of a measurement in the usual decimal notation, it is often necessary to round off to fewer significant figures than originally shown. Rounding off can be done by deleting the unwanted digits to the right of the decimal point (the fractional part of a number) and by substituting zeros for those to the left of the decimal place (the integer part). Three cases can arise: (1) If the deleted or replaced digits represent less than one-half unit in the last required place, no further change is required. (2) If the deleted or replaced digits represent more than one-half unit in the last required place, then this significant figure is raised by one. (Note that, if the significant figure in the last required place is 9, it changes to zero and the digit preceding it is increased by one.) (3) If the deleted or replaced digits represent exactly one-half unit in the last required place, a recommended convention is to raise this last digit by unity if it is odd but let it stand if it is even. Thus, 31.45 would be rounded to 31.4 but 31.55 would be 31.6. Here are a few examples.

| | Number Rounded To: | | |
Number	4 Significant Figures	3 Significant Figures	2 Significant Figures
4.6495	4.650	4.65	4.6
93.65001	93.65	93.7	94
567851	567900	568000	570000
0.99687	0.9969	0.997	1.0

1-11 SIGNIFICANT DIGITS IN ARITHMETIC OPERATIONS

In arithmetic operations involving measurements, where figures are only approximations, the question of how many significant digits there are in the result becomes important.

In multiplication and division, the factor with the fewest significant figures limits the number of significant digits in the product or quotient. Thus, in multiplying a numerical measurement with five significant figures by another with three significant figures, only the first three figures of the product will be trustworthy, although there may be up to eight digits in the product. For example, if the measurement 895.67 and 35.9 are multiplied, the product is 32,154.553. Only the first three figures in the product—3, 2, and 1—are significant. The number 895.67 represents a measurement between 898.665 and 895.675. The number 35.9 represents a measurement between 35.85 and 35.95. The products of the four possible limiting combinations will differ in all except the first three figures; thus,

$$(895.665)(35.85) = 32109.59025$$

$$(895.665)(35.95) = 32199.15675$$

$$(895.675)(35.85) = 32109.94875$$

$$(895.675)(35.95) = 32199.51625$$

Therefore, the first three figures are the only reliable ones in the product. Similarly, in dividing a measurement with eight significant digits by a measurement with three significant figures, the quotient will have only three significant figures. But, if a measurement is to be multiplied or divided by an exact number or a factor that is known to any desired number of significant digits, a slightly different situation occurs. For example, a total weight could be estimated as the product of a mean weight having five significant digits times 55. However, the 55 is an exact number and could also be validly written as 55.000. The product would thus still have five significant digits. It may be helpful to remember that multiplication is merely repetitive addition and the 55, in this case, means that a measurement is added exactly 55 times. Similarly, if the 55 objects had been weighed, as a group, to five significant digits, dividing by 55 would give a mean weight to five significant figures. In these cases, the significant digits are controlled by the number in the measurement. Another case occurs if a measurement is to be multiplied or divided by a factor such as π or e (base of Naperian logarithms), which are known to any number of significant figures. The number of significant digits in π or e should be made to agree with the number in the measurement before the operation of multiplication or division so that there is no loss in precision.

A good rule in multiplication or division is to keep one more digit in the product or quotient than occurs in the shorter of the two factors. This minimizes rounding-off errors in calculations involving a series of operations. At the end of the calculations, the final answer can be rounded off to the proper number of significant figures.

In addition and subtraction, the position of the decimal points will affect the number of significant digits in the result. It is necessary to align numbers according to their decimal places in order to carry out these operations. The statement that measurements can be added or subtracted when significant digits coincide at some place to the left or right of the decimal point can be used as a primary guide. Also, the number of significant digits in an answer can never be greater than those in the largest of the numbers, but may be fewer. As one example, measurements of 134.023 and 1.5 can be added or subtracted as shown below, since significant digits coincide at some place.

$$
\begin{array}{r}
134.023 \\
+1.5 \\
\hline
135.523
\end{array}
$$

The sum has only four significant figures and should be expressed as 135.5. The last two significant figures of 134.023 cannot be used, since there is no information in the smaller measurement for coinciding positions. It is desirable to take measurements to uniform standards of significant figures or decimal places to avoid discarding a portion of a measurement, as we did in the case of the last two digits of 134.023.

Another example to consider is the addition of a series of measurements, the final total of which may have more figures than any of the individual measurements. The number of significant digits in the total will not exceed the number in the largest measurement. Consider the eleven measurements shown here.

Measurement	Range	
845.6	845.55	845.65
805.8	805.75	805.85
999.6	999.55	999.65
963.4	963.35	963.45
897.6	897.55	897.65
903.1	903.05	903.15
986.9	986.85	986.95
876.3	876.25	876.35
863.2	863.15	863.25
931.2	931.15	931.25
998.1	998.05	998.15
10,070.8	10,070.25	10,071.35

The total, 10,070.8, contains six digits, but only the first four are significant. Each measurement can be thought of as an estimate within a range, as shown in the two right-hand columns. Ths sum of the lesser values is 10,070.25 and that of the larger is 10,071.35. The total value of the sum of the eleven measurements can fall anywhere within these limits. The significant figures are 1, 0, 0, and 7. Beyond this, the digits are unreliable.

2
Linear Measurement

Linear, or length, measurement consists of determining the length of a line from one point to another. Since the configuration of objects vary, a length might be the straight-line distance between two points on an object, or the curved- or irregular-line distance between two points on an object. An example of the latter is the periphery of the cross-section of a tree stem.

Length measurements can be made directly or indirectly. Direct length measurement is accomplished by placing a prototype standard of a defined unit beside the object to be measured. The number of units between terminals is the length. The application of a foot or meter rule is an example of a direct measurement. Indirect linear measurement is accomplished by employing geometry or trigonometry, or by using knowledge of the speed of sound or light.

The varied task of length measurement in forestry includes determination of the height of a tree, the length of a log, the width of a tree crown's image on an aerial photograph, the length of a tracheid under a microscope, the length of the femur of a mountain lion, the length of the boundary of a tract of land, and so on. Keep in mind, however, that measurement techniques are rarely unique to specific measurements. Hence, rather than describe techniques for many different measurement tasks, only tree diameter and length measurements, the specialized forestry measurements most commonly applied, are discussed in this chapter. Forest land measurements are covered in Chapter 3. Information on measurements peculiar to specialized phases of forestry and allied subjects can be obtained from references given throughout the text. However, the reader should be able to understand the broad applications of linear measurement techniques from the information given in this chapter.

2-1 DIAMETER MEASUREMENT

A *diameter* is a straight line passing through the center of a circle or sphere and meeting at each end the circumference or surface. The most common diameter measurements taken in forestry are of the main stem of standing trees, cut portions of trees, and branches. Diameter measurement is important because it is one of the directly measurable dimensions from which tree cross-sectional area, surface area, and volume can be computed.

The use of the word diameter implies that trees are circular in cross section. In many cases, however, the section is somewhat wider in one direction than another, or it may be eccentric in other ways. Since for computational purposes tree cross sections are assumed to be circular, the objective of any tree diameter measurement is to obtain the diameter of a circle with the same cross-sectional area as the tree. This will be further discussed in Section 2-1.2.

The point at which diameters are to be measured will vary with circumstances. In the case of standing trees, a standard position has been established. In the United States, the diameter of standing trees is taken at 4.5 feet above the ground level. This is referred to as diameter breast high and is abbreviated to d.b.h. or dbh. In countries that use the metric system the diameter of standing trees is taken at 1.3 meters above the ground level and is abbreviated as d (IUFRO, 1959). On slopes, dbh is commonly taken from the average ground level occupied by the tree, whereas d is commonly taken from the uphill side. Diameters at other points along the stem are often indicated by subscripts: $d_{0.5h}$ = diameter at half total height; $d_{0.1h}$ = diameter at 0.1 total height; d_6 = diameter at 6 meters above ground level. Diameters should be qualified as o.b. (outside bark) or i.b. (inside bark). However, when this designation is omitted from breast height measurements (dbh or d), as it often is, the measurement is assumed to be outside bark.

In the United States, tree diameters are generally recorded in inches. In countries where the metric system is used, diameters are generally recorded in centimeters (occasionally in millimeters).

In measuring at breast height in the field (dbh or d), we recommend the following standard procedures.

- When trees are on slopes, measure 4.5 feet (for dbh) or 1.3 meters (for d) above the ground on the uphill side of the tree.

- When a tree has a limb, a bulge, buttresses, or some other abnormality at breast height, measure diameter above the abnormality; strive to obtain the diameter the tree would have had if the abnormality had not been present.

- When a tree consists of two or more stems forking below breast height, measure each stem separately. When a tree forks at or above breast height, measure it as one tree. If the fork occurs at breast height, or slightly above, take the diameter measurement below the enlargement that is caused by the fork.

- When a tree has a paint mark to designate the breast height point, assume that the point of measurement is at the top of the paint mark.

2-1.1 Bark Thickness

Whether *inside bark* (i.b.) or *outside bark* (o.b.) diameter measurements are taken depends on the purpose for which the measurements are made.

When a rule, or scale stick, is used on a cut section, it is simple to measure diameter inside bark (d.i.b.) or outside bark (d.o.b.). One simply measures an appropriate line, or lines, on the section along which the bark is intact. If both d.i.b. and d.o.b.

are measured, bark thickness is one-half of the difference between them. If one measures d.o.b. and bark thickness, d.i.b. equals d.o.b. minus twice the bark thickness.

Bark thickness on standing trees can be determined with a bark gauge. The Swedish bark gauge is commonly used. The cutting edge of this instrument is a half circle that is dull on one side so that the instrument can be driven through the soft bark, but not through the wood. A sliding cross arm is provided. When this arm is pressed against the bark, one can read the bark thickness on a scale without removing the instrument. Mesavage (1969) found that sizable errors could occur in measuring bark thickness due to the unevenness of the bark surface and failure to reach the wood with the cutting edge. He found that such errors could be practically eliminated by measuring radially from the wood surface to the line of a diameter tape wrapped tautly around the tree (o.b.). The required measurement can be obtained with a Swedish bark gauge or, particularly when the bark is thick and tough, by boring to the wood surface with a brace and bit and obtaining the distance from tape to wood with a small ruler.

2-1.2 Effect of Eccentricity of Tree Cross Sections

Although the cross section of the woody parts of trees approaches circular form, the shape is often not circular. However, for computational purposes, the cross section is assumed to be circular. Consequently, the objective of any tree diameter measurement is to obtain the diameter of a circle with the same cross-sectional area as the tree. (We will refer to this as the "true" diameter.)

When a tree cross section is elliptical, one might measure the major and minor diameters, d_1 and d_2, and obtain the average diameter from the arithmetic mean of d_1 and d_2. This would, however, overestimate the "true" diameter of the cross section. If the periphery of the elliptical cross section was measured with a tape, and the periphery divided by π (this assumes the periphery is a circle), the "true" diameter of the cross section would also be overestimated.

So how does one obtain the "true" diameter of an elliptical cross section? One should use the geometric mean (geometric mean $= \sqrt{d_1 d_2}$).

Unfortunately, tree cross sections often depart from elliptical form as well as from circular form. Consequently, for practical purposes, the best practice is to take the arithmetic average of the long and short "diameters," or axes, or if it is not feasible to secure the long and short "diameters," to take the arithmetic average of two diameters perpendicular to each other.

2-1.3 Instruments for Measuring Tree Diameters

In this section, we describe the most commonly used instruments for measuring tree diameter at breast height. These instruments may also be used to measure diameters at other points on the tree, although there are other instruments that are more efficient for this purpose (Section 2-1.4).

Calipers. Calipers are often used to measure tree diameters when the diameters are not over 24 inches (about 60 centimeters). Calipers of sufficient size to measure large trees, or those with high buttresses (as in tropical regions), are awkward to carry and handle, particularly in dense undergrowth. Figure 2-1 shows three examples of calipers. One form (a), which may be constructed of metal or wood, consists of a graduated beam with two arms perpendicular to it. One arm is fixed at the origin of the scale and the other arm slides. When the beam is pressed against the tree and the arms are closed, the tree diameter can be read on the scale. Another form (b) consists of a set of fixed arms. The graduations are calibrated so that, when the fork is placed on the tree, the points of tangency indicate the tree diameter. The Finnish parabolic caliper (c) consists of a parabolic scale radiating from a common origin (Wiljamaa, 1942; Vuokila, 1955). Numerous other caliper-type instruments have been devised, including ones that provide for the automatic recording of measurements for data processing.

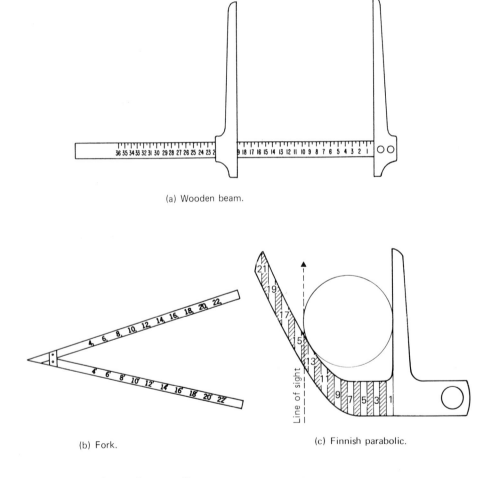

(a) Wooden beam.

(b) Fork.

(c) Finnish parabolic.

Fig. 2-1 *Calipers for tree diameter measurement.*

20   LINEAR MEASUREMENT

For an accurate reading, the beam of the caliper shown in Fig. 2-1a must be pressed firmly against the tree with the beam perpendicular to the axis of the tree stem and the arms parallel and perpendicular to the beam. It is, of course, important that all types of calipers be held perpendicular to the axis of the tree stem at the point of measurement.

Rapid measurement of trees is possible with calipers, particularly when only one measurement is taken per tree. Eccentric trees may be treated as explained in Section 2-1.2.

Diameter Tape. The diameter of a tree cross section may be obtained with a flexible tape by measuring the "circumference" of the tree and dividing by π ($D = C/\pi$). The diameter tapes used by foresters, however, are graduated at intervals of π units (inches or centimeters), thus permitting a direct reading of diameter. These tapes are accurate only for trees that are circular in cross section. In all other cases, the tape readings will be slightly too large, because the circumference of a circle is the shortest line that can encompass any given area.

The diameter tape is convenient to carry and, in the case of eccentric trees, requires only one measurement. Although it is slower to use than other diameter measuring instruments, the time element is generally not important. Care must be taken that the tape is correctly positioned at the point of measurement, that it is kept in a plane perpendicular to the axis of the stem, and that it is set firmly around the tree trunk.

Biltmore Stick. The Biltmore stick, which can hardly be classed as a measuring instrument, is an aid in estimating diameter at breast height. It consists of a straight rule, normally 24 to 36 inches long, that is held perpendicular to the axis of the tree stem. By holding the stick (Fig. 2-2) so that the 0-point of the graduation at one end of the stick lies on the line EA that is tangent to the tree cross section at A, the diameter of the tree can be read at the intersection at the other end of the stick on

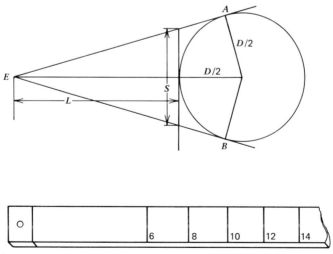

Fig. 2-2 *Biltmore stick.*

the line *EB* that is tangent to the tree cross section at *B*. The distance, *L*, from the eye, *E*, to the tree is usually 25 inches; however, it may have some other value. The graduations, *S*, of the stick for different values of *D* and *L* are obtained from the following formula.

$$S = \sqrt{\frac{D^2L}{L + D}} \qquad (2\text{-}1)$$

Inaccuracies in estimating diameters with the Biltmore stick are due to (1) the difficulty of holding the stick exactly *L* inches from the eye, (2) the failure to keep the eye at breast height level, (3) the failure to hold the stick at breast height level, and (4) the eccentricity of tree cross sections (the Biltmore stick is correct only for circular cross sections).

Sector Fork. Bitterlich's sector fork (*Visiermesswinkel*), Fig. 2-3, is similar in principle to the Biltmore stick (Bitterlich, 1959). It determines diameter from a sector of a circular cross section. One side of the sector is a metal arm; one side is a line of sight. The line of sight intersects a curved scale on which diameters or cross-sectional areas are printed. It is not necessary to hold the instrument at a fixed distance from the eye because a sighting pin fixes the line of sight for any distance. The instrument is especially suited for measuring trees with diameters of less than 50 centimeters (about 20 inches).

Because many trees are eccentric in cross section, all instruments for measuring tree diameter will in the long run give results that average too large. However, when an accurate determination of cross-sectional area is of prime importance, the calipers will give the best results. But when different people caliper the same irregular tree,

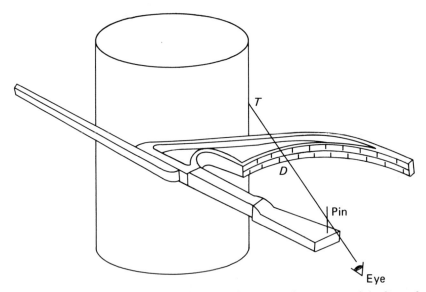

Fig. 2-3 *Sector fork (Visiermesswinkel). (T is the point of tangency of the line of sight; D is the diameter reading on the scale.)*

22   LINEAR MEASUREMENT

there will always be some variation in the measurements. A good part of this variation results because the tree is not always calipered in the same direction. Since this element of the variation does not affect measurements with the diameter tape, tape measurements are more consistent. Such consistency is important in growth studies, when the same trees are remeasured at intervals. Then, the actual diameters of the trees are less important than the changes in diameter during the period between measurements. These changes will be accurately determined by the diameter tape. Errors due to eccentricity will appear in both measurements and will not significantly affect the difference between them. In any case, whenever repetitive diameter measurements are made to determine growth, it is desirable to mark the position of measurement with a scribe or paint mark.

When the average of a number of diameter measurements, d, is required, one might use the arithmetic mean. However, if the primary interest is to obtain an average for the calculation of cross-sectional area and volume, then the quadratic mean, Q, is more appropriate.

$$Q = \sqrt{\frac{\sum\limits_{i=1}^{n} d_i^2}{n}} \tag{2-2}$$

The quadratic mean gives the diameter of the tree of arithmetic average cross-sectional area.

2-1.4 Upper-Stem Diameters

Tree stem diameters above breast height are often required to estimate form or taper and to compute the volume of sample trees from the measurement of diameters at several points along the stem. Upper-stem diameters can be obtained by climbing a tree and using the instruments described in Section 2-2.3, or by mounting calipers on a pole (Ferree, 1946), or by mounting a diameter tape on a pole (Godman, 1949). When the caliper or tape is mounted on a pole, diameters are usually taken at the top of the first log (about 17 feet) to determine Girard form class (Section 8-6.2).

The most practical dendrometers[1] for measuring upper-stem diameters employ optical means that allow diameters to be determined from the ground at some distance from a tree. Optical dendrometers may be classified (Grosenbaugh, 1963a; Smith, 1970) as optical forks, optical calipers, fixed-base rangefinders, and fixed-angle rangefinders.

Optical forks. An optical fork employs a fork angle on which the lines are tangent to the tree cross section at the level of the diameter measurement and on which the

---

[1] There is considerable ambiguity in the use of the term *dendrometer*. In one usage, the term denotes an instrument that takes diameters at points above breast height. The Society of American Foresters (1971), however, collaborating with the International Union of Forest Research Organizations, defines a dendrometer as "Generally, any instrument for measuring the dimensions of trees or logs . . . ," or "More specifically, any instrument for measuring or estimating the diameters of trees or logs."

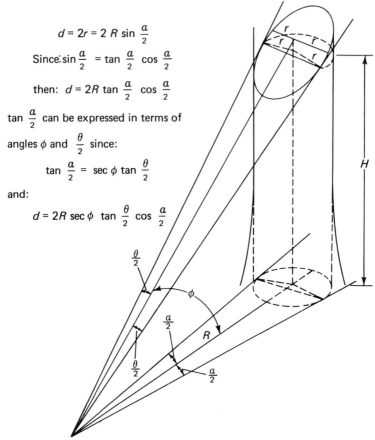

The following equations appear in the figure:

$$d = 2r = 2R \sin \frac{a}{2}$$

Since: $\sin \frac{a}{2} = \tan \frac{a}{2} \cos \frac{a}{2}$

then: $d = 2R \tan \frac{a}{2} \cos \frac{a}{2}$

$\tan \frac{a}{2}$ can be expressed in terms of

angles ϕ and $\frac{\theta}{2}$ since:

$$\tan \frac{a}{2} = \sec \phi \tan \frac{\theta}{2}$$

and:

$$d = 2R \sec \phi \tan \frac{\theta}{2} \cos \frac{a}{2}$$

Fig. 2-4 *Principle of the optical fork.*

vertex is at the observer's eye. The basic geometry is shown in Fig. 2-4. Note that the cos (a/2) can be approximated by using cos (θ/2).

Optical forks vary in design and use, depending on whether the fork angle is fixed or variable or whether it is coupled to the vertical angle. If the fork angle is fixed, the distance from the tree must be varied. If it is varied, a fixed distance from the tree may be used. When the fork angle is coupled to the vertical angle, the need to measure it and introduce it in subsequent calculations is eliminated.

There are several instruments and methods based on the optical fork principle. These include the Spiegel relaskop, the Tele-relaskop with 8 power magnification (Bitterlich, 1978), the use of terrestrial photogrammetry (Ashley and Roger, 1969), and the use of a transit (Leary and Beers, 1963). With the transit neither the vertical angle φ nor the angle θ/2 need be measured (Fig. 2-4). The horizontal angle is formed on the plates of the transit when one sights to the selected upper-stem diameter.

Upper-stem diameter is then

$$d = 2R \sin \frac{a}{2}$$

or for small angles,

$$d = R \sin a$$

A modification of the engineer's transit, the transit dendrometer, has been developed and placed on the market by an American forestry supply company.

Optical Calipers.   Basically, an optical caliper is made up of a graduated scale, two penta prisms, and a sighting tube that may, or may not, have an optical system to produce magnification (Fig. 2-5a). (The sighting tube may be omitted, if magnification is not used.) One of the penta prisms is fixed in place at the zero mark on the scale; the other is mounted so it may be moved along the scale at right angles to a direct line of sight to the tree. The observer sights over the fixed prism to the point of tangency on the left side of the tree cross section and, at the same time, through both prisms to the right side of the tree cross section. The sliding prism is moved along the scale until the line of sight through the prism is tangent to the right side of the tree cross section (i.e., until the "picture" shown in Fig. 2-5b is obtained).

Since the measurements obtained with the optical caliper are based on the assumption that the lines of sight to the tree are parallel, the instrument is, in effect, a "long-armed" caliper. When the instrument is properly oriented and sighted, diameters may be read directly from the scale for any cross section on the tree stem or on a limb. Accuracy is limited by the degree of parallelism of the lines of sight and by the limitations of the human eye. Without magnification, if parallelism obtains, diameters may be obtained by the average individual within a range of ± 0.1 inches at 50 feet, ± 0.2 inches at 100 feet, and so on. With magnification, accuracy will, both theoretically and practically, increase directly as the power of the optical system increases.

For a penta prism the deviation, D, of a light ray will be $D = 2(90° - A)$, where A is the main prism angle (Fig. 2-5). Thus, if A equals 45 degrees, a deviation of 90 degrees will be produced. This deviation will be obtained even if the line of sight is not perpendicular to the prism face which it enters. Consequently, in an optical caliper, collimation and adjustment are simple. If total reflection prisms (two angles of 45 degrees and one of 90 degrees) or mirrors are used, collimation and adjustment are difficult. Several instruments of this type have been devised. The Wheeler penta prism caliper (Wheeler, 1962) is the one most commonly used. This instrument is available with a base line that will permit measurement of diameters up to 34 inches. The instrument may be mounted on a tripod to obtain greater precision.

Fixed-Base Rangefinders.   The principle of the fixed-base rangefinder is embodied in the Barr and Stroud dendrometer (Hummel, 1951; Jeffers, 1956; Grosenbaugh, 1963a). From a fixed base the optics of the system are manipulated so that an angle is varied to measure the required distances (Fig. 2-6).

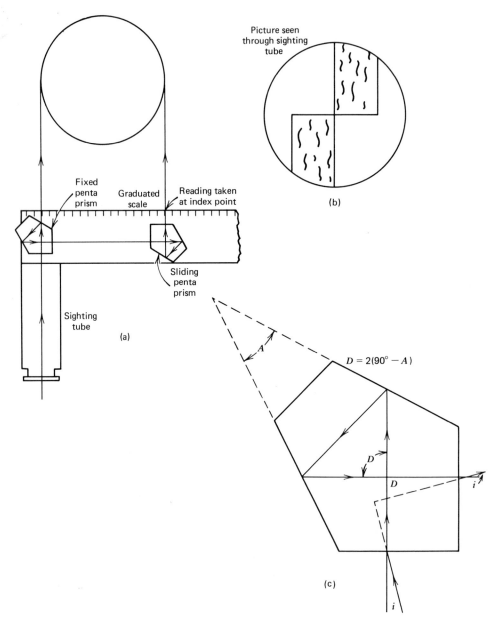

Fig. 2-5 *The optical caliper.*

Fixed-Angle Rangefinders. The principle of the fixed-angle rangefinder is embodied in the Breithaupt Todis dendrometer (Eller and Keister, 1979). From a fixed angle the optics of the system are manipulated, so that the baseline is varied to measure the required distances.

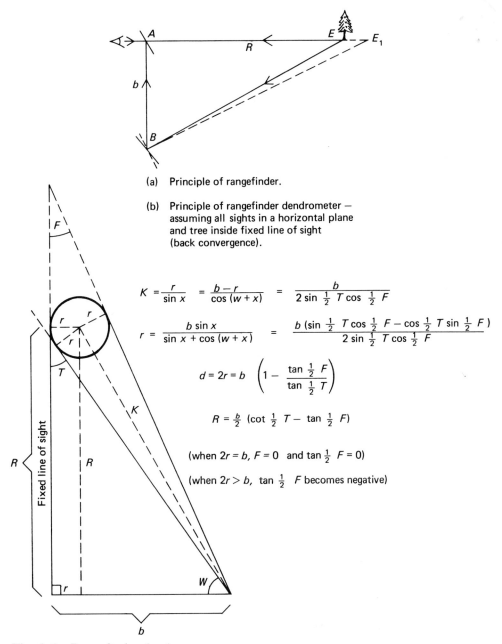

(a) Principle of rangefinder.

(b) Principle of rangefinder dendrometer —
assuming all sights in a horizontal plane
and tree inside fixed line of sight
(back convergence).

$$K = \frac{r}{\sin x} = \frac{b - r}{\cos (w + x)} = \frac{b}{2 \sin \frac{1}{2} T \cos \frac{1}{2} F}$$

$$r = \frac{b \sin x}{\sin x + \cos (w + x)} = \frac{b (\sin \frac{1}{2} T \cos \frac{1}{2} F - \cos \frac{1}{2} T \sin \frac{1}{2} F)}{2 \sin \frac{1}{2} T \cos \frac{1}{2} F}$$

$$d = 2r = b \left(1 - \frac{\tan \frac{1}{2} F}{\tan \frac{1}{2} T}\right)$$

$$R = \frac{b}{2} (\cot \frac{1}{2} T - \tan \frac{1}{2} F)$$

(when $2r = b$, $F = 0$ and $\tan \frac{1}{2} F = 0$)

(when $2r > b$, $\tan \frac{1}{2} F$ becomes negative)

Fig. 2-6 *Rangefinder dendrometer.*

In the foregoing discussion the optical caliper has been given the most space because this instrument is one of the best for measuring upper-stem diameters and because it can be constructed by a novice. Optical components are available from optical supply

DIAMETER MEASUREMENT 27

companies at reasonable prices. However, if the student desires to delve deeper into the construction and use of other optical devices, we suggest Smith (1970), one of the best books available on the subject.

2-1.5 Precise Measurement of Diameter Change

Precise measurement of minute changes in diameter may be required in research. Such changes, which may be for periods as short as an hour, cannot be detected with calipers or diameter tapes. Consequently, dendrometer bands, dial-gauge micrometers, recording dendrographs, and transducers are used to measure minute changes. Dendrometer bands, as described by Liming (1957), Bower and Blocker (1966), and Yocom (1970), consist of aluminum or zinc bands with vernier scales. The band is placed around the stem of a tree and held taut with a spring. Changes in diameter as small as 0.01 inches can be read.

A dial-gauge micrometer was first described by Reineke (1932). This instrument utilized a stationary reference point, a hook screwed into the xylem. The distance from this fixed point to a metal contact glued to the bark is measured with the micrometer. Changes in diameter as small as 0.001 inches can be read. Daubenmire (1945) modified this instrument by inserting three screws into the xylem as the fixed reference point.

A precision dendrograph was devised by Fritts and Fritts (1955). The instrument consists of a pen on an arm bearing on a fixed point on the tree stem. The pen records diameter changes to 0.001 inches on a chart mounted on the drum of an eight-day clock. Phipps and Gilbert (1960) designed an electrically operated dendrograph similar in principle to the dial-gauge micrometer. A potentiometer was fixed to a tree by screws anchored in the xylem. A movable shaft was fixed to the outer bark; any displacement in the shaft was measured by a change in electrical resistance and recorded on a continuous strip chart. Imprens and Schalck (1965) used a variable differential transformer rather than a potentiometer in a similar instrument.

A transducer was designed by Kenerson (1973). The transducer uses a linear motion potentiometer fixed to an invar plate. When the device is fixed to a tree stem, changes in the stem diameter move the potentiometer shaft.

2-1.6 Crown Diameter

Crown diameter measurements are useful to estimate diameter at breast height and therefore tree volume. They are best made on vertical aerial photographs (Section 13-5). Field determination is difficult because of the irregularity of a tree crown's outline. The usual field technique is to project the perimeter of the crown vertically to the ground and to make "diameter" measurements on this projection. Most of the instruments used to achieve the vertical projection incorporate a mirror, a right-angle prism, or a penta prism; they may be hand-held or staff-mounted. Instruments that employ the penta prism are recommended because they do not invert or revert images, and because slight movement of the prism will not affect the true right angle of reflection. Instruments for determining the vertical projection have been described by Husch (1947), Nash (1948), Raspopov (1955), and Shepperd (1973).

Crown diameters measured on vertical aerial photographs are more clearly defined than those measured on the ground, although crown diameters measured on aerial photographs are smaller than those measured on the ground, because parts of the crown are not resolved on photographs. However, the photo measurement is probably a better measure of the functional growing space of a tree and is better correlated with tree and stand volume.

2-2   HEIGHT MEASUREMENT

Height is the linear distance of an object normal to the surface of the earth, or some other horizontal datum plane. Aside from land elevation, tree height is the principal vertical distance measured in forestry.

At the outset one should understand that "tree height" is an ambiguous term unless it is clearly defined. Thus, a logical classification of height measurements that can be applied to standing trees of either excurrent or deliquescent form is shown in Fig. 2-7. Here are the pertinent definitions.

Total height is the distance along the axis of the tree stem between the ground and the tip of the tree. (In determining this height the terminals—i.e., the base and tip of the tree—can be more objectively determined than other points on the stem. It is often difficult, however, to see the tip of a tree in closed stands and to determine the uppermost limit of a large-crowned tree.)

Bole height is the distance along the axis of the tree stem between the ground and the crown point. (As shown in Fig. 2-7, the crown point is the position of the first crown-forming branch. Therefore, bole height is the height of the clear, main stem of a tree.)

Merchantable height is the distance along the axis of the tree stem between the ground and the terminal position of the last usable portion of the tree stem. (The position of the upper terminal is somewhat subjective. It is taken at a minimum top diameter or at a point where branching, irregular form, or defect, limit utilization. The minimum top diameter will vary with the intended use of the timber and with market conditions. For example, it might be 4 inches for pulpwood and 8 inches for sawtimber.)

Stump height is the distance between the ground and the basal position on the main stem where a tree is cut. (A standard stump height, generally about a foot, is established for volume table construction and timber volume estimation.)

Merchantable length is the distance along the axis of the tree stem between the top of the stump and the terminal position of the last usable portion of the tree stem.

Defective length is the sum of the portions of the merchantable length that cannot be utilized because of defect.

Sound merchantable length equals the merchantable length minus the defective length.

Crown length is the distance on the axis of the tree stem between the crown point and the tip of the tree.

2-2.1 Tree Height and Length Measurements

Generally speaking, the techniques and instruments devised for general height measurement may be applied to tree-height measurement. However, instruments must be economical, light, portable, and usable in closed stands.

Total tree heights may be estimated from measurements made on aerial photographs. The use of photo height measurements is discussed in Chapter 13.

The heights of short trees can be measured directly with an engineer's self-reading level rod, a graduated pole, or similar devices. The heights of many taller trees can be measured directly, or accurately estimated, with the aid of sectional or sliding poles made of wood, fiber glass, or lightweight metal. Although measurement with poles is slow, they are often used to measure height on continuous forest inventory plots where high accuracy is desired and merchantable length is less than 70 feet.

Most height measurements of tall trees are taken indirectly with hypsometers (*hypso,* meaning height, plus *meter,* meaning an instrument for measuring). Hypsometers are based on the relation of the legs of similar triangles or on the tangents of angles. A comprehensive summary of hypsometers has been compiled by Hummel (1951). Patrone (1963) and Pardé (1961) describe the more popular European hypsometers. (Note that terms such as altimeter and clinometer are also applied to instruments that are used to measure height.)

2-2.2 Hypsometers Based on Similar Triangles

The Christen, Klausner, Merritt, Chapman, and JAL hypsometers are examples of instruments of this type. Of these, only the Christen and Merritt hypsometers are in general use in the United States. Consequently, only these two instruments will be described in this section.

Christen Hypsometer. This hypsometer consists of a scale about 10 inches long (Fig. 2-8). To use it, a pole (usually 5 or 10 feet long) is held upright against the base of the tree, or a mark is placed on the tree at a height of 5 or 10 feet above the ground. The hypsometer is then held vertically at a distance from the eye such that the two inside edges of the flanges are in line with the top and base of the tree. It may be necessary for the observer to move closer to or farther from the tree to accomplish this, but except for this, the distance from the tree is immaterial. The graduation on the scale that is in line with the top of the pole, or the mark, gives the height of the tree. The following proportion gives the formula for graduating the instrument.

$$\frac{A'C'}{AC} = \frac{A'B'}{AB} \tag{2-3}$$

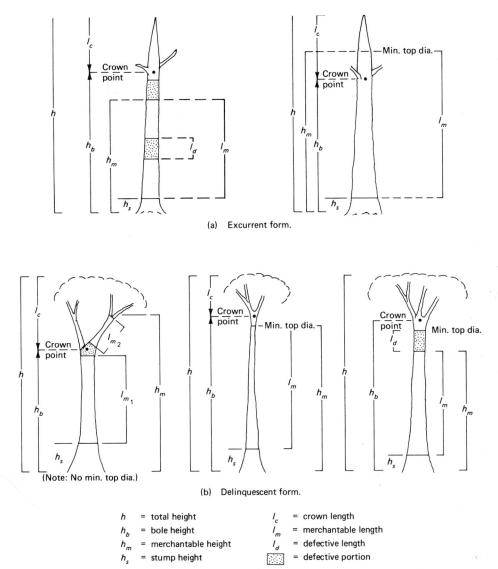

(a) Excurrent form.

(b) Delinquescent form.

(Note: No min. top dia.)

h	= total height	l_c	= crown length
h_b	= bole height	l_m	= merchantable length
h_m	= merchantable height	l_d	= defective length
h_s	= stump height	▨	= defective portion

Fig. 2-7 *Tree height and stem length classification.*

For a given length of instrument $A'B'$ and a given pole length or mark height AC, the graduations $A'C'$ can be obtained by substituting different values of height AB in the equation.

Merritt Hypsometer.   This simple instrument, which is often combined with the Biltmore stick, is a convenient aid in estimating the number of logs in a tree. It consists of a straight graduated stick that is held vertically at a predetermined distance, usually

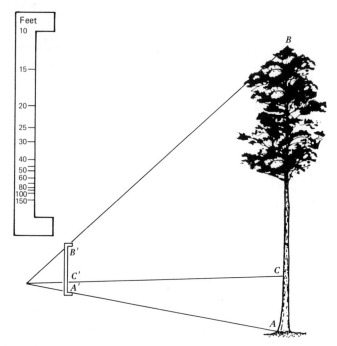

Fig. 2-8 *Christen hypsometer.*

25 inches, from the eye. (If the stick is held 25 inches from the eye along a horizontal line, then the distance to the tree should be measured on the horizontal; if the stick is held 25 inches from the eye along a line to the lower end of the stick, as shown in Fig. 2-9, then the distance to the tree should be measured on the slope.) The observer must stand at a predetermined distance, such as 50 feet, from the tree. Then the height of the tree may be read on the stick at the point where the line of sight to the top terminal on the tree intersects the scale. In Fig. 2-9,

$$\frac{A'B'}{AB} = \frac{EA'}{EA} \tag{2-4}$$

Now, if EA' is 25 inches and EA is 50 feet, for each 16 feet of height in AB the length of $A'B'$ will be 8.00 inches. The stick may thus be graduated in 8-inch intervals with the successive graduations marked 1, 2, 3 (indicating 16-foot logs). When merchantable length is measured, as it often is with the Merritt hypsometer, one should remember that the line of sight EA (Fig. 2-9) must be to the top of the stump.

Although the Christen hypsometer may be used to measure any type of height, it is practical only for total height measurements. Furthermore, an examination of Fig. 2-8 shows a crowding of scale graduations at the bottom of the scale. This makes the instrument unreliable for the determination of the height of tall trees.

In case of the Merritt hypsometer, it is difficult to hold the stick in a vertical position exactly 25 inches from the eye. Small deviations in orientation result in considerable

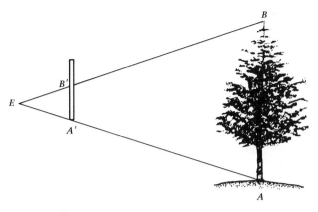

Fig. 2-9 *Merritt hypsometer.*

errors in height readings. Therefore, this instrument should be used only to make rough checks on ocular height estimates.

2-2.3   Hypsometers Based on Tangents of Angles

Although numerous hypsometers of this type have been developed, the basic principle is the same for all of them. One first sights to the upper terminal of the height desired and takes a reading; then one sights to the base of the tree, or the top of the stump, depending on the height desired, and takes a second reading. Figure 2-10a illustrates the situation, with the distance D and the angles α_1 and α_2 known. Then, if the vertical arc is graduated in degrees, the total height of the tree may be calculated as follows.

$$\tan \alpha_1 = \frac{BC}{D}$$

from which we obtain

$$BC = D \tan \alpha_1$$

Similarly,

$$CA = D \tan \alpha_2$$

Since the height of the tree AB is $BC + CA$,

$$AB = D(\tan \alpha_1 + \tan \alpha_2) \tag{2-5}$$

On steep ground a situation as shown in Fig. 2-10b might occur. In this case, the height of the tree AB is $BC - CA$, and

$$AB = D(\tan \alpha_1 - \tan \alpha_2)$$

Hypsometers often are provided with percentage arcs and topographic arcs. The percentage arc is based on an angular unit represented by the ratio of 1 unit vertically to 100 units horizontally. It also means for any vertical angle α

$$\tan \alpha = \frac{\text{percent } \alpha}{100}$$

Thus, when using a percentage arc (Fig. 2-10a),

$$AB = \frac{D}{100} (\text{percent } \alpha_1 + \text{percent } \alpha_2)$$

The percentage arc simplifies height calculations, particularly when readings are taken at such distances as 100, 75, 50, and 25 feet from a tree.

The topographic arc (topo α) is based on an angular unit represented by the ratio of 1 unit vertically to 66 units horizontally. It also means for any vertical angle α

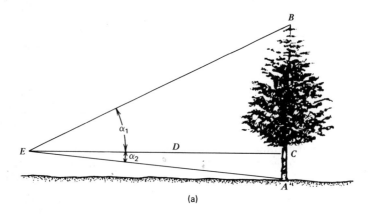

(a)

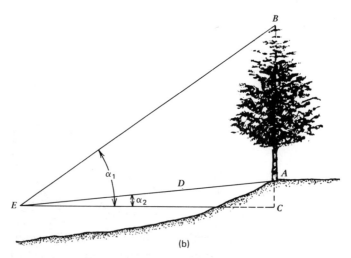

(b)

Fig. 2-10 *Measuring tree height with hypsometers based on the tangents of angles.*

$$\tan \alpha = \frac{topo \; \alpha}{66}$$

Thus, when using a topographic arc (Fig. 2-10a),

$$AB = \frac{D}{66} (topo \; \alpha_1 + topo \; \alpha_2)$$

More details on the topographic arc are given in Chapter 3. However, the topographic arc is as convenient to use as the percentage arc if readings are taken at such distances as 99, 66, and 33 feet from a tree.

The relationship between topographic and percentage readings is as follows

$$Percent \; \alpha = \frac{100 \; topo \; \alpha}{66}$$

Hypsometers may also be provided with arcs based on angular units represented by the ratio of 1 unit vertically to 15, 20, 25, 30, 45, 60, or other, units horizontally. One should note that scales of this type (and this includes percentage and topographic scales) can be used with any units: feet, yards, meters, and so on. One only has to read the scale in the same unit as one measures the baseline. However, since scale numbers usually represent convenient base distances for tree measurement with a particular unit, the unit to use is generally indicated.

Tree heights might be measured with a transit. However, the transit is too clumsy and too slow for practical forest work. Furthermore, the precision is greater than necessary. Consequently, lighter, simpler, and more economical instruments are used. The most commonly used hypsometers in the United States are the Abney level (Calkins and Yule, 1935), the Haga altimeter (Wesley, 1956), the Blume-Leiss altimeter (Pardé, 1955), and the Suunto clinometer. The Spiegel relaskop (Daniel, 1955) is also used for height measurement.

Abney Level. This instrument consists of a graduated arc mounted on a sighting tube about 6 inches long (Fig. 2-11a). The arc may have a degree, percentage, or topographic scale. When the level bubble, which is attached to the instrument, is rotated while a sight is taken, a small mirror inside the tube makes it possible to observe when the bubble is horizontal. Then the angle between the bubble tube and the sighting tube may be read on the arc.

Haga Altimeter. This instrument consists of a gravity-controlled, damped, pivoted pointer and a series of scales on a rotatable, hexagonal bar in a metal, pistol-shaped case (Fig. 2-11b). The six regular American scales are 15, 20, 25, 30, percentage, and topographic. Sights are taken through a gun-type peep sight; the indicator needle is locked by squeezing a trigger, and the observed reading is taken on the scale. A rangefinder is available with this instrument.

Blume-Leiss Altimeter. This instrument is similar in construction and operation to the Haga altimeter, although its appearance is somewhat different (Fig. 2-11c). The five regular metric scales are 15, 20, 30, and 40. A degree scale is also provided. All scales can be seen at the same time. A rangefinder is incorporated into the instrument.

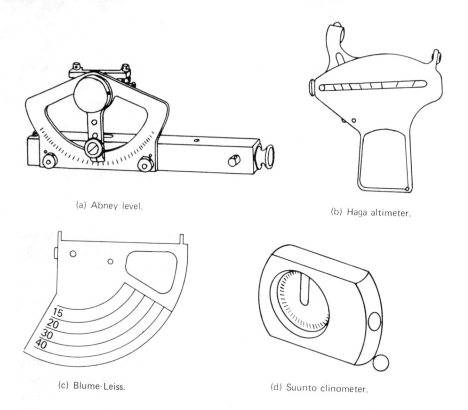

(a) Abney level.

(b) Haga altimeter.

(c) Blume-Leiss.

(d) Suunto clinometer.

Fig. 2-11 *Examples of hypsometers based on trigonometric principles.*

Suunto Clinometer. This instrument is a hand-held device housed in a corrosion-resistant lightweight alloy body (Fig. 2-11d). The scale is supported by a jewel-bearing assembly, and all moving parts are immersed in a damping liquid inside a hermetically sealed plastic capsule. The liquid dampens undue scale vibrations. The instrument is held to one eye and raised or lowered until the hairline is seen at the point of measurement. At the same time, the position of the hairline on the scale gives the reading. Due to an optical illusion, the hairline seems to continue outside the frame and can be observed at the point of measurement. The instrument has provisions for only two scales. However, in the United States the instrument is available with the following scale combinations: percentage and degree, percentage and topographic, topographic and degree. It is also available with a rangefinder.

Generally speaking, hypsometers based on the tangents of angles are more accurate than hypsometers based on similar triangles. The Abney level, the Haga altimeter, the Blume-Leiss Altimeter, and the Suunto clinometer are similar in accuracy. The Abney level, however, is slower to use, and large vertical angles are difficult to measure, because of the effect of refraction on observations of the bubble through the tube from beneath. This makes the Abney level difficult to use in tall timber that is so dense that the tops cannot be seen from a considerable distance. A choice among the other three instruments is largely a matter of personal preference.

Curtis and Bruce (1968) have shown how a clinometer that measures slope in percent can be used in conjunction with a pole of fixed length to measure height without measuring the distance from the observer to the tree. The principle is similar to that of the Christen hypsometer, an instrument specifically designed to eliminate the need to measure the distance from the observer to the tree. Using the Curtis-Bruce method, a pole of known length (between one-fourth and one-fifth of tree height) is placed against the tree. Then slopes are read from any convenient point to tree top α_t, to pole top α_p, and to tree base α_b. When p is the pole height, tree height h is

$$h = p \, \frac{\alpha_t - \alpha_b}{\alpha_p - \alpha_b}$$

Bell and Gourley (1980) point out that this method simplifies locating a point in a dense stand where the tree tip is visible.

When the distance of the observer from the tree is to be determined, a suitable method of determining distance should be selected. With the Merritt hypsometer, which is not highly accurate, pacing is appropriate; but with the other, more accurate, instruments, a tape or a rangefinder should be used.

2-2.4 Special Considerations in Measuring Heights with Hypsometers

It is difficult to accurately measure the height of large, flat-crowned trees, such as the oaks, elms, maples, and so on. There is a tendency to overestimate their height. Consequently, total height determinations for such trees are of little value and are seldom taken.

In general, the *optimum viewing distance* for any hypsometer is the distance along the slope equal to the height to be measured. This rule of thumb, which is adapted from Beers (1974), should be used with discretion. For example, if one were using a Suunto clinometer with a percentage scale to determine the merchantable height of a tree estimated to be about 56 feet, it would be logical to use a viewing distance of 50 feet.

Since all hypsometers assume trees are vertical, trees leaning away from an observer will be underestimated and trees leaning toward an observer will be overestimated. This error will be minimized if measurements are taken such that the lean is to the left or right of the observer. If one must measure leaning trees, one should determine the point on the ground where a plumb bob would fall if suspended from the tip of the tree. Then height should be measured from this point to the tip. The measured height is multiplied by the secant of the angle of lean ϕ to obtain the correct height (Fig. 2-12). The correction will be small except for abnormally leaning trees.

2-2.5 Ocular Estimates of Height

The measurement of tree height with an accurate hypsometer is slow and expensive. Consequently, it is customary to make ocular estimates whenever precision is not essential, such as for commercial timber inventories where none of the required values

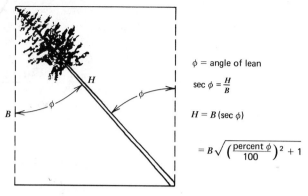

ϕ = angle of lean

$$\sec \phi = \frac{H}{B}$$

$$H = B (\sec \phi)$$

$$= B \sqrt{\left(\frac{\text{percent } \phi}{100}\right)^2 + 1}$$

Fig. 2-12 *Measurement of the height of a leaning tree.*

are precisely determined, and where the large number of trees measured makes precision of individual measurements unimportant. Good ocular estimates can be obtained by an experienced person. Ocular estimates are, however, subject to serious errors because of sudden changes in the timber type or the weather. Furthermore, most people do not make reliable estimates at the start of the day or after a rest. Consequently, estimates should be checked frequently by instrumental measurements. Although the Merritt hypsometer is often used for this purpose, we recommend that one use the Haga altimeter, the Blume-Leiss altimeter, the Suunto clinometer, or a similar lightweight hypsometer.

2-3 EXPRESSIONS OF STAND HEIGHT

It is often necessary to use a single value to characterize the height of a forest stand. The principal applications of such values are to determine stand volume from a stand volume equation and to determine site index from a curve or an equation. The following guidelines describe the more important methods that have been proposed to obtain the average height of a stand.'

1. Measure and average the heights of all of the trees, or a sample of the trees, regardless of their size or relative position in the stand. (This method is excessively time consuming.)

2. Measure and average the heights of the dominant trees, or of the dominant and codominant trees. (The selection of a dominant or a codominant tree is subjective. Consequently, different observers often obtain different results. This height expression is commonly used in determining the site index in North America.)

3. Measure and average the heights of a fixed number of the largest trees (normally, largest in diameter, although it could be largest in height) per unit area, usually 100 per acre. (This height expression is used in Great Britain and New Zealand.)

4. Determine the average height that will yield the correct volume when using the following equation.

$$V = \bar{h}Gf \qquad (2\text{-}6)$$

where

V = volume per unit area
$\underline{G}$ = total basal area per unit area
$\bar{h}$ = average height of the stand
f = form class of the stand

The best known average height of this type is Lorey's mean height h_L.

$$h_L = \frac{n_1 g_1 h_1 + n_2 g_2 h_2 + \cdots + n_z g_z h_z}{G} = \frac{\sum\limits_{i=1}^{z} n_i g_i h_i}{\sum\limits_{i=1}^{z} n_i g_i} \qquad (2\text{-}7)$$

where

n_i = number of trees in a diameter class
g_i = average basal area of a diameter class
h_i = average height of the trees in a diameter class

Lorey's mean height is derived from equation 2-6 by considering that the stand is made up of a series of diameter classes.

$$Gh_L f = n_1 g_1 h_1 f_1 + n_2 g_2 h_2 f_2 + \cdots + n_z g_z h_z f_z$$

By assuming a constant form factor (i.e., $f_1 = f_2 = \cdots = f_z = f_i$), equation 2-7 is obtained.

It should be noted that the arithmetic average height of the trees selected in horizontal point sampling (Chapter 14) yields Lorey's mean height (Kendall and Sayn-Wittgenstein, 1959).

In addition to the above, several expressions of average stand height have been developed that utilize a height-diameter curve constructed from sample-tree data representative of the stand. After one has prepared the curve, one decides what average diameter is representative of the stand and reads the height of a tree of this diameter from the curve. For example, the diameter of the tree of average basal area might be used (symbolized h_g); the arithmetic mean diameter might be used (symbolized h_d); the median diameter might be used (symbolized h_{dM}); or the diameter of the tree of median basal area might be used (symbolized h_{gM}).

3
Forest Surveying

Generally speaking, resource managers do not conduct many original property surveys. However, they often retrace old lines, locate boundaries (of burns, clearcuts, properties, etc.), run cruise lines or transects, and so on. Consequently, this chapter will emphasize the fundamental operations involved in doing these jobs: determining the direction and length of lines in wildlands. For the most part, information available in elementary surveying texts such as Brinker and Wolf (1977) and Moffitt and Bouchard (1975) will not be included.

3-1 MEASUREMENT OF HORIZONTAL DISTANCES

In surveying, the distance between two points is commonly meant to be the horizontal distance. Although slope distances are sometimes measured in the field, these distances are subsequently reduced to their horizontal equivalents, since horizontal distances are the ones used in the preparation of maps and in the computation of areas. The method used in obtaining the distance between two points is largely determined by the accuracy required.

3-1.1 Pacing

Where approximate results are sufficient, as in making tree-height measurements with certain hypsometers, in reconnaissance work, and in some types of cruising, distance can be obtained by pacing. Although the engineer thinks of a pace as a single step, the forester, the wildlifer, and others who work in wildlands define a pace as a *double step*. Thus, as used here, a pace is two steps.

To graduate the pace we recommend that one use a measured line of 20 chains (or 400 meters) that has been laid out in terrain of the type to be encountered. Stakes should be set on the line at 0, 5, 10, 15, and 20 chains (or 0, 100, 200, 300, and 400 meters). One should walk as naturally as possible four times over the line, record the number of paces between each pair of stakes, and compute the average number of paces between each pair of stakes and for each trip. From this one can get a good picture of the consistency of the pacing and can determine the average pace for the terrain.

The pace should be graduated to meet the varying conditions encountered: wooded slopes of 0 to 10 percent, wooded slopes of 10 to 20 percent, open level woods, and so forth. (One should not use the average pace data given in tables for individuals of various heights, or try to adopt a convenient length of pace. In wildland pacing one should walk in one's own way.) However, it must be recognized that on slopes over 30 percent, in swamps, and in slashings, it is impossible to pace accurately. It is advisable to staff pace on steep slopes and to "rough chain" in swamps and in slashings.

To staff pace one uses a 4.125-foot (or 1.25-meter) staff. This gives 16 staffs to a chain (or 16 staffs to 20 meters). The use of the staff is simple. *Traveling uphill:* while holding the staff horizontal, locate the ground position of the rear end of the staff by plumbing by eye and of the forward end by contact with the ground. *Traveling downhill:* while holding the staff horizontal, locate the ground position of the rear end by contact with the ground and of the forward end by plumbing by eye.

With foot or staff pacing, an experienced pacer should consistently attain an accuracy of 1/100 or better.

3-1.2 Steel Tapes

The types of tapes commonly used by resource managers in the United States to measure horizontal distances are the 100-foot steel tape and the steel topographic trailer tape. When one desires to measure distances in meters, 30 and 50 meter tapes are convenient to use.

We will not discuss measurements with the 100-foot tape or with metric tapes because this subject is covered in all standard surveying texts. We will discuss measurements with the topographic trailer tape (Section 3-2.) because this subject is not adequately covered in surveying texts or mensuration manuals.

3-1.3 Electronic and Optical Distance Measurement

Electronic distance-measuring instruments are used for measuring distances in all types of surveying work. Electro-optical equipment was first introduced in 1947; microwave equipment in 1957. In general, microwave instruments are capable of producing good medium range (1 to 30 miles) measurements. Short-range measurements (under 1 mile) are best made with electro-optical instruments. Consequently, when the resource manager has occasion to use an electronic distance-measuring instrument, it will most likely be an electro-optical instrument. With the more sophisticated instruments, in a matter of seconds one can obtain the slope distance and vertical angle electronically and automatically compute the horizontal and vertical distances. Typical accuracy is plus or minus 5 millimeters plus 5 parts per million (Tomlinson and Burger, 1978). Although the line of sight must be free of leaves, shrubs, and other obstacles, only a small opening is required.

Although it is not likely that electronic distance-measuring instruments will replace tapes and foot pacing for distance measurement on cruise lines, sample plots, and some boundaries, in rough, wooded wildlands, such equipment is already being used by resource managers for road surveys, boundary surveys, and so on, in many forested areas. And it will gain wider acceptance in the years ahead.

The alternative of measuring horizontal distances by taping and electronic methods is to use *optical methods*. There is a family of instruments, methods, and procedures classified as optical distance measurements. Some have limited practical application; some are expensive; some are complicated. But some have great potential for resource managers.

Optical distance-measuring devices include fixed-base rangefinders, fixed-angle rangefinders, tacheometers for use with theodolites, distance-measuring wedges for use with theodolites, subtense bars, plane table alidades, and miscellaneous instruments. Resource managers will find fixed-base rangefinders useful for rapid check of plot dimensions, for rapid distance measurement, for sketch mapping, and for other applications where high accuracy is not required. They will also find that the subtense bar and the distance-measuring wedge have possibilities for quick and accurate distance measurement (see Curtis, 1976).

3-2 THE TOPOGRAPHIC TRAILER TAPE

In spite of technological developments, the topographic trailer tape will remain for many years an important tool in the resource manager's tool kit. In rough, wooded wildlands, where great accuracy is not required, it permits fast, cheap measurement of boundary lines, cruise lines, and traverse legs, as well as an inexpensive method of obtaining elevations. (Measurements can easily be made to the precision generally required for wildlands.) It also provides an efficient method of measuring sample plot dimensions, although it is not often used for this purpose.

The topographic trailer tape is graduated in chains and links. A chain is 66 feet; a link is 1/100 of a chain or 0.66 foot. There are three tabs on a trailer tape: one at 0 links, one at 100 links (one chain), and one at 200 links (two chains). Beyond the two-chain tab there is about one-half chain of tape that trails the body of the tape. The trailer is used to convert slope distance to horizontal distance.

The topographic trailer tape was designed to be used with a clinometer, such as the Abney level, that has a topographic arc. The topographic arc (topo arc) is graduated in an angular unit that represents 1 unit vertically to 66 units horizontally. Thus, a topo reading of plus 17 indicates a vertical rise of 17 feet per 66 feet, or 17 feet per chain.

Since a slope distance is greater than its horizontal equivalent, if we measure two chains along a slope we must add a correction to obtain two chains of horizontal distance (Fig. 3-1).

The trailer on the topographic trailer tape carries corrections for converting a slope distance of two chains to two chains of horizontal distance. These corrections are on the top of the tape. Corrections for converting a slope distance of one chain to one chain of horizontal distance are on the underside of the tape beyond the one-chain tab. It is simple to apply these corrections. For example, if a topo reading of "15" is obtained for a slope distance of two chains, and the trailer is let out to the "15" graduation, the slope distance becomes the hypotenuse of a right triangle whose

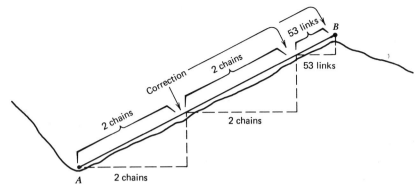

Fig. 3-1 *Applying slope corrections.*

horizontal leg is two chains. If a topo reading of "20" is obtained for a slope distance of one chain, the proper correction is applied when the tape is let out to the "20" graduation on the underside of the tape beyond the one-chain tab.

When it is necessary to measure slope distances less than one chain or between one and two chains, tables are used to obtain the necessary corrections. For example, Table 3-1 shows the amount that must be added per chain of slope distance on different grades to obtain one chain of horizontal distance. Computation of the table values is simple.

$$\text{Correction per chain} = \sqrt{1.0^2 + \left(\frac{\text{topo reading}}{66}\right)^2} - 1$$

Thus, for a topo reading of 33, the correction per chain of slope distance is 0.118 chain (11.8 links). For other slope distances on the same grade the correction equals the slope distance times 0.118 chain. Thus, if a slope distance is 0.73 chain, the correction will be 0.118 times 0.73, or 0.086 chain (8.6 links). When 0.73 + 0.086 = 0.816 chain (81.6 links) is measured along the slope, a horizontal distance of 0.73 chain will be obtained. For a distance of 1.50 chains, the correction would be 0.118 times 1.50, or 0.177 chain (17.7 links).

In applying slope corrections with the trailer tape on irregular topography, the head chainer must stop short of every major break in topography. After the correction is applied, the head chainer must be on the "break." For example, assume it is desired to measure the horizontal distance between points *A* and *B*, a distance between four and five chains (Fig. 3-1). From *A* the tape is drawn up the slope for two chains. Then the clinometer reading is taken and the correction applied by letting out the trailer. This is repeated for another two chains. To measure the remaining distance of less than one chain, the head chainer must estimate the correction and stop short of point *B* by the estimated amount. When the correction is applied, the head chainer should be at *B*, or reasonably close to it.

If there were stakes at points *A* and *B* (Fig. 3-1), the distance could be measured as described above. However, to obtain the correction for the final distance of 53

Table 3-1
Correction Factors to Add to a One-Chain Slope Distance on Different Grades to Obtain a One-Chain Horizontal Distance[1]

Topo Reading	Correction (chains per chain)	Topo Reading	Correction (chains per chain)	Topo Reading	Correction (chains per chain)	Topo Reading	Correction (chains per chain)
1	0.000	31	0.105	61	0.362	91	0.703
2	.000	32	.111	62	.372	92	.716
3	.001	33	.118	63	.382	93	.728
4	.002	34	.125	64	.393	94	.740
5	.003	35	.132	65	.404	95	.753
6	.004	36	.139	66	.414	96	.765
7	.006	37	.146	67	.425	97	.778
8	.007	38	.154	68	.436	98	.790
9	.009	39	.162	69	.447	99	.803
10	.011	40	.169	70	.458	100	.815
11	.014	41	.177	71	.469	101	.828
12	.016	42	.185	72	.480	102	.841
13	.019	43	.194	73	.491	103	.854
14	.022	44	.202	74	.502	104	.866
15	.026	45	.210	75	.514	105	.879
16	.029	46	.219	76	.525	106	.892
17	.033	47	.228	77	.537	107	.905
18	.037	48	.236	78	.548	108	.918
19	.041	49	.245	79	.560	109	.931
20	.045	50	.255	80	.571	110	.944
21	.049	51	.264	81	.583		
22	.054	52	.273	82	.595		
23	.059	53	.283	83	.607		
24	.064	54	.292	84	.619		
25	.069	55	.302	85	.631		
26	.075	56	.311	86	.643		
27	.080	57	.321	87	.655		
28	.086	58	.331	88	.667		
29	.092	59	.341	89	.679		
30	.098	60	.351	90	.691		

[1] Correction factors for other than one-chain slope distances are obtained by multiplying the correction factor by the slope distance in chains.

links, it might require several trials before the head chainer would end on the stake. Consequently, it is preferable to measure the slope distance to stake B and to convert this distance to the horizontal equivalent by the formula: $H = S(\cos A)$, where H is the horizontal distance, S is the slope distance, and A is the angle the slope makes with the horizontal.

Since $\cos A = \cos [\text{arc tan (topo reading/66)}]$, it is convenient to prepare a table that gives $\cos A$ for various topo readings.

3-2.1 Field Operations

Before commencing field work, the height of instrument (HI) must be determined for the clinometer. This is determined by the rear chainer, the clinometer operator, by sighting with the instrument set at "0" along level ground to the head chainer. The point at which the line of sight strikes the head chainer is the HI. All sights should be made to this point. (If elevations are being measured, it is also desirable for the head chainer to take clinometer readings.)

The head chainer, who generally determines the direction of a line with a staff compass (some head chainers use a good hand-held, liquid-filled compass), commonly ties the rawhide at the zero end of the tape to the back of his or her belt so both hands will be free. If it is necessary to move off the line, as is often required in cruising, a loop should be tied in the rawhide at the zero end of the tape and slipped over the shoulder. Then, the loop can be removed from the shoulder when it is necessary to move off the line.

The head chainer should pace when two chain "shots" can be taken so he or she will know to stop when approximately two chains have been covered.

On reaching a station, the head chainer's procedure should be: (1) stand parallel to the line of travel; (2) hold the tape at waist height in the center of the body and plumb the zero point of the tape to the ground by eye (the rear chainer should hold the tape in the same manner and plumb the approximate tape mark to the ground by eye); (3) at the command "stick" from the rear chainer, make a scuff on the ground

Line	Azimuth (degrees)	Station (chains)	Distance (chains)	Topo reading +	Topo reading −	Correction[1] (chains)	Elevation difference (feet)	Elevation (feet)
1-2	73	0.00						789.0
			2.00	30			60.0	
		2.00						849.0
			1.00	32			32.0	
		3.00						881.0
			2.00	28			56.0	
		5.00						937.0
			0.75	26		0.056	19.5	
		5.75						956.5
			2.00		24		−48.0	
		7.75						908.5
			1.63		23	0.096	−37.5	
		9.38						871.0
2-3	17	0.00						871.0
			2.00	10			20.0	
		2.00						891.0

[1] Corrections for distances of one or two chains are on the tape and are not shown.

Fig. 3-2 *Rear chainer's notes.*

to mark the station and reply "stuck" to the rear chainer; (4) turn and face the rear chainer and stand erect while he or she takes the clinometer reading; and (5) move up, pull the tape taught, and make a new scuff on the ground if the rear chainer calls "correction."

Before moving, the head chainer should wait until the rear chainer is ready to move. When the rear chainer begins to move, the head chainer should also move, but at a slower rate because, when the rear chainer reaches the new station, he or she stands about 4 feet back of the station and lets the tape slide through the hand so that the two-chain tab can be located when it comes along. As the tab passes the rear chainer's hand, "chain" is called out to stop the head chainer before the tab goes by the station.

As explained, after the clinometer reading is taken the correction on the tape is applied if the "shot" is one or two chains, but if the distance is less than one chain, or between one and two chains, the correction is calculated from a table of correction factors (Table 3-1). After the correction is applied, if elevations are being measured the rear chainer should take another reading, because there might be a slight change in the slope after the correction has been applied. The rear chainer should keep field notes in the form shown in Fig. 3-2. However, if elevations are not measured, the last two columns may be omitted.

3-3   ACCURACY OF TAPED MEASUREMENTS

An accuracy of at least 1/300 can be attained with the topographic trailer tape–clinometer combination. The 100-foot tape used in combination with plumb bobs and chaining pins will give an accuracy of better than 1/1000.

It should be kept in mind that the precision required in any particular case depends on the use to be made of the results. For example, if data are being secured for the preparation of a map to be drawn to a scale of 1 inch = 1000 feet, distances need to be measured only to the nearest 20 feet, because distances on the map cannot be plotted closer than 0.02 inches. Generally, this would require an accuracy of less than 1/300.

As another example, assume that the area of a piece of land about 10 acres in size is required to an accuracy of 0.05 acre. Since the desired accuracy of the area determination is about 1/200, the sides of the tract must be measured to an accuracy of 1/400.

3-4   OBTAINING THE DIRECTION
OF A LINE BY COMPASS

The compass was one of the most important surveying instruments in the early centuries of the modern era for determining the direction of a line. It is not, however, an instrument of precision; results of great accuracy are not to be expected in compass surveys. On the other hand, it has the advantage of speed, economy, and simplicity. It is useful in retracing lines of old surveys established by compass, in running all types of boundary lines, in maintaining the direction of cruise lines, and so forth.

Many cases also arise in which the resource manager must depend on the compass for direction.

The direction of a line is generally indicated by the angle between the line and some line of reference (i.e., by an azimuth or a bearing). The line of reference is generally a *true meridian* or a *magnetic meridian.*

The axis on which the earth rotates is an imaginary line cutting the earth's surface at two points: the north geographic pole and the south geographic pole. The *true meridian* at any place is the great circle drawn on the earth's surface passing through both poles and the place.

If a magnetized steel bar, such as a compass needle, is allowed to rotate on a pivot, the bar will take very nearly the same direction at any given place. The direction of the line so indicated is the *magnetic meridian* at the place. Although the magnetic meridian has the general direction of the true meridian, except in a few places, the magnetic meridian through a point on the earth's surface does not coincide with the true meridian at that point.

Many people assume that the magnetized bar is acted on by the earth's poles, and that one end will point toward the north magnetic pole, and one toward the south magnetic pole. This, however, is not a correct statement. Although the magnetized bar points in the general direction of the magnetic pole over most of the earth, the bar is actually turned so that it points in the direction of the horizontal projection of the earth's magnetic field at the point of observation (the earth acts like a great spherical magnet) and not toward the magnetic poles.

3-4.1 Magnetic Declination

The magnetic azimuth or bearing of a line can be obtained by direct reading of a compass. The true azimuth or bearing can be approximated by converting magnetic readings to true readings by applying the magnetic declination for the location in question. Consequently, a thorough understanding of magnetic declination is important for proper use of the compass.

The *magnetic declination D* is the angle between the true meridian and the magnetic meridian; it is considered east if magnetic north is east of true north, and west if magnetic north is west of true north (Fig. 3-3). Declination is often called *variation of the compass*, or simply *variation*. When it is desired to attach a sign to declination, east declination is considered positive, and west negative.

The declination at any point can be measured with the compass when the true direction of a line is known, or it can be obtained from charts distributed by the Branch of Distribution, U.S. Geological Survey, 1200 South Eads Street, Arlington, Virginia 22202. The use of charts, while not as accurate, is the method most commonly used. On these charts lines called *isogonic lines* are drawn through points where the magnetic declination is the same. The *agonic line* passes through points where the declination is zero.

At any specific point, the earth's declination is continually changing. It is convenient to classify these changes as *daily variation, irregular change, secular change,* and *annual variation.*

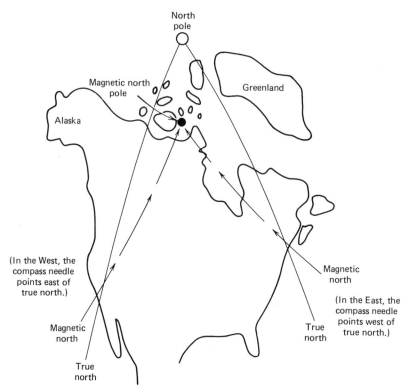

Fig. 3-3 *The compass needle does not point true north.*

Daily variation is a fairly systematic departure of the declination from its daily mean value. This repeats itself with fair regularity day after day. The amount of departure, however, depends on the time of day, the season, and other factors not wholly understood. During the summer season in the United States, the north end of the compass needle points on the average 12 minutes more to the west at 1 P.M. than at 8 A.M. A line run northerly 1000 feet by compass at 8 A.M. would end 3 feet to the right of where it would end if run at 1 P.M.

Superimposed on the regular daily variation, there are usually *irregular changes*. When they become large, we say there is a *magnetic storm*. These are associated with sunspots and are characterized by auroral displays and pronounced disturbances to radio wave transmission. In the United States, the change in declination in the course of a magnetic storm may, on rare occasions, be more than four degrees. Since the amplitude of *daily variation* and *irregular changes* is not predictable for any one day, these changes are not considered in compass surveys.

In general, the average value of declination changes from one year to the next, and the change usually continues in one direction for many years. This is called *secular change*. The amount of change in one year is called the *annual variation*. Unfortunately, there is sometimes an abrupt, unpredictable change in the rate of secular change. Thus, secular change can be determined only by observations at magnetic observatories. Information from such observations are available for each state from the earliest date of valid observations to the present, at 10-year intervals up to 1900, and at 5-year intervals thereafter. This information, which is often needed for rerunning old survey lines and reestablishing corner markers, may be obtained from the National Geophysical and Solar-Terrestrial Data Center (complete address given at close of this section). Requests should specify the point of interest, preferably giving latitude and longitude.

In most regions the above changes are gradual enough so that one can use the same declination throughout an area for an entire season. But in some regions *local disturbances* cause large differences within small areas—sometimes several degrees within a hundred feet. If these disturbances are caused by human interference, they are called *artificial disturbances*; otherwise, they are called *natural disturbances*.

Iron, steel, or direct-current electricity near the compass can also cause *artificial disturbances*. Unless precautions are taken, the observer's clothing may contain iron in belt buckles, zippers, glasses, and so on. The magnitude of this effect diminishes rapidly with increasing distance.

Natural disturbances of several degrees are usually the result of deposits of magnetite. Other ores and geological formations cause smaller irregularities. However, even in undisturbed regions, minor irregularities are the rule. Almost anywhere, the declination at two points 100 feet apart differs by a few minutes. Such disturbances are responsible for many of the shortcomings of the compass as a surveying instrument.

A compass survey should be made with the compass in good condition. Consequently, one should follow the suggestions in the instruction manual for the specific compass. Although a compass may be in good operating condition, it may have an appreciable *index correction;* that is, there may be an angle between the real magnetic north and the direction shown by the compass. This correction for a specific compass may be determined by observation at a magnetic station (i.e., a marked point on the ground where the magnetic and true meridians have been determined accurately). Several thousand of these stations have been established throughout the United States. The National Geophysical and Solar-Terrestrial Data Center (EDS/NOAA, Boulder, Colorado 80302) will provide the site description of the nearest magnetic station, if one specifies the location of interest.

Thoughtful study of this discussion should make one fully aware of the precautions that must be taken to attain reasonable accuracy in compass surveys, as well as the limitations of compass surveys.

3-5   THE U.S. PUBLIC LAND SURVEYS

The U.S. rectangular surveying system was devised to establish legal subdivisions for describing and disposing of the public domain under the general land laws of the United States. The system has been used in 30 states. It has not been used in the older states along the Atlantic seaboard and in a few states where the lands were in private hands before the federal government was founded. Many of the land holdings in states that do not use the U.S. rectangular surveying system are described by "metes and bounds" (*metes* means to measure or to assign by measure; *bounds* refers to a boundary); the length and direction of each side of the boundary are determined and the corners monumented. The descriptions of old surveys of this type are often vague and the corners difficult or impossible to relocate.

3-5.1 Units of Measurement

In the United States the *chain* (often called "Gunter's chain") is the unit of linear measure for the survey of public lands. A chain is 66 feet long and is divided into 100 links, each 0.66 foot or 7.92 inches long. In deeds of private lands, distances are sometimes recorded in poles, rods, or perches: 1 pole = 1 rod = 1 perch = 25 links = 16.5 feet. The unit of area for the survey of public lands is the acre. One *acre* equals 43,560 square feet or 10 square chains. Since 1 mile equals 80 chains, 1 square mile equals 640 acres.

3-5.2 The U.S. Rectangular Surveying System

The Ordinance of 1785 established the rectangular surveying system. The system provides for townships 6 miles square, containing 36 sections 1 mile square. In any given region the survey begins from an *initial point*. Through this point is run a meridian, called the *principal meridian*, and a parallel of latitude, called the *baseline*. With the establishment of an initial point, the latitude and longitude of the point is determined by accurate astronomical methods. Monuments are placed on the principal meridian and on the baseline at intervals of 40 chains. Since the baseline is a parallel of latitude, it is run as a curve with cords 40 chains in length (Fig. 3-4).

The next step in the subdivision of the district being surveyed is to run *standard parallels* or *correction lines*. These lines, which are parallels of latitude, are established in the same manner as the baseline. They are located at intervals of 24 miles north and south of the baseline and extend to the limits of the district being surveyed. Standard parallels are numbered as First, Second, Third, etc., Standard Parallel North, or South (Fig. 3-4).

The survey district is next divided into tracts approximately 24 miles square by means of guide meridians. These lines are true meridians that start at points on the baseline, or standard parallels, at intervals of 24 miles east and west of the principal meridian, and extend north to their intersection with the next standard parallel (Fig. 3-4). Because of the convergence of the meridians, the distance between these lines will be 24 miles only at the starting points. At all other points the distance between them will be less than 24 miles. Guide meridians are numbered as First, Second, Third, etc., Guide Meridian East, or West. Note that two sets of monuments are found on the standard parallels. The monuments that were set when the parallel was first located are called *standard corners*. They govern the area north of the parallel. The second set, found at the intersection of the parallel with the meridians from the south, is referred to as the *closing corners*. They govern the area south of the parallel.

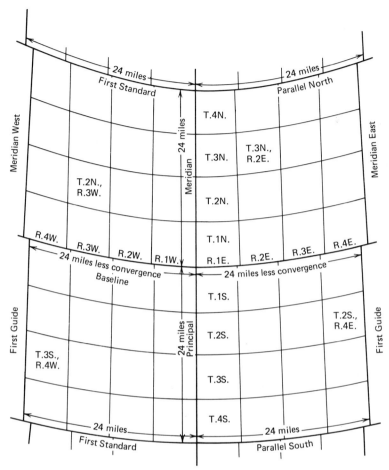

Fig. 3-4 *Standard parallels and guide meridians.*

The townships of a survey district are numbered meridionally into ranges and latitudinally into tiers from the principal meridian and the baseline of the district. As illustrated in Fig. 3-4, the third township south of the baseline is in tier 3 south. Since the word township is frequently used instead of tier, any township in this tier is often designated as township 3 south. The fourth township west of the principal meridian is in range 4 west. By this method of numbering, any township is located if its tier, range, and principal meridian are given, as Township 3 south, Range 4 west, of the Fourth Principal Meridian. This is abbreviated T. 3 S., R. 4 W., 4th P.M.

In subdividing a township into *36 sections*, the aim is to secure as many sections as possible that will be 1 mile on a side. To accomplish this, the error due to convergence of meridians is thrown as far to the west as possible by running lines parallel to the east boundary of the township, rather than running them as true meridians. Errors in linear measurements are thrown as far to the north as possible by locating monuments at intervals of 40 chains along the lines parallel to the east boundary of the

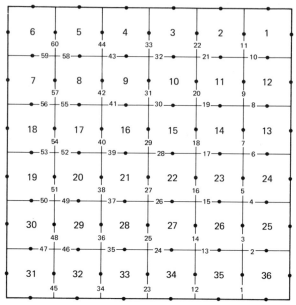

Fig. 3-5 *Subdivision of township.*

township, all the accumulated error falling in the most northerly half-mile, which may be more or less than 40 chains in length.

The system now used in numbering the sections of a township was established in 1796. This numbering system and the most recent order in which the lines are run to subdivide the township into sections (indicated by the numbers on the lines) are shown in Fig. 3-5. This system of subdividing a township throws the errors due to survey and losses from convergence into the extreme north and west sections. Other sections, however, may contain more or less than 640 acres due to survey errors. Nevertheless, the established boundaries are final, regardless of errors made in the original survey.

If any of the monuments of an original survey are missing, it is necessary for surveyors to know the methods used in that original survey as well as the principles that have been adopted by the courts, if they are to restore the missing corners correctly. (Procedures for relocating original survey lines and corners are given in the U.S. Department of Interior's "Manual of Instructions for the Survey of the Public Lands of the United States.")

After all the original monuments have been found or any missing ones have been replaced, the first step in the subdivision is the location of the center of the section. Regardless of the location of the section within the township, this point is always at the intersection of the line joining the east and west quarter-section corners with the line joining the north and south quarter-section corners. By locating these lines on the ground, the section is divided into quarter sections containing approximately 160 acres each.

The method of dividing these quarter sections into 40-acre parcels depends on the

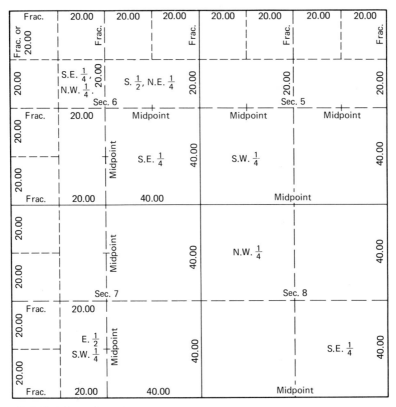

Fig. 3-6 *Subdivision of sections.*

position of the section within the township. For any section except those along the north and west sides of the township, the subdivision is accomplished by bisecting each side of the quarter section and connecting the opposite points by straight lines. The intersection of these lines is the center of the quarter section.

The method of subdividing the sections along the north and west sides of the township is shown in Fig. 3-6. The corners on the north–south section lines are set at intervals of 20.00 chains, measured from the south, the discrepancy being thrown into the most northerly quarter-mile. Similarly, the monuments on the east–west lines are set at intervals of 20.00 chains, measured from the east, the discrepancy being thrown into the most westerly quarter-mile.

The rectangular system of subdivision provides a convenient method of describing a piece of land that is to be conveyed by deed from one person to another. If the description is for a 40-acre parcel, the particular quarter of the quarter section is first given, then the quarter section in which the parcel is located, then the section number, followed by the township, range, and principal meridian. Thus, the 40-acre parcel labeled in Sec. 6 of Fig. 3-6 can be described as the S.E. $\frac{1}{4}$, N.W. $\frac{1}{4}$, Sec. 6, T. 3 S., R. 4 W., 4th P.M. The legal descriptions of other parcels appear in Fig. 3-6.

4
Time Measurement

Time is utilized in forestry (1) to denote the position in a continuum of time at which some event took place, (2) to measure the duration of a given event, and (3) to determine the speed or rate at which an event or physical change occurred.

The accurate measurement of time by establishing accurate time standards poses difficult technological problems. But any measure of time is ultimately based on counting the cycles of some regularly recurring phenomenon and accurately measuring fractions of a cycle. The earth rotates on its axis at a very nearly constant rate, and the angular positions of celestial bodies can be precisely determined. Thus, astronomical observations provide a good method of measuring time. The period of rotation of the earth with respect to the sun is the solar day, which is the basis of *solar time*. Since the length of the true solar day varies seasonally, true solar time, as measured by a sundial, does not move at a constant rate. Therefore, the mean solar day, with a length equal to the annual average of the actual solar day, was introduced as the basis of mean solar time. Mean solar time moves at an almost constant rate and is the basis of civilian time, which is kept by clocks. However, in reality, the earth's rotation is being slightly braked by tidal and other effects, so that even mean solar time is not strictly uniform. The ultimate standard for time is provided by the natural frequencies of vibration in atoms and molecules. Atomic clocks, based on masers and lasers, lose only about 1 second over periods of thousands of years. Therefore, the fundamental time unit that has been internationally accepted is the atomic second (Section 1-4.1).

4-1 TIME MEASUREMENT METHODS

Time-measuring instruments can be classified as astronomical, mechanical, and atomic. Astronomical instruments are used to obtain accurate star positions from which to calculate precise times. Mechanical or atomic instruments that indicate time are referred to as clocks. Clocks consist of (1) a part that vibrates or a part in which some repetitive event occurs at equal time intervals, and (2) a counter that indicates the total interval count by hands and a dial, or by digital display. A quartz-crystal clock is based on the frequency of vibrations of a quartz crystal. An atomic clock uses energy changes within atoms to produce regular waves of electromagnetic radiation that are counted.

A chronograph is a timepiece used to mark the exact instant of an occurrence.

Chronographs are illustrated by weather instruments (e.g., barographs, thermographs, hygrographs, and recording rain gauges) and instruments such as the dendrograph which records changes in the diameter of a tree with the passage of time. The typical chronograph includes a stylus and a drum that rotates with the passage of time. The stylus traces a line on a chart on the drum that indicates changes at given times.

The rate of disintegration of radioactive elements provides another means of measuring time; in particular, it provides a means of dating both natural and manufactured objects through analysis of their content of carbon-14, a radioactive isotope of carbon. Since 1947, when the method was first invented by W. F. Libby, impressive strides have been made in the use of Libby's method, which is based on measuring the number of beta particles emitted by the carbon-14 present in a given sample as it decays. In 1978, measurements with a precision for determining carbon-14 content to better than 0.2 percent (or 16 years) were reported. However, in 1977, a new technique was discovered that promises to revolutionize the field of radiocarbon dating by allowing direct detection of carbon-14 atoms, that is, radiocarbon dating with accelerators. The new method can analyze samples less than a milligram in size, a thousand times smaller than those required for conventional radiocarbon dating, and measurements can be made in hours rather than days for very old samples. Accelerator dating could be useful not only in the fields of anthropology and archaeology but in geology, where the possibility of analyzing smaller samples would allow researchers to focus on a single layer or stratum, a thing they cannot do with conventional dating. Also, accelerator radioisotope analysis has great potential in biology (Keilson and Waterhouse, 1978).

Dendrochronology provides another means of dating archaeological and geological events as far back as 3000 years and arranging them in order of occurrence. Basically, it is the study of growth rings in living trees and aged woods to establish a time sequence in the dating of past events. Dendrochronology has been used primarily for the analysis of growth rings from trees on sites where precipitation is the most limiting climatic factor affecting tree growth. Tree response is indicated by the width of annual rings—narrow rings when precipitation is low; wide rings when precipitation is high. Since climatic variations tend to occur over rather large regions, characteristically narrow rings in two or more trees can be matched, provided the trees grew at the same time. Recognition of this fact led to cross-dating, the foundation of dendrochronology. But it has been demonstrated that tree rings from selected trees may be affected by air pollution, as well as moisture conditions. Thus, dendrochronological studies may be useful in determining the date and spacial extent of air pollution.

4-2   TREE AGE AND GROWTH DETERMINATION

The age of a tree is the length of time that has elapsed since the germination of the seed or the budding of the sprout.

In certain species, branch whorls can be used to determine age. Each season's height growth starts with the bursting of the bud at the tip of the tree; this lengthens to form the leader. The circle of branchlets that grows out at the base of the leader marks the height of the tree at the very start of the season's growth. This process is repeated the following year, and a new whorl appears to mark the beginning of that season's growth. A count of these branch whorls thus gives the age of the tree. It is only in certain coniferous species, however, that whorls are well defined, and in old trees of these species the evidence of the former whorls can seldom be distinguished.

Annual rings afford the best method of determining tree age in temperate regions. Here most timber trees grow in diameter by adding a new layer of wood each year between the old wood and bark. The formation of this layer begins at the start of the growing season and continues through it. The woody tissue formed in the spring (spring wood) is more porous and lighter in color than the woody tissue formed in the summer (summer wood). Thus, the annual growths of the tree appear on a cross section of the stem as a series of concentric rings. A count of the number of rings on a given cross section gives the age of the tree above the cross section. Consequently, if the count is made on a cross section at ground level, the count gives the total tree age. If the count is made on a cross section above ground level, the number of years for the seedling to grow to the height of the cross section must be added to the ring count to obtain total tree age. (The number of years for a seedling to reach stump height varies from 1 year for hardwood sprouts to 20 years for coniferous seedlings growing on dry sites.)

In tropical and subtropical regions that have alternating wet and dry seasons, growth rings similar to those that occur in temperate regions are produced. These rings may be useful for determining tree age. However, in tropical regions that do not have a regular alternation in growing conditions, any rings that may be produced are useless for determining age.

Difficulties may be encountered in making ring counts. In slow-growing trees the rings may be so close together that they are hard to count. In some species, the difference in appearance between the spring and summer wood is not marked, and the rings are indistinct. In addition, abnormal weather during the growing season may lead in certain species to the formation of false growth rings—a dry spell may interrupt the annual growth, and the growth may resume when the rains come. Then, too, defoliation by insects may cause a tree to produce a second set of leaves and an additional growth ring for a given year. False rings, however, are not as clear as true growth rings and often do not extend around the entire circumference of a tree cross section.

Ring counts are often made on a sawed surface such as the stump, although such surfaces are generally too rough to permit the rings to be accurately counted. Consequently, the rings on a smoothed strip extending from the bark to the center of the section should be cut with a sharp knife or plane and viewed with a hand lens to facilitate counting. On standing trees, an *increment borer* may be used (Fig. 4-1). This consists of a hollow cutting bit that is screwed into the tree. The core of wood that is forced into the hollow center of the bit is removed with the extractor. The rings on this core are then counted. Unfortunately, total age of large trees is difficult to obtain with the increment borer because the maximum practical length of borer bits

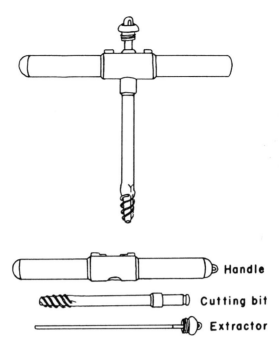

Fig. 4-1 *Increment borer.*

is about 16 inches and because of the difficulty of finding the pith. The chief use of the increment borer is therefore to study the growth of trees for the past few years.

Average radial growth may be determined from several increment cores (cores for this purpose are normally taken at breast height), but usually only one core is taken. When a single core is taken, it should be extracted halfway between the long and short diameters. The length of the core depends on the past period for which the growth measurement is desired. If growth for a past period is needed as a basis of growth predictions, the boring should include the number of rings in the period, usually 5 or 10 years. If growth for the life of the tree is needed, the boring must reach the pith.

4-2.1 Age of Even-Age Stands

Even-aged stands are stands in which all the trees are essentially of the same age. If a stand, such as a plantation, is absolutely even-aged, a count of the rings of a single tree will give the age. But stands originating through natural reproduction, which generally takes 5 to 15 years or more, contain trees of various ages. There are several conceptions of the age of such a stand. The average age of all the trees in the stand, as determined by sampling, is sometimes used. The objection to this concept is that younger trees should not be given the same weight as the large, older trees that make up the major part of the volume of the stand. The following method has thus been suggested: take the age of several sample trees whose volume is the average for the stand. A simpler concept, however, with many advantages, is to consider the age of the stand to be the average age of the dominant and codominant trees (i.e., the largest individuals).

Growth and yield of trees and stands are discussed in Chapters 15, 16, and 17.

4-3 TIME SERIES

A time series is concerned with data for which the essential problem is the analysis of chronological variation. It is important to distinguish between the problems arising in the analysis of a time series and those involving the organization and analysis of materials for which the time factor does not enter. In studying a time series the primary object is to measure and analyze the chronological variations in the value of the variable. Thus, one may study variations in temperature over a period of time, fluctuations in the production of pulpwood, or changes in prices paid for pulpwood. Quite different is the procedure used in the study of standing timber volume at a given time. In this case we might desire to know how much volume for a given forest falls in each of a number of species and diameter classes. Data of this sort, when organized, constitute a *frequency series*, as opposed to a *time* or *historical series*.

Here are some examples of time series that are important in the study of forestry.

1. Economic time series (prices of forest products on successive days, months, or years; exports or imports of forest products at given times).

2. Physical time series (rainfall, temperature, or relative humidity for different times or dates).

3. Marketing time series (sales or production figures for a forest industry over a period of years).

4. Performance time series (changes in the quality or quantity performance for a manufacturing process over time).

A time series is *continuous* if the variable can be measured at every point in time; it is *discrete* if observations can be made only at specific times.

Continuous time series are most readily obtained from observations of a physical variable by a mechanical recorder or chronograph (e.g., temperature by a thermograph, barometric pressure by a barograph, and relative humidity by a hygrograph). Discrete time series are generated if the variable can be observed only at certain times (e.g., volume of timber at rotation age), or if the variable exists only by aggregation over a period (e.g., monthly precipitation, annual paper production).

If a time series can be used to predict the exact future values of the variable of interest, it is *deterministic*. If future values of the variable have a probability distribution that is only partly determined by past values, the time series is *stochastic*. The analysis of time series is covered by Kendall (1973) and Chatfield (1975).

4-4 MOTION AND TIME STUDY

Forestry operations, whether silvicultural treatments, inventories, timber harvests, or lumber manufacturing, can be considered industrial processes. As in all industries, the guiding principle is to carry out the given tasks with the minimum of time, cost, and effort possible. Motion and time studies are used to accomplish these objectives. For a comprehensive treatment of the subject the reader is referred to Barnes (1963). Winer (1961) and Wright (1962) treat the subject as it applies to forest operations.

Motion and time study consists of studying a complete process and dividing it into its fundamental steps. The lengths of time necessary for the individual steps are then observed. Attempts are made to eliminate superfluous steps or motions and to minimize the work and time of the essential steps by improving the human contribution, by changing machines, or by rearranging the sequence of the steps.

A time study ideally obtains the amount of time necessary to accomplish a job, using a given method, under given conditions of work, by a worker possessing a specified amount of skill on the job, when working at a pace that will produce, within a unit of time, a specified physical effect upon the worker. The time obtained is called the *standard time*. But more often a time study obtains the amount of time necessary to accomplish a job, using a given method, under given conditions of work.

Time studies are used for the following purposes.

1. To set production schedules.
2. To determine operating effectiveness.
3. To set labor standards.
4. To compare methods.
5. To determine standard costs.
6. To determine equipment and labor requirements.
7. To provide a basis for setting incentive wages.

To determine the time devoted to a unit of work or a job, two procedures are available: timing and work sampling.

4-4.1 Timing

Time values for a time study are commonly recorded with a stopwatch. Ordinary watches may be used, but the time of the beginning and end of an event must be read to determine the elapsed time. Motion picture cameras and videotape recorders may be used in conjunction with time recording.

In timing a job, or an element of a job, the following occurrences should be noted.

1. A fumble or a false movement.
2. A machine adjustment or repair.
3. Faulty work due to low skill.

4. Faulty work due to poor equipment, the nature of the raw material (e.g., the tree or log), or the ground conditions.

5. Personal time out.

An occurrence should be handled as follows.

1. If it is not a necessary part of the job, such as personal time out, or if it represents work improperly done, the time value should be discarded and have no influence on the final result.

2. If it is inherent to the job, such as "fumbles" with a tangled choker, it should be allowed to remain in the study.

3. If it is an irregular occurrence, such as a machine adjustment or the erratic handling of an occasional log due to log defects, it should be evaluated separately and added to the final time standard in proportion to its rate of occurrence.

It is interesting to note that timing procedures may be combined with the measurement of variables affecting the time required to do a job. For example, one might wish to develop a relationship between dbh, total height, or crown length and the time to fell and buck trees. In the analysis of any logging operation the relationship between such tree factors and time may be sought (FAO, 1976). This relationship can be determined by a simple linear regression analysis using the equation $y = b_0 + b_1 d$, or a multiple regression analysis using the equation

$$y = b_0 + b_1 d + b_2 d^2 + b_3 h + b_4 l$$

where

$$\begin{aligned}
y &= \text{time required for job under study} \\
d &= \text{diameter at breast height} \\
h &= \text{total tree height} \\
l &= \text{crown length} \\
b_i &= \text{regression coefficients}
\end{aligned}$$

Fig. 4-2 shows the results of a time study by Jensen (1940).

4-4.2 Work Sampling

Work sampling consists of taking a series of short-interval observations during a production process at random time intervals and noting which step, or job, in the process is taking place. The observation intervals are short periods, usually 1 minute. (If one used 1-minute intervals for an 8-hour workday, there would be 480 possible time units for the workday.) The proportion of time taken for a given job in the total process p_i is then the ratio of the frequency of the observed number of intervals for a given job n_i to the total number of observations n.

$$p_i = \frac{n_i}{n}$$

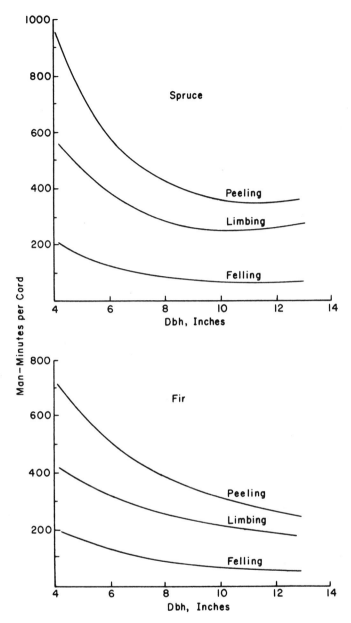

Fig. 4-2 *Production time for commercial clear-cutting by two people: felling, limbing, and peeling spruce and fir pulpwood from trees of different sizes. Based on 597 trees (Jensen, 1940).*

The distribution of n_i will follow a binomial distribution when the sample is large. The number of samples needed should be estimated for each job of the process. The job requiring the largest sample size should dictate the sample size.

Although work sampling is not as accurate as timing, it has some advantages over timing: it is less tedious, less time consuming, and less expensive. Lussier (1961) showed how work sampling could be utilized in forestry and gave an example of its application in determining the time of various steps in a log-hauling operation.

5
Weight Measurement

Weight is being used more and more for scaling traditional forest products (Guttenberg, 1973). In addition, the growing interest in complete tree utilization (roots, stump, branches, etc.), the use of residues from the manufacture of forest products, biomass quantities, fuel quantity in relation to forest fire conditions, and other topics, has increased the use and importance of weight measurement.

The force of attraction that the earth exerts on a body—that is, the pull of gravity on it—is called the *weight* of the body. Weight is often used as a measure of mass; however, the two are not the same. Mass is the measure of the amount of matter present in a body and thus has the same value at different locations. Weight varies depending on the location of the body in the earth's gravitational field (or the gravitational field of some other astronomical body). Since the gravitational effect varies from place to place on the earth, the weight of a given mass also varies.

The distinction between weight and mass is confused by the use of the same units to measure both—the gram, the kilogram, and the pound (Section 1-4).

The basic procedure of weighing at any locality consists of comparing an object of unknown weight with the weight of an object of known mass. Since the force of gravity at the place affects both objects identically, the weight thus determined is a relative value, independent of the gravitational force. Thus, an object would weigh the same at any locality, provided the standard used for comparison was of the correct mass.

5-1 INSTRUMENTATION

The weight (or more precisely, the force) of objects can be measured by any one of five general methods (Doebelin, 1966).

1. By balancing the object against the known gravitational force on a standard mass. (This method employs the well-known weighing machines: equal-arm balances, unequal-arm balances, and pendulum scales.)

2. By measuring the acceleration of a body of known mass to which the unknown force is applied. (This method which uses an accelerometer for force measurement is of restricted application since the force determined is the resultant of several inseparable forces acting on the mass.)

3. By balancing it against a magnetic force developed by the interaction of a current-carrying coil and magnet. (At present this method is used for the determination of weights of no more than 1 gram.)

4. By transducing the force to a fluid pressure and then measuring the pressure. (This method is exemplified by a container filled with a liquid, such as oil, under a fixed pressure. Application of a load increases the liquid pressure that can then be read on a gauge. Instruments of this type can be constructed for measuring large weights.)

5. By applying the force to some elastic member and measuring the resulting deflection. (This method permits the measurement of both static and dynamic loads while the previously described methods are restricted to static or slowly moving loads. By using deflection transducers, the force applied to some elastic element will cause it to move, and this motion can then be transformed to an electrical signal. The various devices differ principally in the form of the elastic element and in the displacement transducer which generates the electrical signal. The movement of the elastic element may be a gross, ocularly perceptible motion, or it may be a very small motion that requires the use of strain gauges to sense the force.)

All of the above methods of weight measurement may be applied to some aspect of forestry. However, for weighing large, bulky quantities of wood, weighing machines have been, and will continue to be, most widely used.

5-2 THE USE OF WEIGHT AS A MEASURE

Weighing may be necessary if there are no practical alternatives; it may be desirable if its advantages are greater than those of alternatives. But basically a decision to use weight measurement will rest on these factors.

1. PHYSICAL CHARACTERISTICS OF THE SUBSTANCE. The volume of pulpwood, coal, soil, seed, fertilizer, and other solids that occur as irregular pieces can be measured by filling a space or container of known volume. However, the ratio of air space to solid material will vary with the shape, arrangement, and compaction of pieces. This can seriously affect the accuracy and reliability of the volume measurement. Therefore, weight is a better measure than volume for solids that occur in irregular pieces.

2. LOGICALNESS OF WEIGHT AS AN EXPRESSION OF QUANTITY. Weight may be the most useful and logical expression of quantity. For example, the weight measurement of pulpwood is supplanting cord and cubic measurement because it can be done rapidly and accurately, and because the derived product, pulp, is expressed in weight. Indeed, when weight is the ultimate expression of quantity, it is logical to apply it consistently from the beginning. (Since transportation charges for forest products are based primarily on weight, weight becomes an even more logical measure of quantity.) Furthermore, a material object may be composed of several components. For example, a tree consists of roots, stem, and crown; each part is composed of additional units. The ratio of any component to the whole, such as bark

to wood in the merchantable stem, can best be expressed in terms of weight. And when the entire tree is of interest (roots, branches, etc.), weight is the most logical measure of quantity.

   3. FEASIBILITY OF WEIGHING. The substance in question must be physically separable from material not relevant to the measurement, or weighing is not feasible. In addition, a weighing device must be available. For example, it might be desirable to determine the weight of the merchantable stem of a standing tree. But unless one felled the tree, separated the merchantable stem, and had a weighing machine available, it would be impossible to directly determine the weight.

   4. RELATIVE COST OF WEIGHING. Weight might be a better expression of quantity than volume, but the costs of weighing might be greater than the costs of volume estimates, and vice versa. The final decision depends on the value of the material to be measured. When the value of material is high, the costlier procedure may be better if it is more accurate.

5-3 FACTORS INFLUENCING WOOD WEIGHT ESTIMATES

It is comparatively easy to determine the gross weight of wood, but estimates of dry weight are more difficult because the gross weight is affected by (1) *density*, (2) *moisture content*, and (3) *bark and foreign material*.

5-3.1 Density

The density of a substance is its mass per unit volume. In the International System of Units (SI) density is expressed in kilograms per cubic meter, or grams per cubic centimeter. In the English system density is expressed in pounds per cubic foot. Wood density is generally expressed as the dry mass of wood substance per unit volume. Since wood density will vary with the moisture content, it is essential to specify the moisture conditions at which volume was determined.

   The density of water under standard conditions is approximately 62.4 pounds per cubic foot, or 1 gram per cubic centimeter (1000 kilograms per cubic meter). These figures are standards to which the densities of solids and liquids are compared. When the density of a substance is divided by the density of water, the quotient is the specific gravity.

$$\text{Specific gravity} = \frac{\text{Density of liquid or solid}}{\text{Density of water}} \qquad (5\text{-}1)$$

$$= \frac{\text{Weight of liquid or solid}}{\text{Weight of equal volume of water}}$$

For example, if the oven-dry weight of white oak is 46.8 lb/ft³ or 750 kg/m³, the specific gravity, sg, is $46.8/62.4 = 750/1000 = 0.75$.[1] Thus, the specific gravity of wood is the ratio of the oven-dry weight of a given volume of wood to the weight of an equivalent volume of water. There are, however, three different specific gravities, depending on the manner of determining wood volume: current-volume specific gravity, sg; dry-volume specific gravity, sg_d; and wet-volume specific gravity, sg_s. The formulas for determining the three types of specific gravity are

$$sg = \frac{W_d}{W_{vg}}$$

$$sg_d = \frac{W_d}{W_{vd}}$$

$$sg_s = \frac{W_d}{W_{vs}}$$

where

W_d = over-dry weight of a sample of green wood (the sample is dried at $103 \pm 2°C$ until it reaches a constant weight)

W_{vg} = weight of water equivalent to the volume of a sample of green wood

W_{vd} = weight of water equivalent to the volume of a sample of oven-dry wood

W_{vs} = weight of water equivalent to the volume of a sample of saturated wood (wood is saturated with free water and bound water)

Unless otherwise specified, the term specific gravity means current-volume specific gravity. Procedures for determining specific gravity are given in Forbes and Meyer (1955) and Besley (1967).

The cell wall substance of woody plants is quite uniform; the specific gravity is about 1.5. Solid wood with no air spaces would thus weigh about 93.6 lb/ft³ or 1500 kg/m³. However, the density of wood never attains this figure because wood is a porous structure. The specific gravities of commercial woods in North America range between 0.3 and 0.8. (For specific gravities of commercial woods in the United States, see *Wood Handbook No. 72, Rev.*, U.S. Forest Products laboratory, 1974.) The range of specific gravities of woods from other parts of the world is somewhat wider. One should note that there is considerable variation in density between trees of the same species.

For most species there is a tendency for the specific gravity to decrease from base to tip of stem. It appears that this variation is small in comparison to the difference

[1] To convert lb/ft³ to kg/m³, multiply by 16.0185. To convert kg/m³ to lb/ft³, multiply by 0.0624280.

between trees. The density of the wood in the cross sections of stems tends to increase from the pith to the cambium. Large variations of specific gravity within individual annual rings have also been found.

5-3.2 Moisture Content

The moisture content of wood varies by species, by location in the tree, and by length of time since cutting. The practical determination of this variability constitutes a major problem in the use of weight as a measure of wood quantity. Moisture occurs in wood as free water in cell cavities and as absorbed water in the cell wall. The condition that exists in a cell when the cell cavity contains no free water and the cell wall is saturated with bound water is known as the *fiber saturation point*. Its value will vary from 27 to 32 percent of the dry weight for most species. It is customary to express the moisture content of wood as a percentage of dry weight. The dry weight is obtained by drying a sample of wood at $103° \pm 2°C$ until a stable weight has been reached.[2] The percentage is calculated from

$$MC_d = \frac{W_w - W_d}{W_d} (100) \qquad (5\text{-}2)$$

where

MC_d = moisture content as a percentage of oven-dry weight
W_w = green weight of wood
W_d = oven-dry weight of wood

In the pulp and paper industry, moisture content is often expressed as a percentage of the wet weight.

$$MC_w = \frac{W_w - W_d}{W_w} (100) \qquad (5\text{-}3)$$

where

MC_w = moisture content as a percentage of wet weight

For example, if the weight of a wood sample before drying is 155 grams, and the weight after drying is 85 grams, the moisture percentages are

$$MC_d = \frac{155 - 85}{85} (100) = 82.3\%$$

$$MC_w = \frac{155 - 85}{155} (100) = 45.1\%$$

---

[2] There are numerous methods of determining moisture content: oven drying, vacuum drying, distillation, electric moisture meters, electric hygrometers, microwave absorption, nuclear magnetic resonance, and nuclear radiation.

Moisture percentages of wood vary with location in standing trees (Besley, 1967; Nylinder, 1967; and Koch, 1972). The moisture percentage in the heartwood of conifers is usually appreciably lower than in the sapwood; in hardwoods there is variation, but it is less pronounced. The wood in the upper portion of conifers usually has a higher moisture percentage than in the lower sections because of the larger proportion of sapwood. Again, there are differences in moisture percentages at varying heights on the stems of hardwoods, but they are less pronounced. There are also variations in moisture content by seasons; studies indicate a higher moisture content in winter and spring than at other times of the year (Marden, Lothner, and Kallio, 1975).

When the tree is cut down and cut into logs, the wood immediately begins to lose moisture. If permitted to air-dry, its moisture content will reach about 12 percent. The rates of drying for logs vary with air temperature, humidity, species, log size, knottiness, method of stacking, presence or absence of bark, and other factors. Besley (1967), Nylinder (1967), and Adams (1971) reported varying weight losses over time for several species in North America and Scandinavia. The results of Adam's study to determine the weight loss of red-oak logs are shown in Fig. 5-1.

5-3.3   Bark and Foreign Material

If only the weight of wood is desired, bark and foreign material such as ice, snow, mud, and rocks must be removed before measurement, or their weight must be deducted from the gross weight to obtain the weight of the wood. Green bark was found to be between 9 and 19 percent of the weight of green, rough (i.e., unpeeled) hardwood pulpwood in Maine (Hardy and Weiland, 1964). Oven-dry bark was found to be between 3.8 and 9.1 percent of the weight of green, rough logs for several species in Minnesota (Marden et al., 1975).

5-4   WEIGHT MEASUREMENT OF PULPWOOD

The use of stacked measure (e.g., cord or stere) has long been used to measure bulk forest products, such as fuelwood and pulpwood (Section 7-3). Although stacked measure is not an accurate measure of the solid cubic contents, when measurements must be made in the woods, particularly at scattered locations, it is often a necessary compromise. However, if measurement can be done on trucks, and weighing facilities are available, weighing provides a cheaper, more accurate, and more objective method of scaling.

The green weight and the moisture content of wood provide good measures of the wood available for pulp. In spite of some problems in obtaining moisture content, weight as a measure of pulpwood quantity has several advantages: (1) It permits an accurate determination of the yield of pulp, which is measured in weight; (2) it

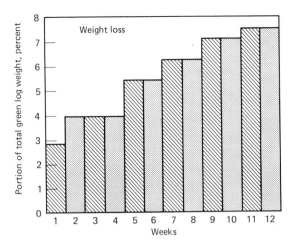

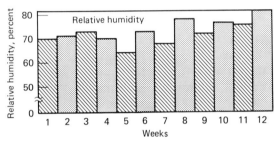

Fig. 5-1 *Cumulative weekly weight loss due to moisture loss for 21 red-oak sawlogs, with associated relative humidity record (Adams, 1971).*

encourages wood suppliers to promptly deliver wood to the mill to reduce weight loss (fresh, green pulpwood is preferred at the mill); (3) it encourages better loading of trucks since there is no advantage in loose piling; (4) it is faster and more economical than volume scaling; and (5) it eliminates personal judgment.

When weight scaling is employed, a dependable procedure for obtaining moisture content is essential. If unpeeled wood is being handled, estimates of bark weight must also be made. There are two ways to deal with these problems.

1. Develop average moisture and bark percentages for green wood at the time of cutting and at intervals of time since cutting.

To use the percentages for reducing fresh, green-wood weight to dry-wood weight, it is best that weighing be done immediately following cutting. In practice, the conversion from green weight with bark to dry weight without bark is often not made. Instead, the wood is weighed immediately after cutting, and prices are based on this weight. Although the quantity of dry wood is not explicitly determined, it is implicit in its effect on the price per unit weight. Some moisture percentages for several species used for pulpwood in the southern and northeastern parts of the United States are shown in Table 5-1. The percentages shown are based on freshly cut green wood without bark. When weighing is done at varying times following cutting, moisture and bark percentages are required at invervals of time since cutting. Percentages for the moisture component are more variable and less reliable than those for fresh green wood since season, weather, method of stocking, and so on, will affect moisture content.

2. Determine current moisture and bark percentages at the time of weighing.

This approach requires that an estimate be made, at the time of weighing, of the contribution of moisture and bark to the total weight. Sampling procedures must be utilized to provide these estimates. When moisture percentages have been determined, the dry weight of wood can be calculated using equation 5-2 or 5-3. The relationship expressed by equation 5-2 is shown in Fig. 5-2. Nylinder (1958) found that sample disks cut 10 centimeters from the ends of logs gave satisfactory estimates of the moisture content of logs. Braathe and Okstad (1967) found that a thin triangular segment of the log cross section cut with a chain chipper, or a sample taken from the log radius with a drill, gave satisfactory estimates of the moisture content of logs. Electrical moisture meters are available but they have not been too satisfactory for moisture determinations of pulpwood or sawlogs.

Since about 1950, weight scaling of pulpwood has been used by more and more of the larger companies in North America and northern Europe. Large platform scales

Table 5-1
Sample Weights and Moisture Contents of Some Speices Used for Pulpwood
(all weights are per cord)

Species	Weights per Cord Green (pounds)			Oven-Dry Weight (pounds), Barked Wood	Percent of Moisture Content, Barked Wood
	Unbarked	Bark	Barked		
Longleaf pine	6374	660	5714	2920	95.7
Shortleaf pine	5669	675	4994	2037	145.1
White ash	5031	795	4236	2992	41.5
Beech	5584	446	5138	3157	62.7
Grey birch	5690	677	5013	2700	85.6
White birch	5731	705	5026	2788	80.3
Yellow birch	6090	786	5304	3013	76.0
Elm	5857	763	5094	2627	93.9
Red maple	5482	720	4762	2877	65.5
Sugar maple	5977	801	5176	3178	62.0

SOURCE: Derived from Taras (1956) and Swan (1959).

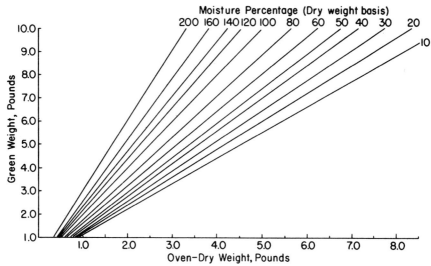

Fig. 5-2 *Relationship of green weight, moisture percentage, and oven-dry weight of wood.*

capable of determining the gross weight of a truck and its load are commonly used. The gross weight of the truck and load minus the weight of the empty truck, the *tare weight*, equals the weight of wood, including bark and moisture. The dry weight of wood can be estimated by applying corrections for moisture and bark.

Since pulpwood quantity in terms of stacked volume may be desired after the wood has been weighed, estimates of weight per cord are available (Table 5-1).

5-5 WEIGHT MEASUREMENT OF SAWLOGS

Weight scaling of sawlogs can be used (1) to estimate the amount of lumber, veneer, etc., in logs, and (2) to estimate the total quantity of wood in logs. In both cases the determination of weight is an intermediate step in the estimation of volume, because the wood-using industries, particularly in North America, customarily measure sawlogs in volume units. Weight scaling of sawlogs, therefore, normally consists of weighing and converting the weight to volume. The same problems with moisture content, bark, and foreign material that pertain to weight scaling of pulpwood pertain to weight scaling of sawlogs.

Generally, weights of entire truckloads are determined and converted to volume equivalents, although weights of individual logs can be determined. Weight scaling in truckload lots is most suitable for logs of a single species, with uniform diameter, uniform length, and uniform quality. This is the reason weight scaling has been most widely used for southern pine in the southeastern United States.

Weight scaling of truckloads eliminates the need for measurement of individual logs, speeds up scaling, and reduces errors of judgment. It also encourages the delivery of freshly cut logs, since price is almost always based on green weight. When logs are variable in size and quality, however, prices must be adjusted for these factors. Consequently, objections have been raised to weight scaling of hardwood logs, because they vary more in size, shape, and quality than softwood logs.

In North America weight scaling of sawlogs has generally been viewed as a more rapid method of obtaining volume estimates than conventional log scaling. Thus, the main efforts have been to develop volume-weight relationships. Relationships of this type can be developed from studies that relate volumes of a sample of sawlogs to their weights (volumes may be estimated from log rules or obtained from mill studies). Guttenberg, Fassnacht, and Siegel (1960) developed a regression equation showing the relationship of the board-feet mill tally to weight in pounds for individual shortleaf and loblolly pine sawlogs in Arkansas and Louisiana (Table 5-2). Yerkes (1966) developed a regression equation showing the relationship of cubic feet to weight in pounds for Black Hills ponderosa pine sawlogs: cubic foot volume $= -2.09 + 0.020W$, where W is total log weight in pounds.

For industrial application of sawlog weight scaling, it is desirable to develop relationships of volume to weight for truckloads of logs (Row and Fasick, 1966; Row and Guttenberg, 1966; Timson, 1974; Adams, 1976; and Donnelly and Barger, 1977). Donnelly and Barger found that the accuracy of conversion from weight to volume was improved by including number of logs per load as an additional independent variable.

In applying truckload weighing, sampling procedures are generally used to determine, or adjust, the conversion factors. The number of truckloads to measure is first determined for an acceptable sampling error. Each load sampled is chosen randomly or systematically, weighed, and scaled log by log for volume. Average volume-weight ratios are then calculated for converting weight to volume for all the truckloads in the operation (Chehock and Walker, 1975; Adams, 1976; and Donnelly and Barger, 1977).

5-6 WEIGHT MEASUREMENT OF PULP

In the United States pulp is usually measured in air-dry tons (English system). The air-dry ton consists of 90 percent, or 1800 lb, of dry pulp, and 10 percent, or 200 lb, of water. The oven-dry ton consists of 2000 lb of oven-dry pulp. The air-dry metric ton consists of 90 percent, or 900 kg, of dry pulp, and 10 percent, or 100 kg, of water. The oven-dry metric ton consists of 1000 kg of oven-dry pulp.

To convert the oven-dry weight of pulp to cubic volume of wood equivalent, the oven-dry weight of pulp is divided by the yield factor to obtain the oven-dry weight of wood. Then this figure is divided by the species density (oven-dry weight per green cubic unit) to give the green volume equivalent. For example, assume that we desire to determine the number of cubic feet of solid green wood of a given species that is

Table 5-2
Board-Foot Lumber Yields from Loblolly and Shortleaf Pine

Log Weight (pounds)	Predicted Green Lumber Yield (board feet)
200	9
600	51
1000	94
1400	134
1800	184
2000	207
2200	231
2600	278
3000	328

SOURCE: Guttenberg et al., 1960.

Based on the equation:

$$\text{Board feet} = [(\text{Weight, lb})/9.88] + [(\text{Weight, lb})^2/254{,}362] - 10.96$$

required to produce 1 air-dry ton (2000 lb) of unbleached kraft pulp. Assuming the yield of oven-dry unbleached kraft pulp is 54 percent of oven-dry wood, and that the oven-dry weight is 25 lb per green cubic foot, then

$$\text{Oven-dry weight of raw material} = \frac{1800}{0.54} = 3333 \text{ lb}$$

$$\text{Green volume equivalent} = \frac{3333}{25.0} = 133 \text{ ft}^3$$

To express the wood requirement in cords, one simply divides the cubic-foot volume by the appropriate number of solid cubic feet per cord (Section 7-4). For example, assume that a cord of unpeeled pulpwood contains 80 solid cubic feet of wood. The cord equivalent for the above example would be: 133/80 = 1.66 cords per air-dry ton of pulp.

5-7 WEIGHT MEASUREMENT OF OTHER FOREST PRODUCTS

The weights of sawn lumber, plywood, veneer, wood chips, and sawmill residues may be desired. Such weights may be determined directly by weighing, or indirectly by estimating volume and converting to weight.

The problem of indirectly determining weight can be illustrated by an example. Assume that we desire to know the weight of a thousand board feet of rough 2 by 4 inch lumber with a moisture content of 18 percent. Since the cubic foot equivalent of 1000 board feet of this lumber is 64.8, and the density of this lumber at 18 percent moisture content is 35.7 lb/ft³, the estimated weight per thousand board feet is (64.8)(35.7) = 2313 ib. The weights of other products can be estimated in a similar manner.

5-8 TREE WEIGHT RELATIONSHIPS

The increasing use of weight to measure forest products has developed a need to estimate the weight of wood in standing trees. Since the early 1960s, studies have been conducted to estimate (1) weight of the merchantable portion of a standing tree, and (2) weight of the complete tree.

Husch (1962) developed tree weight relationships to estimate the oven-dry weight of wood in the merchantable stem of standing white pine trees in southeastern New Hampshire (Fig. 5-3). A similar study of the green weight of red pine in New York yielded the results shown in Table 5-3. Other studies (Schroeder, Taras, and Clark, 1975) yielded the predicted weights of the stem to different top diameters, and the predicted weights of bark residue, sawdust, lumber, and chippable residue.

The green or oven-dry weight of wood in a stand can be estimated in two ways. (1) Obtain the volumes of individual trees from conventional volume tables, or from individual stem measurements, and convert to weight using an appropriate weight-volume relationship. (2) Obtain the weights of individual trees directly (Table 5-3 and Fig. 5-3).

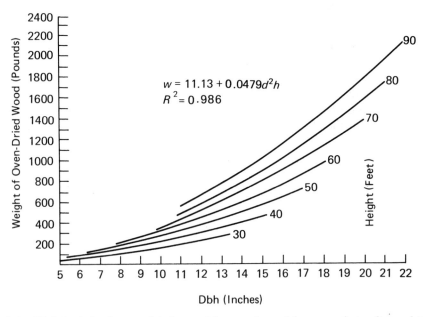

Fig. 5-3 *Weight (w) of oven-dried wood in merchantable stem of standing white pine trees according to dbh (d) and total height (h). Weight includes stem without bark from stump to top dob of 3 inches (Husch, 1962).*

Table 5-3
Weight in Pounds of Green Wood in Merchantable Stem of Red Pine to a 4-Inch Top

	Total Tree Height in Feet					
Dbh in Inches	16	24	32	40	48	56
6	122	168	215	261	307	
8	194	276	358	441	523	605
10		415	543	672	800	929
12			769	954	1139	1324
14			1037	1288	1540	1792
16			1345	1674	2002	2331

Weight $= 29.5872 + 0.16055$ dbh^2h.
$R^2 = 0.946$; $SE = 49.416$.
Line indicates extent of basic data.

SOURCE: Adapted from Cody, 1976.

Young (1964) pointed out that the increasing value of wood as a raw material makes it desirable to recognize the *complete tree*. *Tree biomass*, which is used to denote the total quantity of material in a tree, can best be measured in terms of weight.

In past years only the merchantable stems of trees were taken from the woods.

Table 5-4
Example of Proportional Weights of a Tree in 8 Components

	Weights[1]			
	Red Spruce 12-Inch dbh, 70 Feet Total Height		Red Maple 12-Inch dbh, 70 Feet Total Height	
Tree Component	Pounds	Percent	Pounds	Percent
Roots less than 1 inch in diameter	55	3	62	3
Roots from 1 to 4 inches diameter	115	6	96	5
Roots larger than 4 inches to base of stump	115	6	115	6
Stump—from 6 inches above ground to large roots	109	5	159	8
Merchantable stem from stump to 4 inches upper diameter	1218	60	1224	63
Branches larger than 1 inch diameter	76	3	109	6
Branches smaller than 1 inch diameter including leaves	320	16	118	6
Stem above merchantable portion	20	1	58	3
Total tree	2028	100	1941	100

[1] Weights are based on moisture conditions when freshly cut and include bark.

SOURCE: Young et al., 1964.

Table 5-5
Summary Table Showing Biomass Data by Species Group for
Two-Diameter Intervals from the Bureau of Public Lands Biomass Inventory for All Districts Combined

Species Group	Stems per Hectare			Biomass/Hectare/Metric Tons			Total Biomass/Metric Tons		
	Under 13 cm dbh	Over 13 cm dbh	Total	Under 13 cm dbh	Over 13 cm dbh	Total	Under 13 cm dbh	Over 13 cm dbh	Total
Industrial tree species	7,266	630	7,896	12.15	109.44	121.59	768,290	6,922,983	7,690,973
Nonindustrial tree species	817	16	833	1.23	1.73	2.96	77,918	109,712	187,630
Woody shrubs	3,962	0	3,962	0.69	0.00	0.69	43,932	0	43,932
Total	12,045	646	12,591	14.07	111.17	125.24	890,140	7,032,395	7,922,535

Number of hectares = 25,606; number of plots = 597.
SOURCE: Young, 1978.

This is also true for many present-day operations. The unutilized material in roots, stumps, and branches constitutes a sizable portion of the total volume of cut trees. Young, Strand, and Altenberger (1964) have shown the proportions of the entire tree made up of the various components (Table 5-4).

Taras and Clark (1977) and Crow (1978) have developed equations to express the relationships among weight Y of the individual components of a tree and diameter breast height d and total tree height h. Taras and Clark's model for longleaf pine is

$$\log_{10}Y = b_0 + b_1 \log_{10}(d^2h)$$

Table 5-6
Yield of Dry Wood per Acre in Tons
by Stand Density, Age Class, and Site Index

Age Class	Site Index	Stand Density—Basal Area per Acre (square feet)				
		60	80	100	120	140
	70			9.95	12.15	14.39
15	80		10.76	13.74	16.79	19.88
	90	10.43	14.31	18.28	22.32	26.44
	70	8.66	11.88	15.17	18.53	21.95
20	80	11.97	16.41	20.96	25.60	30.32
	90	15.91	21.82	27.88	34.05	40.33
	70	11.16	15.30	19.55	23.87	28.28
25	80	15.42	21.14	27.00	32.99	39.07
	90	20.50	28.11	35.91	43.87	51.95
	70	13.21	18.12	23.14	28.27	33.48
30	80	18.25	25.03	31.97	39.05	46.25
	90	24.28	33.29	42.52	51.94	61.51
	70	14.91	20.44	26.11	31.89	37.77
35	80	20.59	28.24	36.07	44.06	52.18
	90	27.39	37.55	47.97	58.60	69.40
	70	16.32	22.37	28.58	34.91	41.35
40	80	22.54	30.91	39.49	48.23	57.12
	90	29.98	41.11	52.52	64.15	75.97
	70	17.51	24.00	30.66	37.46	44.36
45	80	24.19	33.17	42.37	51.75	61.29
	90	32.17	44.11	56.35	68.83	81.51
	70	18.53	25.39	32.44	39.63	46.93
50	80	25.59	35.09	44.82	54.75	
	90	34.03	46.66	59.61		

$\text{Log}_e U = 1.0974 \log_e (BA) + 2.4208 \log_e (SI) - \dfrac{25.333}{(A)} - 11.352$, where U = weight yield in tons/acre, BA = basal area, SI = site index, A = stand age. $R^2 = 0.7866$, $s_{y.123} = 0.3484$.

SOURCE: Zobel et al., 1969.

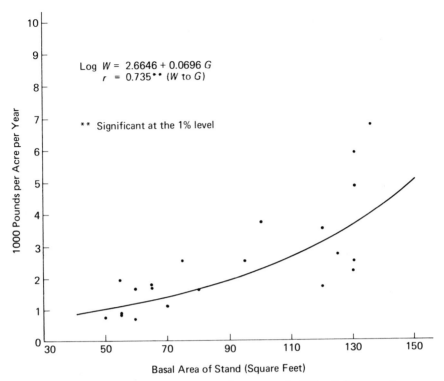

Fig. 5-4 *Growth of loblolly pine expressed as pounds (W) of dry wood per acre per year in relation to basal area of stand in square feet (G) (Zobel et al., 1965).*

Crow's model for tropical hardwoods is

$$\log_e Y = b_0 + b_1 \log_e(d^2h)$$

In recent years some attention has been given to the total productivity of forestland in terms of tree biomass per unit area. Young (1978) has advocated estimation of the total biomass of shrubs and trees per unit area. Table 5-5 gives the partial results of such a biomass inventory of 25,606 hectares in Maine.

Other tree weight studies have concentrated on estimating the weight of the crown components of trees and stands (Boyer, 1968; Olson, 1971; Brown, 1976; and Wartluft, 1977) and the weight of forest fuel for fire hazard determination (Alexander, 1978).

5-9 YIELD AND GROWTH STUDIES

There is some interest in expressing the yield of stands in weight instead of volume. Zobel, Roberds, and Ralston (1969) studied the effects of site, age, and stand density on wood yields in oven-dry weight for loblolly pine (Table 5-6). Maeglin (1967) used weight to express quantity in a study of the effect of spacing on yields of plantation-grown red and jack pine.

Only a limited number of tree and stand growth studies have used weight. An interesting study by Zobel, Ralston, and Roberds (1965) is summarized in Fig. 5-4.

6
Area Measurement

Area is the measure of the size of a surface region, usually expressed in units that are the *square* of linear units—for example, square feet or square meters. In elementary geometry, formulas for the areas of simple plane figures and the surface areas of simple solids are derived from the linear dimensions of these figures. Examples are given in Appendix Tables A-2 and A-3. The areas of irregular figures, plane or solid, can be computed by the use of coordinates and integral calculus. Also, the areas of simple plane figures or of irregular figures can be obtained or closely approximated by the use of dot grids, weighing, or planimeters.

6-1 AREA BY COORDINATES

When coordinates are used, the area of a polygon of any shape can be computed. Given the direction and distance between two points, A and B (Fig. 6-1), the X and Y components of the line can be calculated. The north-south vector is Y and the east-west vector is X. The coordinates of all vertices of Fig. 6-1 can be obtained by addition or subtraction of successive vectors, and the area of the figure, which must be closed, can be calculated by the continuous product method. If the X coordinates are designated as $X_1, X_2, X_3, \ldots, X_n$ and the Y coordinates as $Y_1, Y_2, Y_3, \ldots, Y_n$, then the area of the polygon whose vertices are $(X_1, Y_1), (X_2, Y_2), (X_3, Y_3), \ldots, (X_n, Y_n)$ is

$$\text{Area} = \frac{1}{2}[(X_1Y_2 + X_2Y_3 + \cdots + X_nY_1)$$

$$- (X_2Y_1 + X_3Y_2 + \cdots + X_1Y_n)] \tag{6-1}$$

Using the coordinate values given in Fig. 6-1, the area of the polygon is

$$\text{Area} = \frac{1}{2}\{[(30)(20) + (80)(70) + (150)(110) + (120)(90)$$

$$+ (50)(40)] - [(80)(40) + (150)(20)$$

$$+ (120)(70) + (50)(110) + (30)(90)]\}$$

$$= \frac{1}{2}\{35,500 - 22,800\}$$

$$= 6350 \text{ square units}$$

This method is frequently used to calculate land areas.

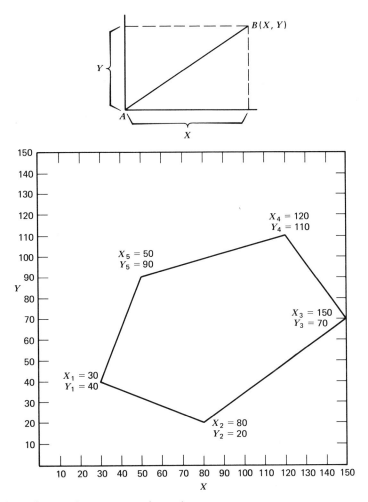

Fig. 6-1 *Coordinates for vertices of a polygon.*

6-2 AREA BY INTEGRATION

If a plane surface is bounded by known functions, its area can be found by solution of the definite integral

$$A = \int_a^b f(X)\, dX \quad \text{or} \quad \int_c^d f(Y)\, dY \tag{6-2}$$

The simplest illustration is the area under a curve plotted from a known function and bounded by the X axis and ordinates at $X = a$ and $X = b$. Areas bounded by intersections of curves or straight lines of known functions can be found in this way by setting up the appropriate $f(X)$ or $f(Y)$ to define the area of interest. In forest mensuration, the method is useful when the function can be shown on a graph and the enclosed area is geometrically related to some variable of interest. One application is the calculation of the total volume of a fully regulated forest from an even-aged yield relationship expressible in equation form. The total volume in a fully regulated forest of rotation age n (or n units or area) would be the area under the curve of the yield function bounded by the ordinates at the initial stage n_0 and rotation n. Curtis (1967) evaluated the area under a curve, setting up the expression $\int_{A_0}^{A}$ (increment function) dA, to obtain cumulative cubic volume growth by age A for Douglas fir.

The plane area of a figure represented in polar coordinates can also be determined by integration. The plane area bounded by a curve $\rho = f(\theta)$ between the radius vector $\theta = \theta_1$ and $\theta = \theta_2$ is given by

$$A = \frac{1}{2} \int_{\theta_1}^{\theta_2} \rho^2 \, d\theta \tag{6-3}$$

Matérn (1958) used several functions in polar form to describe the contour of a tree's cross section.

In the plane area defined by a continuous curve, expressed as $Y = f(X)$, the X axis and the ordinates $X = a$ and $X = b$ can be revolved about the X axis to generate a solid of revolution (Section 7-1). The surface area of this solid, excluding the ends, can be determined from

$$S_x = 2\pi \int_a^b Y \sqrt{1 + \left(\frac{dY}{dX}\right)^2} \, dX \tag{6-4}$$

If the curve is $X = f(Y)$ and it is revolved about the Y axis, the surface area between abscissa values c and d is

$$S_Y = 2\pi \int_c^d X \sqrt{1 + \left(\frac{dX}{dY}\right)^2} \, dY \tag{6-5}$$

Since tree stems approach the forms of solids of revolution, this method is useful in approximating the total surface or cambial area of a tree stem (see Section 6-7).

6-3   AREA BY APPROXIMATE INTEGRATION

When it is impossible, or difficult, to integrate by means of elementary functions, the area of a plane figure may be approximated by the *trapezoidal rule* or by *Simpson's rule*. In applying these rules, we form trapezoids like the ones shown in Fig. 6-2 for which h is equal. Since the area of a trapezoid is the product of one-half the sum of the parallel sides and the altitude, the total area is the sum of the areas of the individual trapezoids.

$$\text{Area} = \left(\frac{1}{2}Y_0 + Y_1 + Y_2 + \cdots + Y_{n-1} + \frac{1}{2}Y_n\right)h \tag{6-6}$$

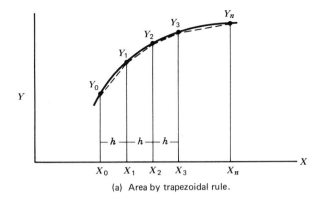

(a) Area by trapezoidal rule.

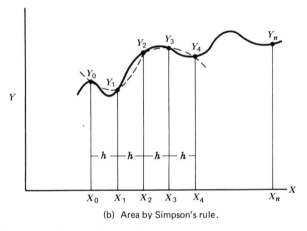

(b) Area by Simpson's rule.

Fig. 6-2 *Area by approximate integration.*

The greater the number of intervals, the smaller h will become, and the closer the sum of the areas of the trapezoids will approach the true value. An example of the application of the trapezoidal rule is given by Davis (1957) for the calculation of the total volume of growing stock from yield-table data for a fully regulated forest. (This procedure was used because no yield function in equation form was available.)

Simpson's rule is used when there is an even number of intervals. It assumes that the terminals of each set of three successive ordinates are connected by arcs of parabolas. The areas of the successive "double parabolic strips" are summed yielding the formula

$$\text{Area} = \frac{h}{3}(Y_0 + 4Y_1 + 2Y_2 + 4Y_3 + \cdots + 2Y_{n-2} + 4Y_{n-1} + Y_n) \qquad (6\text{-}7)$$

6-4 AREA BY DOT GRIDS, LINE TRANSECTS, AND WEIGHING

If a plane figure is drawn on rectangular coordinate paper consisting of uniform squares of known area (e.g., 0.01 in.^{2}), the area of the figure can be estimated by counting the number of squares that fall within the boundaries of the figure. Where boundary lines include only portions of squares, one must estimate these portions and add their total to the total number of whole squares within the boundaries. If the figure represents an area drawn to scale (e.g., a timber type), the area of the figure can be converted to the represented area by computing, for the known scale, the appropriate scale conversion (e.g., acres per square inch).

If a dot is placed in the center of each square on the rectangular coordinate paper and the lines are removed, a dot grid is formed. (Of course, the lines may be retained, if desired.) Each dot now represents an area equal to that of the square. Thus, the area of a plane figure can be estimated by counting the number of dots that fall within the boundaries of the figure and can be converted to the represented area by computing the appropriate scale conversion. Dot grids, which are generally on transparent sheets, can be prepared, or purchased, with any desired number of dots per unit area.

In addition to being used to obtain the area of individual figures, dot grids may be used to obtain area ratios. For example, a dot grid can be placed over a map or an aerial photograph on which numerous forest types are outlined, and the number of dots within each type counted. Then the ratio of the number of dots in each type to the total number of dots on the map or photograph can be computed. The total area of the map or photograph can then be multiplied by the ratios to obtain the type areas. (Sections 9-4.1 and 13-5.1 give more information on applications of dot grids; Section 6-5 compares dot grids with planimeters.)

An analogous procedure for determining the area of plane figures is the line-transect method. This method is used primarily to determine area ratios. For example, equally spaced lines are drawn on a map or an aerial photograph on which numerous forest types are outlined, and the lengths of lines within each type measured. Then, the ratios of line lengths in each type to total line length on the map or photograph can be computed. These ratios can be used in the same manner as ratios computed by use of dot grids.

Weighing can also be employed to estimate the area of a plane figure drawn on paper, or other material, of uniform weight per unit area. The procedure is simple. The ratio of area to unit weight is developed for the paper, or other material. Then, the figures are cut out and weighed. Area is determined by multiplying weight of figure by area per unit weight.

6-5 AREA BY PLANIMETER

The *planimeter,* which was invented in 1854 by Jacob Amsler, is a mechanical device for measuring the area of plane figures. It is used to obtain areas on maps, photographs, drawings, diagrams, and so on. Over the years Amsler's polar planimeter has been greatly improved, and other forms, such as the rolling disc planimeter, have been developed. In addition, computing planimeters with built in calculators that can be programmed for any scale ratio are available.

The polar planimeter (Fig. 6-3) consists of a *pole arm, BP,* a *tracing arm, TB,* and a *carriage.* The carriage furnishes bearings for a vertical measuring wheel *W* that revolves on a horizontal axis. When set in position the instrument rests on three points, the fixed point *P,* the measuring wheel *W,* and the tracing point *T.* As *T* is moved about, the instrument pivots about *P* and the wheel revolves. To determine an area one simply moves the tracing point *T* once around the boundary of a figure (Fig. 6-3) and reads the resulting movement of the measuring wheel, which is graduated for this purpose. In effect,

$$\text{Area} = 2\pi r(TB)n \tag{6-8}$$

where

r = radius of measuring wheel W
TB = length of planimeter tracing arm
n = algebraic sum of rotations of measuring wheel W

[Note that $2\pi r(TB)$ is a constant for a given planimeter. Thus, the measuring wheel is graduated in terms of n times this constant.]

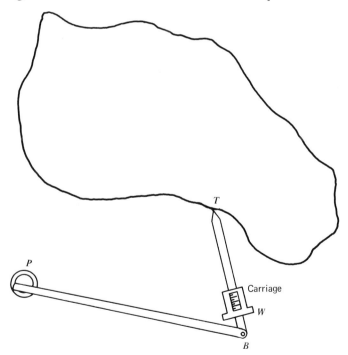

Fig. 6-3 *Area measurement with planimeter.*

There are no significant differences in the accuracies of dot grids and planimeters, assuming dot grids with a suitable number of dots per unit area are used. However, measurements taken with dot grids take longer than those done with planimeters. Consequently, if many figures must be measured and manpower is limited, a planimeter is economically superior to a dot grid. Indeed, if a large number of figures must be measured day after day, a computing planimeter may be a good investment, because measurements can be made almost twice as fast as with an ordinary planimeter.

6-6 LAND AREA MEASUREMENT

Area determinations of the earth's surface are usually for planar projections of the irregular land surface. However, in some cases, such as for watershed management, total surface area may be required.

Areas of the earth's surface may be expressed as the square of any linear unit. Different units, however, have become standard in different countries. For example, the *acre* is the common land area unit in the United States; the *hectare* and the *are* are the common land area units in countries that use the metric system. Converting factors for these, and other area units, are given in Appendix Table A-1.

Some of the important techniques for obtaining measurements to determine land areas are given in Chapters 3 and 13. Additional details can be found in Brinker and Wolf (1977) and Moffitt and Bouchard (1975).

6-7 TREE CROSS-SECTIONAL AREA MEASUREMENT

The cross-sectional areas of planes cutting the stem of a tree normal to the longitudinal axis of the stem are often desired. If the cross section of a tree is taken at breast height, it is called the *basal area*. The total basal area of all trees, or of specified classes of trees, per unit area (e.g., per acre or per hectare) is a useful characteristic of a forest stand. For example, basal area is directly related to stand volume and is a good measure of stand density (Chapter 17).

When a tree cross section (for either a standing tree or a cut section) is circular, as it is often assumed to be, its area can be computed from its diameter or circumference; thus,

$$g = \frac{\pi}{4} d^2$$

and since $d = c/\pi$,

$$g = \frac{c^2}{4\pi}$$

where g = tree cross-sectional area; d = diameter of cross section; and c = circumference of cross section.

In American forest practice, diameter d is commonly expressed in inches, and cross-sectional area in square feet (ft²). Consequently, it is convenient to express g (in square feet) as a function of diameter d in inches.

$$g(\text{ft}^2) = \frac{\pi d^2}{4(144)} = 0.005454 d^2$$

In countries that use the metric system, diameter d is commonly expressed in centimeters, and cross-sectional area in square meters (m²). In this case it is convenient to express g (in square meters) as a function of diameter d in centimeters.

$$g(\text{m}^2) = \frac{\pi d^2}{4(10000)} = 0.00007854 d^2$$

Unfortunately, cross sections of tree stems often are not circular. Thus, when they are assumed to be circular, errors in determining cross-sectional area may result. The problems of determining the diameter of a circle with the same cross-sectional area as a tree cross section (i.e., "true" diameter) are discussed in Section 2-1.2. For most purposes, the geometric mean of the long and short diameters, d_1 and d_2, of a section will give the most accurate results (geometric mean $= \sqrt{d_1 d_2}$), although for practical purposes a satisfactory practice is to take the arithmetic average of the long and short "diameters," or axes; or if it is not feasible to secure the long and short "diameters," to take the arithmetic average of two diameters perpendicular to each other.

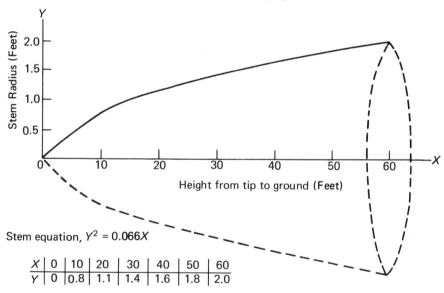

Stem equation, $Y^2 = 0.066X$

X	0	10	20	30	40	50	60
Y	0	0.8	1.1	1.4	1.6	1.8	2.0

Surface area

$$S_x = 2\pi \int_0^{60} \sqrt{0.066X} \sqrt{1 + \left(\frac{0.066}{2\sqrt{0.066X}}\right)^2} \, dX$$

Fig. 6-4 *Surface area of a tree stem of paraboloid form.*

6-8 TREE BOLE SURFACE AREA

The exterior surface area of the stem of a tree approximates the cambial surface (i.e., the area under bark). This area represents the surface on which the wood substance accumulates and is therefore useful in the estimation of tree and stand growth. For tree stems that assume the shape of a geometric solid, the surface area can be computed by the calculus or by the appropriate formula (Appendix Table A-3). For example, assume the form of a tree approximates the paraboloid generated by revolving the equation $Y^2 = 0.066X$ about the X axis (Fig. 6-4). In this equation Y is the radius of the stem at a given point (e.g., at the stump), and X is the distance from the tree tip to the given point. The surface of the tree between $Y = 2$ feet (i.e., $X = 60$) and the tip of the tree (i.e., $X = 0$) can be determined by using equation 6-4 and evaluating the following.

$$S_x = 2\pi \int_0^{60} \sqrt{0.066X} \sqrt{1 + \left(\frac{0.066}{2\sqrt{0.066X}}\right)^2} \, dX$$

$$= 2\pi \int_0^{60} \sqrt{0.066X + 0.001089} \, dX$$

$$= 2\pi \left[\frac{2(0.066X + 0.001089)^{3/2}}{(3)(0.066)}\right]_0^{60}$$

$$= 2\pi \left[10.10 \, (0.066X + 0.001089)^{3/2}\right]_0^{60}$$

$$= 159.25\pi \text{ square feet}$$

$$= 500.3 \text{ square feet}$$

In this example the same value would have been obtained if the formula for the surface area of a paraboloid (Appendix Table A-3) had been used. For more complex stem equations the analytical procedure used above is appropriate. However, recent advances in programmable calculator technology make numerical approximation of complex integrals an attractive alternative.

Lexen (1943) has shown that bole surface area can be approximated by using Huber's formula or Smalian's formula and substituting circumferences for cross-sectional areas and summing the surface areas for all sections in a tree. Bole surface area can also be obtained by plotting on rectangular coordinate paper the circumferences for several points along the stem and planimetering the area below the curve drawn through these points.

Swank and Schreuder (1974) tested sampling methods to estimate foliage, branch, and stem surface areas. Hann and McKinney (1975) developed prediction equations for the surface area of four species in the southwestern United States by using formulas for the surface areas of a cylinder, cone, and paraboloid.

6-9 LEAF AREA AND CROWN COVERAGE

The surface area of the foliage of forest trees is a useful measure for the study of precipitation interception, light transmission through forest canopies, forest litter accumulation, soil moisture loss, and transpiration rates. Cable (1958) developed a regression equation to estimate surface area of ponderosa pine fascicles from their oven-dry weight. Madgwick (1964) and Drew and Running (1975) also studied methods of determining surface area of conifer needles. Cummings (1941) and Wargo (1978) studied methods of determining surface area of broadleaf trees.

Crown coverage, a measure of relative density, is the proportion of the area of a forest stand covered by tree crowns. It has been used principally to estimate the volume of timber per unit area from aerial photographs. Thus, crown coverage is usually estimated on vertical aerial photographs. (Indeed, crown coverage estimates are more easily made on vertical aerial photographs than on the ground.) Fine dot grids and crown density scales (Section 13-5) allow rapid estimation on vertical aerial photographs.

The "moosehorn" crown closure estimator can be used to estimate crown closure from the ground (Garrison, 1949). While this instrument is held in a vertical position, the operator sights through a side aperture and counts the number of dots on a transparent surface on the top of the instrument that fall on tree crowns. Then, the ratio of dots on crowns to total number of dots gives an estimate of the proportion of the area covered by crowns.

The "spherical densiometer" developed by Lemmon (1957) can also be used to estimate crown closure from the ground. The instrument employs a spherical mirror that reflects a large area. A grid of squares is engraved on the mirror or on a transparent overlay. The ratio of squares on crowns to total number of squares gives an estimate of the proportion of the area covered by crowns.

7

Volume Measurement

Volume is the measure of solid content or capacity, usually expressed in units that are cubes of linear units, such as cubic meters and cubic feet, or in units of dry and liquid measure, such as bushels, gallons, and liters (Appendix Table A-1).

Volume has been, and will continue to be, the most widely used measure of wood quantity. This chapter discusses the principles of volume measurement and the volume units that are used in forest mensuration; Chapters 8 and 9 discuss the estimation of tree and log volumes.

7-1 METHODS OF DETERMINING VOLUME

Solid objects may assume the form of polyhedrons, solids of revolution, and solids of irregular shape. The standard formulas shown in Appendix Table A-3 can be used to compute the volumes of polyhedrons, such as cubes, prisms, and pyramids, and the volumes of solids of revolution, such as cones, cylinders, spheres, paraboloids, and neiloids, and frustums of cones, cylinders, spheres, paraboloids, and neiloids. These formulas, which are of particular importance in forest mensuration, are summarized in Table 8-1.

The volume of a solid with known cross sections can also be obtained by summation of the volumes of cross-sectional slices of the solid. A sum of such volumes approximates the volume of the solid. We take the integral that is suggested by this sum to be the volume

$$V = \int_a^b A(X)\, dX$$

in which the volume of a slice is expressed as a function $A(X)$ of X, times the thickness dX, where the thickness approaches zero. This procedure can be used for solids of any shape for which the cross-sectional area can be expressed as a function.

Whenever the profile of a tree can be expressed by an equation, the formula for the stem volume obtained by rotating the graph of the equation $Y = f(X)$ about the X axis (between $X = a$ and $X = b$) may be written

$$V = \pi \int_a^b Y^2\, dX$$

In the integral the radius of the solid of revolution Y must be expressed as a function of X. The cylinder is formed by rotating a rectangle around one side; a cone by rotating a right triangle around the vertical leg; a sphere by rotating a semicircle around the diameter; and a paraboloid by rotating a second-degree polynomial around its axis. The generalized formulas for the volumes of the solids shown in Appendix Table A-3 can be derived from the integral shown above by substituting the appropriate $f(X)$ for Y. For solids of revolution formed by curves of other shapes, simple formulas such as these are not available; integration is required to obtain volume.

To illustrate the procedure, we will use Fig. 6-4 and will compute the volume of the stem shown in the figure between a 1-foot stump and an upper limit that is 10 feet from the top. The volume would then be

$$V = \pi \int_{10}^{59} 0.066X \, dX$$

Integrating and evaluating yields

$$V = \left[\frac{\pi 0.066X^2}{2} \right]_{10}^{59} = \left[0.104X^2 \right]_{10}^{59}$$

$$V = (0.104)(59^2) - (0.104)(10^2)$$

$$= 351.6 \text{ cubic feet}$$

Since this solid is the frustum of a paraboloid, equation 8-5 in Table 8-1 will yield the same result.

The most accurate method of measuring the volume of an irregularly shaped solid is by measuring the volume of water that it will displace. For example, the cubic volume of any part of a tree may be found by submerging it in a tank in which the water displacement can be accurately read. But to use such a tank, which is termed a *xylometer,* to measure the volume of a tree, it is necessary to fell the tree and cut it into sections that are small enough to fit into the tank.

A graphic method has some advantages over other methods for measuring the volume of solids that have circular cross sections but that vary in diameter along an axis normal to the cross sections. To use the method to obtain the volume of trees, one must have diameter measurements at intervals along the stem, preferably both inside and outside bark. It is most convenient to have measurements that were taken at regular intervals; however, if sufficient diameter measurements are taken so that the taper of the tree is accurately depicted, the measurements may be taken at any chosen interval.

When suitable taper measurements have been obtained for a given tree, the cross-sectional area or diameter squared, preferably both inside and outside bark, should be plotted over height on cross-sectional paper for each cross section for which measurements were taken. Then, the points should be connected by smooth lines to give a profile that is analogous to that of one side of a longitudinal section taken through the center of the tree. A separate graph is prepared for each tree stem.

On the graph it is useful to label diameters at important points, such as top of stump, breast height, log ends, and merchantable limit of the stem, and to record pertinent information on species, locality, observers, date, and so forth. Volumes for the entire stem, or for any of its sections, either inside or outside bark, can be obtained by measuring the required area on the graph and applying the appropriate converting factor. For example, in Fig. 7-1 diameter squared is plotted over length for a section of a tree. The area under the curve can be obtained by using a planimeter or a dot grid, and converted to cubic volume by multiplying by cubic feet per square inch represented by the graph. Specifically,

$$\text{Converting factor} = AL$$

where

A = cross-sectional area per inch of Y axis
L = length per inch of X axis

In the example,

$$\text{Converting factor} = \frac{\pi d^2}{4(144)}L$$

$$= 0.005454(20)(1)$$

$$= 0.109 \text{ cubic feet}$$

And since the area under the curve of the section is 27.21 square inches, the volume of the section is 27.21(0.109) or 2.97 cubic feet.

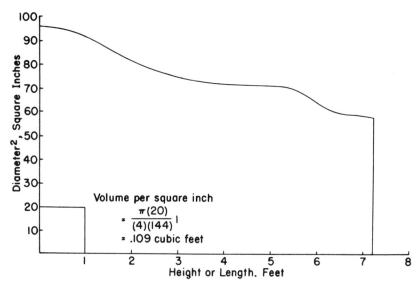

Fig. 7-1 *Graphical estimation of the volume of a section of a tree.*

7-2 VOLUME UNITS USED TO MEASURE WOOD

In the United States the volume of trees and logs is sometimes expressed in cubic feet; however, it is more common to express volume in board feet. In countries that use the metric system, the volume of trees and logs is generally expressed in cubic meters.

Another unit that should be mentioned is the *cunit*. In 1923, the *Pulp and Paper Magazine of Canada* ran a contest for the most suitable name for a unit of 100 cubic feet of solid wood. The winner was C. W. Halligan of the News Print Service who suggested the name cunit. The cunit was used in Canada until the adoption of the metric system in the early 1970s. On the West Coast of the United States, and in a few other places, the cunit is accepted as 100 cubic feet of solid wood, 100 cubic feet of sawdust, and 100 cubic feet of pulp chips.

The *Hoppus foot* is a cubic volume unit that has been used in the United Kingdom, India, Australia, and New Zealand. The formula to determine Hoppus feet, often called the quarter-girth formula, is

$$\text{Hoppus feet} = \left(\frac{C}{4}\right)^2 \frac{L}{144} \tag{7-1}$$

where

C = girth (i.b. or o.b.) in inches at center of log
L = length in feet of log

The Hoppus formula gives 78.5 percent of the actual cubic volume of a log, and a Hoppus foot is considered equivalent to 10 board feet. To obtain cubic volume a divisor of 113 is used in place of 144 in the quarter-girth formula.

The *Blodgett foot,* which is occasionally employed in the northeastern United States, is defined as a cylindrical block of wood 16 inches in diameter and 1 foot in length, equivalent to 1.4 cubic feet. Originally, the Blodgett foot was to be converted to board feet at a ratio of 10 board feet to 1 Blodgett foot. Now, however, factors between 8.7 and 9.4 to 1 are often used.

The *board foot,* which is still widely used in the United States to measure lumber and logs, is correctly defined as a piece of rough green lumber 1 inch thick, 12 inches wide, and 1 foot long, or its equivalent in dried and surfaced lumber. Thus, a board foot is a nominal volume unit. For example, if you visited a lumber yard to obtain some ponderosa pine yard lumber, you would find that the standard finished size of a kiln-dried 1 by 12 inch board would be $\frac{3}{4}$ by $11\frac{1}{4}$ inches; a kiln-dried 2 by 4 inch board would be $1\frac{1}{2}$ by $3\frac{1}{2}$ inches; and so on. However, the board foot volume of a board would be computed from the nominal thicknesses and widths, not the exact dimensions, by the following formula.

$$\text{Board feet} = \frac{W \times T \times L}{12} \tag{7-2}$$

where

W = nominal width in inches
T = nominal thickness in inches
L = actual length in feet

When log and lumber volumes in commercial transactions, inventory reports, and so forth, involve large quantities of material, the volumes are normally given in thousands of board feet. Then the abbreviation M.b.f. (thousand board feet) or M.b.m. (thousand feet, board measure) is used. For example, 11,234,000 board feet = 11,234 M.b.f.

Although the board foot is the standard unit of lumber measurement in the United States, lumber that is thinner than 1 inch in thickness is usually sold on the basis of surface measure. Finish lumber, such as trim and molding, is sold by the linear foot.

Although there is an inexactness in the measurement of the board-foot volume of lumber, board-foot log rules (Chapter 9) showing the estimated number of board feet of lumber that can be sawed from logs of given lengths and diameters are frequently used to estimate the contents of logs and trees. This constitutes an attempt to estimate, before processing, the amount of lumber in logs and trees. This is rather unique. Few raw materials are measured in terms of the finished product. It is somewhat like trying to measure the yield of a field of corn in terms of boxes of corn flakes.

The number of board feet that can be produced from a log depends primarily on its length and diameter. There are unfortunately many other factors that affect the yield of a log. Among these are:

1. Efficiency of the sawmill machinery and, in particular, the thickness of the kerf cut by the various saws.
2. Efficiency of the crew, particularly of the sawyers, edgers, and trimmers.
3. Market conditions. (When markets are good, efforts will be made to utilize small pieces, and the output will be raised. And when the ratio of thick lumber cut to 1-inch lumber cut increases, output will be raised.)
4. Amount of defect and taper in logs.

7-3   VOLUME UNITS FOR STACKED WOOD

A *cord* is a unit of measure that expresses the volume of stacked wood. The need to measure relatively small pieces of low value, rough wood, stacked at scattered locations in the woods gave rise to the cord. Cord measurement can be done rapidly and economically.

A *standard cord* is a pile of stacked wood 8 feet long and 4 feet high that is made up of 4 foot pieces. (The standard cord occupies 128 cubic feet.) The actual solid wood content is generally less than 100 cubic feet and varies by species, method of stacking, form of wood, length and diameter of wood, bark thickness of wood, and other factors.

A *long cord* measures 8 feet long and 4 feet high and is made up of pieces longer than 4 feet. Long cords commonly consist of pieces 5 feet long, or 5 feet 3 inches long, and occupy 160 or 168 cubic feet, respectively.

A *short cord* or *face cord* is a unit smaller than the standard cord and is usually used to measure fuelwood that is cut less than 4 feet long. A *rick* of fuelwood is often considered to be an 8 by 4 foot pile made up of 12-inch pieces (4 ricks per standard cord), or an 8 by 4 foot pile made up of 16-inch pieces (3 ricks per standard cord).

A *pen,* which is used in the southern United States, consists of sticks stacked to a height of 6 feet in the form of a square enclosure with two sticks to a layer. Five pens of 4-foot sticks are assumed to equal a standard cord; five pens of 5-foot sticks are assumed to equal a long cord of 160 cubic feet.

The *stere* or metric cord, which is used in countries that employ the metric system, is a pile of stacked wood 1 meter long and 1 meter high that is made up of 1-meter pieces. (The stere occupies 1 cubic meter.)

All of the above cord units, it should be emphasized, differ from the volume units discussed in Section 7-2 because they are not measures of individual pieces in terms of solid-wood contents.

7-4 VOLUME UNIT CONVERSION

Board foot-cubic foot conversions for logs vary by log rules and log and tree sizes. (Board foot-cubic foot ratios for sawn lumber are seldom used because they are confusing and illogical.) Table 7-1 illustrates the nature of these variations. But, in general, the ratios increase rapidly from small to large diameters and level off once the larger diameters have been reached. The reasons for these variations are explained in Chapter 9. Although any flat conversion factor should be used with care, a "rule-of-thumb" factor in rather general use is 6 board feet (Scribner log rule) per cubic foot.

Cubic foot-cubic meter conversions can be easily accomplished by using an exact mathematical relationship: cubic feet = cubic meters (35.3145); cubic meters = cubic feet (0.028317). (Appendix Table A-1 gives these conversion factors.)

Cubic foot-cord conversions vary so greatly that the closest approximation for "rule-of-thumb" figuring appears to be from 60 to 94 cubic feet per standard cord for unpeeled (rough) wood, depending on species, method of stacking, size of wood, and bark thickness. The average bark volume will run from 10 to 20 percent of unpeeled volume for most species. But good working averages are between 75 and 85 solid cubic feet per standard cord for green, unpeeled southern pine, between 80 and 90 solid cubic feet per standard cord for green, unpeeled Douglas fir or western hemlock, and between 73 and 85 solid cubic feet per standard cord for green, unpeeled hardwoods.

Board foot-cord conversions are unreliable and are used infrequently. However, a converting factor often used is: 1 thousand board feet (Scribner log rule) equals 2 standard cords of unpeeled cordwood. For more accurate conversion, however, con-

Table 7-1
Board–Foot–Cubic–Foot Ratios

(a) For 16-Foot Logs

Diameter Inside Bark, Small End (inches)	Volume[1] (cubic feet)	Board Feet per Cubic Foot (Ratios)		
		International $\frac{1}{4}$-Inch	Doyle	Scribner
10	11	5.9	3.3	5.4
20	39	7.4	6.6	7.2
30	84	8.0	8.0	7.8
40	147	8.2	8.8	8.2

(b) For Standing Trees (Conifers in Lake States)[2]

dbh Outside Bark (inches)	Board Feet per Cubic Foot by Number of 16-Foot Logs (Ratios)			
	1	2	3	4
10	4.7	5.0	—	—
20	3.4	5.1	6.2	6.4
30	3.0	4.8	5.8	6.1
40	—	—	5.6	6.0

[1] Volume by Smalian's formula; taper allowance, 2 inches per 16 feet.

[2] Gevorkiantz, 1950. Board-foot volumes according to International $\frac{1}{4}$-inch rule.

sideration must be given to diameter and length of logs, bark thickness, and other factors. For example, one might obtain a variation in yield, per thousand board feet, of from 1.8 cords with logs of 30-inch diameter and larger, to 3.3 cords with logs of 6-inch diameter.

Cubic meter-stere conversions are analogous to cubic foot-cord conversions. On a percentage basis the solid cubic contents of a stere can be expected to vary as does the cord. For example, if averages are between 73 and 85 solid cubic feet per standard cord for green, unpeeled hardwoods—that is, between 57 and 66 percent of the 128 cubic feet occupied by the cord is solid wood—averages would be between 0.57 and 0.66 solid cubic meters per stere for green, unpeeled hardwoods.

8

Cubic Volume and Measures of Form

It is convenient to think of the tree as consisting of four parts: *roots, stump, stem,* and *crown.*

The *roots* are the underground part of the tree that supply it with nourishment. The *stump* is the lower end of the tree that is left above the ground after the tree has been felled. The *stem* is the main ascending axis of the tree above the stump. (Trees that have the axis prolonged to form an undivided main stem, as exemplified by many conifers, are termed *excurrent;* trees that have the axis interrupted in the upper portion due to branching, as exemplified by many broadleaved species, are termed *deliquescent.*) The *crown* consists of the primary and secondary branches growing out of the main stem, together with twigs and foliage.

The most important portion of a tree, in terms of usable wood, is the stem. In the past, since the roots and stumps of trees were less often utilized, little attention was given to the determination of their volume. The picture is changing, however, because of the increasing interest in complete tree utilization. There has been, and continues to be, a need to determine the volume of crowns because crown-volume data may be used to describe fuel hazards, to estimate volume of material left from line-clearing operations, and to determine volume of pulpwood and fuelwood in crowns. It is interesting to note that the cubic volume of the merchantable stem (to a 4-inch top diameter) of the average tree is about 55 to 65 percent of the cubic volume of the complete tree, excluding foliage, and that the cubic volume of the crown of the average tree, excluding foliage, is about 15 to 20 percent of the cubic volume of the complete tree, excluding foliage, for softwoods and 20 to 25 percent for hardwoods.

Since the stump, stem, and branches are all covered with bark, they are, when utilized, peeled in the woods or hauled, bark and all, to the mill where the bark is removed before the manufacturing process begins. Thus, it is necessary to determine, at one time or another, unpeeled volume and bark volume.

8-1 DIRECT DETERMINATION OF THE CUBIC VOLUME OF TREE PARTS

Direct volume determinations of parts of trees are usually made on sample trees to obtain basic data for the development of relationships between the various dimensions of a tree and its volume (Chapter 9). Relationships of this type are used to estimate the volumes of other standing trees. In the past sample-tree measurements were often taken on trees cut in harvesting operations. But volume relationships developed from such measurements may lead to bias because they may not be representative of all the trees in a stand. Thus, there is a growing tendency to take measurements on a representative sample of standing trees.

The direct determination of the volume of any part of a tree involves clearly defining the part of the tree for which volume is to be determined and carefully taking measurements in accordance with the constraints imposed by the definition. For example, for purposes of measurement we might include the portion of the stem above a fixed-height stump to a minimum upper-stem diameter outside bark, or on stems that do not have a central tendency, to the point where the last merchantable cut can be made. For roots we might include roots larger than some minimum diameter; for tops we might include the branches and the tip of the stem to some minimum diameter outside the bark.

Generally speaking, the tree must be felled and the limbs cut into sections before one can directly determine crown volume. To directly determine root volume, the roots must be lifted from the ground and the soil removed. But of course, stem and stump volume may be directly determined on either standing or felled trees.

The stems and stumps of trees closely resemble certain geometrical solids. Thus, their cubic volume may be determined by formulas (Section 8-2). It is not feasible, however, to determine the volume of roots by formulas because they do not, either in their entirety or in portions, closely resemble any known geometrical solids. Practically speaking, this is also true of crowns.

Thanks to the reasonably regular form of stems, graphical techniques may be used to obtain the cubic volume of the stem and stump of a standing or felled tree (Section 7-1). But again this technique is unsuitable for determining the volume of roots or crowns, because for each set of roots, or for each crown, an inordinate number of diameter measurements and an excessive number of graphs are required to obtain satisfactory results. The displacement method (Section 7-1), on the other hand, can be used to obtain accurately the cubic volume of any part of the tree. It is applicable only to cut trees, however, and it is relatively slow and expensive.

To use formulas or graphical techniques to determine the volume of a stem, it is necessary to have diameter measurements at various intervals along the stem. Both inside and outside bark measurements are desirable. Measurements may be made on down or standing trees (Chapter 2). On standing trees, when it is convenient to obtain only outside bark measurements, inside bark diameters, and thus inside bark volumes, may be calculated by use of bark factors (Section 8-3).

8-2 DIRECT DETERMINATION OF CUBIC VOLUME BY FORMULAS

The stems of excurrent trees are often assumed to resemble neiloids, cones, or paraboloids—solids that are obtained (Fig. 8-1) by rotating a curve of the general form

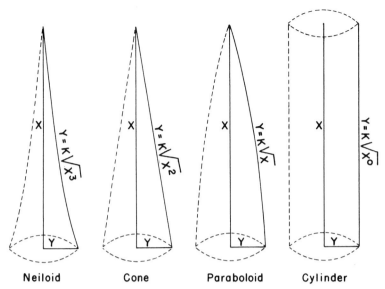

Fig. 8-1 *Solids of revolution descriptive of tree form.*

$Y = K\sqrt{X^r}$ around the X axis. As the form exponent r changes in this equation, different solids are produced. When r is 1, a paraboloid is obtained; when r is 2, a cone; when r is 3, a neiloid; and when r is 0, a cylinder. But the stems of excurrent trees are seldom cones, paraboloids, or neiloids; they generally fall between the cone and the paraboloid. The merchantable portion of stems of deliquescent trees, on the other hand, are assumed to resemble frustums of neiloids, cones, or paraboloids (occasionally cylinders). But they generally fall between the frustum of a cone and the frustum of a paraboloid.

It is more realistic, however, to consider the stem of any tree to be a composite of geometrical solids (Fig. 8-2). For example, when the stem is cut into logs or bolts, the tip approaches a cone or paraboloid in form, the central sections approach frustums of paraboloids, or in a few cases frustums of cones or cylinders, and the butt log approaches the frustum of a neiloid. Although the stump approaches the frustum of a neiloid in form, for practical purposes it is considered to be a cylinder.

Formulas to compute the cubic volume of the solids that have been of particular interest to mensurationists are given in Table 8-1. Newton's formula is exact for all the frustums we have considered. Smalian's and Huber's formulas are exact only when the solid is the frustum of a paraboloid.[1] For example, if the surface lines of a tree section are more convex than the paraboloid frustum, Huber's formula will overestimate the volume while Smalian's formula will underestimate the volume. But if the surface lines of a tree section are less convex than the paraboloid frustum, as they often are, Smalian's formula will overestimate the volume and Huber's formula will

---

[1] Newton's formula is attributed to Sir Isaac Newton. Prodan (1965) claims that Huber's formula came into use in 1785 and Smalian's formula in 1804.

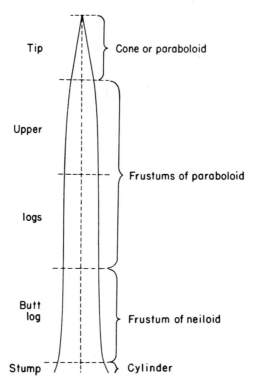

Fig. 8-2 *Geometric forms assumed by portions of a tree stem.*

underestimate the volume. Assuming Newton's formula gives correct volume values, it can be shown by subtracting Newton's formula first from Smalian's formula and then from Huber's formula that the error incurred by Smalian's formula is twice that incurred by Huber's formula and opposite in sign. In a study by Young, Robbins, and Wilson (1967) on 8- and 16-foot softwood logs of 4- to 12-inch diameters, volumes calculated by Newton's, Smalian's, and Huber's formulas were compared with volumes determined by displacement. Average percent errors of about 0 percent were obtained for Newton's formula, +9 percent for Smalian's formula, and −3.5 percent for Huber's formula. They also found that there were no significant errors for any of the three formulas on 4-foot bolts. In a study by Miller (1959) on 16-foot hardwood logs of 8- to 22-inch diameters, volumes calculated by the three formulas were compared with volumes determined by graphical techniques. Average percent errors of about +2 percent were obtained for Newton's formula, +12 percent for Smalian's formula, and −5 percent for Huber's formula.

It should now be apparent that in calculating cubic volume of trees and logs, mensurationists should select their methods carefully. Unless one is willing to accept a rather large error, Smalian's formula should not be used unless it is possible to measure the sections of the tree in 4-foot lengths. For 8- or 16-foot long, Newton's or Huber's formula will give more accurate results.

Table 8-1
Equations to Compute Cubic Volume of Important Solids

Geometrical Solid	Equation for Volume V (cubic units)		Equation Number
Cylinder	$V = A_b h$		(8-1)
Paraboloid	$V = \frac{1}{2}(A_b h)$		(8-2)
Cone	$V = \frac{1}{3}(A_b h)$		(8-3)
Neiloid	$V = \frac{1}{4}(A_b h)$		(8-4)
Paraboloid frustum	$V = \frac{h}{2}(A_b + A_u)$	(Smalian's formula)	(8-5)
	$V = h(A_m)$	(Huber's formula)	(8-6)
Cone frustum	$V = \frac{h}{3}(A_b + \sqrt{A_b A_u} + A_u)$		(8-7)
Neiloid frustum	$V = \frac{h}{4}(A_b + \sqrt[3]{A_b^2 A_u} + \sqrt[3]{A_u^2 A_b} + A_u)$		(8-8)
Neiloid, cone, or paraboloid frustum	$V = \frac{h}{6}(A_b + 4A_m + A_u)$	(Newton's formula)	(8-9)

h = height.
A_b = cross-sectional area at base.
A_m = cross-sectional area at middle.
A_u = cross-sectional area at top.

Newton's formula will give accurate results for all sections of the tree except for butt logs with excessive butt swell. For such butt logs, Huber's formula will generally give better results. Either the paraboloidal formula (equation 8-2, Table 8-1) or the conic formula (equation 8-3, Table 8-1) is appropriate to determine the volume of the tip. The cylindrical formula (equation 8-1, Table 8-1) is normally used to compute the volume of the stump, although the stump actually approaches the neiloid frustum in form.

Newton's and Huber's formulas cannot, of course, be applied to stacked logs, because it is not possible to measure middle diameters. However, these two formulas are as well suited as Smalian's formula to determine the volume of unstacked logs or standing trees.

Newton's formula may be used to compute the volume of the merchantable stem or of the total stem. If the sections are of the same length, the procedure can be summarized in a single formula. To illustrate, consider a stem with diameters in inches from top of stump to a point where the last merchantable cut will be made, d_0, d_1, d_2, d_3, d_4, d_5, and d_6, located at intervals of h feet. To give each section three diameters, the volume is computed by sections of $2h$ length. Thus, with Newton's formula the volume (in cubic feet) V is

$$V = \frac{2h}{6}(0.005454)(d_0^2 + 4d_1^2 + d_2^2) + \frac{2h}{6}(0.005454)(d_2^2 + 4d_3^2 + d_4^2)$$

$$+ \frac{2h}{6}(0.005454)(d_4^2 + 4d_5^2 + d_6^2)$$

$$= h(0.001818)(d_0^2 + 4d_1^2 + 2d_2^2 + 4d_3^2 + 2d_4^2 + 4d_5^2 + d_6^2)$$

$$= 0.003636h\left(\frac{d_0^2}{2} + 2d_1^2 + d_2^2 + 2d_3^2 + d_4^2 + 2d_5^2 + \frac{d_6^2}{2}\right)$$

(The constant 0.005454 comes from the expression: cross-sectional area in square feet $= [\pi/4(144)]d_i^2 = 0.005454d_i^2$. If the metric system is used, d_i will be in centimeters and h in meters, and to obtain volume in cubic meters the constant will come from the expression: cross-sectional area in square meters $= [\pi/4(10,000)]d_i^2 = 0.00007854d_i^2$.) This formula may be extended for as many sections as desired, provided there is an odd number of diameters, that is, an even number of sections of h length.

If the number of diameters measured is even, the last interval of h cannot be computed by Newton's formula, because it will have only two end diameters. Thus, its volume must be found by Smalian's formula and added to the previous formula. For eight diameters, or seven intervals of h length, this yields

$$V = 0.003636h\left(\frac{d_0^2}{2} + 2d_1^2 + d_2^2 + 2d_3^2 + d_4^2 + 2d_5^2 + \frac{5d_6^2}{4} + \frac{3d_7^2}{4}\right)$$

Grosenbaugh (1948) described a systematic procedure using this method.

The volume of the merchantable stem can also be calculated with good accuracy using Smalian's formula if the stem is divided into short sections. To illustrate, consider a stem with diameters in inches, from top of stump to a point where the last mer-

Table 8-2
Sample Tree Data for Computation of Volume by Height Accumulation by 2-Inch Taper Steps and 4-Foot Unit Heights

	dob Taper Steps					
dbh	10	8	6	4	2	Sum
9.5	1	3	4	2	0	10
7.7		1	5	2	0	8
10.5	1	5	3	2	0	11
$L = 2$		9	12	6	0	29
$H = 2$		11	23	29	29	94
$H' = 2$		13	36	65	94	210

chantable cut will be made, $d_0, d_1, d_2, \ldots, d_n$, located at intervals of h feet along the stem. Then, according to Smalian's formula,

$$V = 0.005454h \left(\frac{d_0^2}{2} + d_1^2 + d_2^2 + \cdots + d_{n-1}^2 + \frac{d_n^2}{2} \right)$$

To adapt Huber's formula to the computation of merchantable stem volumes, the diameter measurements are taken at the midpoints of the sections. Thus, when diameter measurements $d_{m_1}, d_{m_2}, \ldots, d_{m_n}$ are taken at the midpoints of sections of h length, Huber's formula yields

$$V = 0.005454h (d_{m_1}^2 + d_{m_2}^2 + \cdots + d_{m_n}^2)$$

8-2.1 Determination of Volume by Height Accumulation

The height accumulation concept was conceived and developed by Grosenbaugh (1948, 1954), who stated that the system can be applied by selecting tree diameters above breast height in diminishing arithmetic progression, say 1- or 2-inch taper intervals, and estimating, recording, and accumulating tree height to each successive diameter. The system uses diameter as the independent variable instead of height, is well adapted to use with electronic computers, and permits segregation of volume by classes of material, log size, or grade. But since optimum log lengths for top log grades depend on factors other than diameter, the best grades may not be secured.

To apply the system one must know the number of sections L in some unit height between taper steps, the cumulative total H of L values, the cumulative total H' of H values, and, if inside bark volume is desired, the mean bark factor k (Section 8-3). This requires that one collect the following sample tree data: dbh to nearest 0.1 inch, length of merchantable stem between successive taper steps to nearest unit height, and measurements to compute the mean bark factor. For example, if one used 2-inch taper steps and 4-foot unit heights (a feasible practice), dbh would be rounded to the nearest even inch, and the first unit height L—the 4-foot section between stump height and breast height—would be 1. The unit heights to each taper step might be estimated, but to obtain acceptable accuracy, an instrument, such as the Spiegel relaskop, should be used.

Table 8-2 gives sample tree data needed to compute the volume V of a number of trees, or of individual trees, by height accumulation. Very simply

$$V = A(\Sigma H') + B(\Sigma H) + C(\Sigma L)$$

Volume coefficients A, B, and C for cubic feet are given in Table 8-3 for 2-inch taper steps, 4-foot unit heights, and mean dib/dob ratios. Grosenbaugh (1954) also gives coefficients for 1-inch taper steps and 1-foot unit heights, coefficients for determination of board-foot volume, formulas to calculate coefficients for other cases, and the theory of height accumulation.

Table 8-3
**Height-Accumulation Coefficients, *A*, *B*, and *C*,
to Compute Cubic-Foot Volume by 2-Inch Taper Steps,
4-Foot Unit Heights, and Various Mean dib/dob Ratios**

Mean Ratio dib/dob	Volume Coefficients for Cubic Feet		
	A	*B*	*C'*
1.00*	0.175	0	0.0291
0.95	0.158	0	0.0263
0.90	0.141	0	0.0236
0.85	0.126	0	0.0210

[1] When computing volume outside bark, use coefficients for ratio of 1.00.

SOURCE: Grosenbaugh, 1954.

For the trees given in Table 8-2, the total cubic foot volume, inside bark, for a mean dib/dob ratio of 0.90 is

$$V = 0.141(210) + 0(94) + 0.0236(29) = 30.3 \text{ ft}^3$$

Similarly, individual-tree cubic-foot volume is

$$V_{9.5} = 0.141(75) + 0(33) + 0.0236(10) = 10.8 \text{ ft}^3$$

$$V_{7.7} = 0.141(46) + 0(23) + 0.0236(8) = 6.7 \text{ ft}^3$$

$$V_{10.5} = 0.141(89) + 0(38) + 0.0236(11) = \underline{12.8} \text{ ft}^3$$

$$\text{Total} \qquad 30.3 \text{ ft}^3$$

Enghardt and Derr (1963) found that computation of cubic volume of young even-aged stands of southern pine by height accumulation required less time and effort than conventional methods and gave satisfactory accuracy for research purposes. Indeed, the potentialities of this unique system have been generally overlooked.

8-3 THE BARK FACTOR AND DETERMINATION OF BARK VOLUME

Average bark volume will run from 10 to 20 percent of unpeeled volume for most species. But we often need to know bark volume more accurately than this to determine peeled stem or log volume from unpeeled stem or log volume, the quantity of bark residue that will be left after the manufacturing process has been completed, and, in cases where the bark has value, the quantity of bark available. We can, of course, compute unpeeled stem or log volume from outside bark diameter, peeled stem or log volume from inside bark diameter, and take the difference to secure a good estimate of bark volume. But the bark-factor method, which deserves more consideration than most foresters choose to give it, is easier to apply and gives sufficiently accurate results for most purposes. Let us look at this method more closely.

Bark thickness, which must be accurately determined to obtain reliable bark factors, may be determined as described in Section 2-1.1. The accuracy of bark measurements is increased if single-bark thickness is measured at two or more different points on a given cross section of the stem and if the average single-bark thickness b computed. Then, diameter inside bark d_u may be computed from diameter outside bark d from equation: $d_u = d - 2b$. When d_u is plotted as a function of d, the relationship will be linear, or close to linear, with a Y intercept of 0, or close to 0. (Figure 8-3 shows this relationship for white oak for the cross section at breast height.) Consequently, it is reasonable to assume that the prediction equation for this relationship may be written in the general form $d_u = kd$. Since the regression coefficient k is normally determined at stump or breast height, we will call it the *lower-stem bark factor*. Such bark factors range from 0.87 to 0.93, varying with species, age, and site. But since the major portion of the variation can be accounted for by species, it is reasonable and convenient to assume that the ratio will be constant for a given species. It is also convenient to assume that this bark factor will remain the same, for a given species, at all heights on the stem. But for many species, upper-stem bark factors are often not the same as lower-stem bark factors. To account for this one might develop multiple regression equations to predict upper-stem bark factors from such variables as tree age, tree dbh, height above ground of cross section for which bark factor is desired, bark factor at breast height, and diameter outside bark at cross section for which bark factor is desired. In this day of computers and programmable calculators, although these equations are fairly complex, their use is not limited because of computation time, but because of the time required to measure the independent variables. Consequently, the common practice is to assume that the bark factor is the same at all heights on the stem. However, whether we assume that the bark factor is the same at all heights on the stem, or that the bark factor will be different at different heights on the stem, once the bark factor has been obtained, the method of using it to obtain bark volume is the same. We can illustrate the method by assuming that k is the same at all heights on the stem.

An average value of k, to be reliable, should be based on 20 to 50 bark-thickness measurements and corresponding diameter-outside-bark measurements. By the method of least squares, k is determined so that the sum of squared deviations of the individual d_u's (Y's) about the fitted regression line is a minimum. In this case, where we assume that when $X = 0$ then $Y = 0$, it is appropriate to use the following equation to determine k.

$$k = \frac{\Sigma \, dd_u}{\Sigma \, d^2}$$

When the variation of the dependent variable is proportional to the independent variable, as it generally is when d_u is plotted over d, Meyer (1953) has shown that the following formula will give the same results.

$$k = \frac{\Sigma \, d_u}{\Sigma \, d}$$

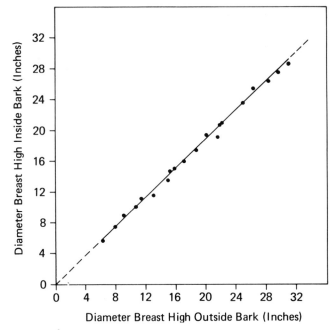

Fig. 8-3 *Relationship between corresponding diameters inside and outside bark of white oak (see Table 8-4).*

For the trees listed in Table 8-4 and plotted in Fig. 8-3, the first formula gives

$$k = \frac{7179.62}{7628.93} = 0.941$$

The second formula gives

$$k = \frac{341.7}{363.3} = 0.941$$

Agreement of these two formulas will not always be this good. But, for calculating k, the second formula will, for practical uses, always give satisfactory results.

The bark thickness b, corresponding to an average value of k, can be determined as follows for any diameter d.

$$b = \frac{1}{2}(d - d_u)$$

And since $d_u = kd$, then

$$b = \frac{1}{2}(d - kd)$$

$$= \frac{1}{2}d(1 - k)$$

(8-10)

Thus, for a white oak cross-section of 14.0 inches in diameter, we may estimate the bark thickness to be

$$b = \frac{1}{2}(14.0)(1 - 0.941)$$

$$= 0.413 \text{ inches}$$

The average value of k can also be used to obtain cubic bark volume V_b and cubic volume inside bark V_u from cubic volume outside bark V for a given stem section. If diameter outside bark at the middle of the section d_m, diameter inside bark at the middle of the section d_{mu}, and section length L are all in the same units, we have

$$V = \frac{\pi d_m^2}{4}(L) \qquad \text{and} \qquad V_u = \frac{\pi d_{mu}^2}{4}(L)$$

And since

$$d_{mu} = kd_m$$

Table 8-4
Diameter and Bark Measurements of 20 White Oak Trees

d dbh Outside Bark (inches)	$2b$ Double Bark Thickness (inches)	d_u dbh Inside Bark (inches)
30.8	2.2	28.6
14.7	1.1	13.6
24.8	1.2	23.6
20.0	0.8	19.2
21.7	1.1	20.6
7.9	0.4	7.5
15.8	1.1	14.7
12.7	1.3	11.4
18.6	1.0	17.6
17.0	1.0	16.0
26.1	0.8	25.3
28.1	1.7	26.4
10.6	0.6	10.0
6.1	0.4	5.7
29.4	1.6	27.8
15.1	0.8	14.3
21.5	2.3	19.2
22.0	1.2	20.8
11.4	0.7	10.7
9.0	0.3	8.7
Sum 363.3	21.6	341.7

$\Sigma d \cdot d_u = 7179.62.$

$\Sigma d^2 = 7628.93.$

then

$$V_u = \frac{\pi}{4}(kd_m)^2(L) = k^2V \qquad (8\text{-}11)$$

Finally, since

$$V_b = V - V_u$$

$$V_b = V(1 - k^2) \qquad (8\text{-}12)$$

$$V_b(\%) = (1 - k^2)100 \qquad (8\text{-}13)$$

When V_u is determined by equation 8-11, it will be theoretically correct. However, V_b determined by equation 8-12 will be greater than the actual value. This is because V_b includes air spaces between the ridges of the bark. Indeed, a study by Chamberlain and Meyer (1950) shows that the difference in volume between stacked-peeled and stacked-unpeeled cordwood is, on the average, 80 percent of the volume given by equation 8-12. One might not expect this result. But it comes about because in a stack of wood the ridges of the bark of one log will mesh with the ridges of another log, and because the weight of the logs will compress the bark. Thus, for practical purposes, we can rewrite equation 8-12 to give the bark volume of cordwood in stacks V_{b_s}.

$$V_{b_s} = 0.8V(1 - k^2) \qquad (8\text{-}14)$$

8-4 METHODS OF STUDYING STEM FORM

There are wide variations in the form of the main stem of trees due to variations in the rates of diminution in diameter from the base to the tip. This diminution in diameter, known as taper, which is a fundamental reason for variation in volume, varies with species, diameter breast high, age, and site.

In a definitive study, Larson (1963) discussed the biological concept of stem form by a comprehensive review of the literature. In their studies of stem form, mensurationists have looked for pure expressions of stem form that are independent of diameter and height. But pure expressions have not been discovered; they are probably fictions. Nevertheless, the methods that have been developed for studying stem form have been useful. They may be considered under four headings: *form factors, form quotients, form point,* and *taper tables, curves,* and *formulas.*

8-4.1 Form Factors

A form factor is the ratio of tree volume to the volume of a geometrical solid, such as a cylinder, a cone, or a cone frustum, that has the same diameter and height as the tree. (The diameter of the geometrical solid is taken at its base; the diameter of the tree is taken at breast height.) A form factor is different from other measures of form in that it can be calculated only after the volume of the tree is known. In formula form, the form factor f is

$$f = \frac{\text{Volume of tree}}{\begin{array}{c}\text{Volume of geometrical solid of}\\ \text{same diameter and height}\end{array}} \qquad (8\text{-}15)$$

Early in the nineteenth century it was recognized that the form of tree stems approached that of the solids discussed in Section 8-2. But it was also recognized that there were many variations in form, and that a tree rarely was of the exact form of one of these solids. Thus, the form factor was conceived as a method of coordinating form and volume. That is to say, the main objective of the early work was to derive factors that would be independent of diameter and height, and by which the volume of standard geometrical solids could be multiplied to obtain the tree volume. For example, the ratio of the volumes of a paraboloid to a cylinder is 0.5 when the base diameter and the height of the two solids are equal; the volume of the paraboloid is obtained by multiplying the volume of the cylinder by 0.5.

The *cylindrical form factor*, which has been most commonly used, may be expressed by the equation

$$f = \frac{V}{gh} \qquad (8\text{-}16)$$

where

V = volume of tree in cubic units
g = cross-sectional area of cylinder whose diameter equals tree dbh
h = height of cylinder whose height equals tree height

The usefulness of form factors to estimate the volume of trees of variable form is limited. However, there are some uses for the rapid approximation of volume as well as the volume of trees of little form variation. Belyea (1931) discussed form factors and their uses at some length.

8-4.2 Form Quotients

A form quotient is the ratio of a diameter measured at some height above breast height, such as one-half tree height, to diameter at breast height. In formula form the form quotient q is

$$q = \frac{\text{Diameter above breast height}}{\text{Diameter at breast height}} \qquad (8\text{-}17)$$

Next to diameter breast high and height, form quotient is the most important variable that can be used to predict the volume of a tree stem. Thus, it may be used as the third independent variable in the construction of volume tables (Chapter 9).

The original form quotient (Schiffel, 1899) took diameter at one-half total tree height $d_{0.5h}$ as the numerator, and diameter breast high d as the denominator. This was termed the *normal form quotient q*

$$q = \frac{d_{0.5h}}{d} \tag{8-18}$$

For this form quotient, as tree height decreases, the position of the upper diameter comes closer to the breast height point until, for a tree whose height is double breast height, they coincide. To eliminate this anomaly, Jonson (1910) changed the position of the upper diameter to a point halfway between the tip of the tree and breast height, $d_{\frac{1}{2}(h+4.5)}$, and called the ratio the *absolute form quotient* q_a.

$$q_a = \frac{d_{\frac{1}{2}(h+4.5)}}{d} \tag{8-19}$$

The absolute form quotient is a better measure of stem form than the normal form quotient. However, it is not a pure expression of stem form. It is not independent of diameter and height and it varies within a given diameter-height class for a given species. For most species absolute form quotients diminish with increasing diameters and heights, varying between 0.60 and 0.80. (Absolute form quotient is 0.707 for a paraboloid, 0.500 for a cone, and 0.354 for a neiloid when diameter d is taken at the base. These values hold irrespective of the diameter and height of the solid.)

In determining the normal or absolute form quotient, the two diameters may be taken either outside or inside bark. Although it is difficult to obtain an accurate upper-stem diameter inside bark, the ratio of the inside bark measurements is a better index of form than the outside bark measurements, because variable bark thicknesses do not then distort the ratio.

At this point it would be well to discuss the term *form class*. Originally the term was applied to a class, as in a frequency table of absolute form quotients. For example, form classes with class intervals of 0.05 have often been laid out as follows: 0.575–0.625, 0.625–0.675, 0.675–0.725, and 0.725–0.775. The midpoints of these classes, 0.60, 0.65, 0.70, and 0.75, were used to name the classes, and a tree falling into a particular class was said to have the form quotient of the midpoint of the class. Tree-volume computations were classified by absolute form class as well as by diameter and height. Form class may be related to the density of the stand (Jonson, 1911) or to *form point* (Fogelberg, 1953), so it may be determined indirectly.

In North America the term form class has been used in a different sense. Girard (1933), in the course of work in the U.S. Forest Service, developed a form quotient for use as an independent variable in volume table construction. This measure, termed *Girard form class*, q_G, is the percentage ratio of diameter inside bark at the top of the first standard log d_u to diameter breast high, outside bark. When 16-foot logs are taken as the standard, the upper diameter $d_{u17.3}$ is taken at the height of a standard 1-foot stump plus 16.3 feet. Thus,

$$q_G = \frac{d_{u17.3}}{d} (100) \tag{8-20}$$

This measure of form has three advantages over normal and absolute form quotients:

1. The top of the first log is close enough to the ground so that the diameter at that point may be accurately estimated or measured.
2. The reference diameter is near enough to the ground to give a measure of butt swell.
3. Bark thickness is taken into account and its effect on taper partially eliminated.

As mentioned in Chapter 9, Girard form class is a useful form quotient that has been widely employed in U.S. forest practice.

Efforts to develop form quotients that can be computed from more accessible diameters have led to a number of form quotients of the type advocated by Maass (1939). Maass's form quotient is the ratio of the diameter at 2.3 meters above the ground to diameter at 1.3 meters above the ground. Unpublished studies by Miller (1952), however, of similar form quotients indicated that this quotient, whether measurements are inside or outside bark, is too variable for trees of the same species, diameter, and height to be of practical use.

8-4.3 Form Point

Form point is the percentage ratio of the height to the center of wind resistance on the tree, approximately at the center of gravity of the crown, to the total tree height.

It was hypothesized by Jonson (1912) that the development of the form of a tree stem, as exemplified by the absolute form quotient, depends on the mechanical stresses to which the tree is subjected. These stresses come from the dead weight of the stem and crown, and the wind force. The wind force, it was concluded, is the most important stress, and "causes" the tree to "construct" its stem in such a way that the relative resistance to fracture or shearing will be the same at all points on the longitudinal axis of the stem. Thus, the main determinative of stem form is the focal point of wind force, or the point of greatest resistance to wind bending. Since the crown offers most of the resistance, it will be the location of the focal point of the wind force. Thus, the point where the wind resistance is the greatest is approximately at the center of gravity of the crown; it will vary with the size and shape of the crown. Except as they affect the size and shape of the crown, such tree characteristics as diameter, height, species, and age, and such things as the site factors, do not, so it is claimed, affect stem form. The greater the form point, the more nearly cylindrical will be the form of the tree.

In cases where the form point has been used, the average form points for the various diameter classes of a stand are obtained by sampling and, with these values as the independent variables, form classes are read from curves or tables (Fogelberg, 1953). Also, with the form point of an individual tree as the independent variable, the form class for an individual tree may be read from a curve or table.

It should be emphasized that the main value of the form point is to predict the absolute form quotient; there appears to be a good correlation between form quotient and form point. There are, however, some serious limitations to the use of form point. The focal point of wind resistance within the crown of any tree varies with the density of the crown and the position of the crown in the stand. Thus, the point is difficult to locate; two estimators will differ considerably on their selection.

8-4.4 Taper Tables, Curves, and Formulas

If sufficient measurements of diameters are taken at successive points along the stems of trees, one can prepare average taper tables that give a good picture of stem form. The ultimate purpose of all taper tables is to portray stem form in such a way that the data can be used in the calculation of stem volume or in the construction of volume tables. There are several ways of preparing taper tables (Chapman and Meyer, 1949), but the method of greatest utility is to average upper-log taper rates inside bark for standard log lengths according to diameter breast high and merchantable height in standard logs (Table 8-5).

Since taper tables may be expressed in curve form, it is logical to express a taper curve by a mathematical function. And it is logical to use the formula to obtain the volume of the solid of revolution. Many formulas have been developed, but as Grosenbaugh (1966) said in his comprehensive study of stem form:

> Many mensurationists have sought to discover a single simple two-variable function involving only a few parameters which could be used to specify the entire tree profile. Unfortunately, trees seem capable of assuming an infinite variety of shapes, and polynomials (or quotients of polynomials) with degree at least two greater than the observed number of inflections are needed to specify

Table 8-5
Average Upper-Log Taper Inside Bark (inches) in 16-Foot Logs

dbh (inches)	2-Log Tree 2nd Log	3-Log Tree 2nd Log	3rd Log	4-Log Tree 2nd Log	3rd Log	4th Log
10	1.4	1.2	1.4			
12	1.6	1.3	1.5	1.1	1.4	1.9
14	1.7	1.4	1.6	1.2	1.5	2.0
16	1.9	1.5	1.7	1.2	1.6	2.1
18	2.0	1.6	1.8	1.3	1.7	2.2
20	2.1	1.7	1.9	1.4	1.8	2.4
22	2.2	1.8	2.0	1.4	2.0	2.5
24	2.3	1.8	2.2	1.5	2.2	2.6
26	2.4	1.9	2.3	1.5	2.3	2.7
28	2.5	1.9	2.5	1.6	2.4	2.8
30	2.6	2.0	2.6	1.7	2.5	3.0

SOURCE: Mesavage and Girard, 1946.

variously inflected forms. Furthermore, coefficients would vary from tree to tree in ways that could only be known after each tree has been completely measured. Thus, explicit analytic definition of tree form requires considerable computational effort, yet lacks generality. . . . Each tree must be regarded as an individual that must be completely measured, or else as a member of a definite population whose average form can only be estimated by complete measurement of other members of the population selected according to a valid sampling plan. . . . Hence, polynomial analysis may rationalize observed variation in form after measurement, but it does not promise more efficient estimation procedures.

Thus, it appears unwise to derive complicated relationships to characterize tree form. This notion is corroborated by Kozak and Smith (1966) who, after studying the multivariate techniques of Fries (1965) and Fries and Matern (1965), concluded that the use of simple functions, sorting, and graphical methods is adequate for many uses in operations and research.

9
Log Rules and Volume Tables

A *log rule* is a table or formula that gives the estimated volume of logs of specified diameters and lengths. A *volume table* is a tabulated statement of the average volumes of trees by one or more tree dimensions. In the United States most log rules, and a majority of volume tables, give volumes in board feet of lumber, although tables that give volume in cubic units are available. In nations where the International System of Units is used, log rules and volume tables usually give volumes in cubic meters.

9-1 BOARD-FOOT LOG RULES

One might think that it would be easy to prepare a board-foot log rule that would be universally applicable. This is not the case, however, because of variations in the dimension of lumber produced from logs, in the equipment used to saw logs, in the skill of operators, in computer programs used to saw logs, and in the logs. All of these things affect the amount of lumber obtained from any run of logs.

In the early years of the lumber industry in the United States and Canada, there were a number of independent marketing areas, and no industrial organization or governmental agency had control over the measurement of lumber or logs. As a result, different areas devised different rules to fit specific operating conditions. As a consequence, as Freese (1973) points out in his excellent report on log rules, "In the United States and Canada there are over 95 recognized rules bearing about 185 names." (Most of these "over 95 recognized rules" have long since been forgotten. Only five or six rules are important in present-day use. There are perhaps half a dozen others that may be encountered in certain localities.)

Board-foot log rules are used to estimate the contents of logs. This constitutes an attempt to estimate, before processing, the amount of lumber in logs. Thus, we must distinguish between the measurement of the board-foot contents of sawn lumber, that is, *mill tally*, and the estimation of the board-foot contents of logs, that is, *log scale*. The board-foot mill tally, though not exact, is a well-defined unit; the board-foot log scale is an ambiguous unit. Thus, the amount of lumber sawed from any run of logs rarely agrees with the scale of the logs. This variation O_v may be expressed in board feet.

$$O_v \text{ (in board feet)} = \text{Mill tally} - \text{Log scale}$$

114

When O_v is positive, a mill has produced an *overrun;* when O_v is negative, a mill has produced an *underrun*.

Foresters, however, have typically expressed O_v as a percentage of log scale.

$$O_v \text{ (in percent)} = \left(\frac{\text{Mill tally} - \text{Log scale}}{\text{Log scale}}\right) 100$$

$$= \left[\left(\frac{\text{Mill tally}}{\text{Log scale}}\right) - 1\right] 100 \tag{9-1}$$

where the ratio Mill tally/Log scale is the overrun ratio, that is, the number of board-feet mill tally per board-foot log scale.

In the construction of all known board-foot log rules, three basic methods have been used: (1) mill study, (2) diagram, and (3) mathematical.

For the log rules in present-day use, generally the yield of logs V in board feet is estimated in terms of lumber 1-inch thick, from average small-end diameter inside bark D in inches, and log length L in feet.

9-1.1 Mill-Study Log Rules

In this method of constructing log rules a sample of logs is first measured on the log deck. Then, as each log is sawed, the boards are measured to determine the board-foot volume of the log. The log rule is prepared by relating board-foot yields, the dependent variable, with log diameters and lengths, the independent variables. The problem may be solved graphically or by the method of least squares.

A rule of this type should give good estimates for mills that cut timber with peculiar characteristics, or for those that use specific milling methods. The method, however, has never been widely used. The Massachusetts log rule, one of the few rules constructed by this method that is still in use, was based on 1200 white pine logs. This rule was constructed for round- and square-edged boards sawed from small logs (¼-inch saw kerf). Some boards over 1-inch thick were included, so the values are slightly high for 1-inch boards.

9-1.2 Diagram Log Rules

The procedure for the construction of a diagram log rule is simple.

1. Draw circles to scale to represent the small ends of logs of different diameters inside bark. Assume logs are cylinders of a specific length, such as 8 feet.
2. Use definite assumptions on saw kerf and shrinkage, and board width, and draw boards (rectangles) 1-inch thick within the circles.
3. Compute the total board-foot content for each log diameter.
4. Determine the board-foot contents of other log lengths by proportion.

When a diagram log rule is prepared for any given log length, it will be found that increases in volume, from one diameter to the next, will be slightly irregular. These irregularities may be eliminated by preparing a freehand curve, or a regression equation, to predict volume from diameter for each length.

The Scribner log rule, the most widely used diagram log rule, was first published in 1846 by J. M. Scribner, a country clergyman. The rule was prepared for 1-inch lumber with a ¼-inch allowance for saw kerf and shrinkage. The minimum board width is unknown. The original table gave board-foot contents for logs with scaling diameters (diameters inside bark at small end) from 12 to 44 inches, and with lengths from 10 to 24 feet. Log taper was not considered. A few years after the original rule was published, Scribner modified the rule by increasing the slab allowance on larger logs. This is the rule in use today. The Scribner rule gives a relatively high overrun (up to 30 percent) for logs under 14 inches. Above 14 inches the overrun gradually decreases and flattens out around 28 inches to about 3 to 5 percent.

A regression equation was prepared from the original Scribner table by Bruce and Schumacher (1950). The equation, which gives volume in board feet V, in terms of scaling diameter in inches D and log length in feet L, is

$$V = (0.79D^2 - 2D - 4)\frac{L}{16} \tag{9-2}$$

Some so-called Scribner tables contain values based on this equation. Values in these tables differ slightly from the original Scribner values because Scribner did not smooth the values he obtained from his diagrams. Note that the Scribner values in Table 9-1 are from the original Scribner table.

Since calculating machines were not generally available in the nineteenth century, scalers found the adding of long columns of figures laborious. Consequently, the Scribner rule was often converted into a *decimal rule* by dropping the units and rounding the values to the nearest 10 board feet. Thus, 114 board feet was written 11, and 159 board feet was written 16.

Because the original Scribner rule did not give values for logs less than 12 inches in diameter, a number of lumber companies extrapolated to derive volumes for small logs. Finally, the Lufkin Rule Company prepared three tables using different assumptions to extend the rule to cover small logs. They published these as decimal rules, and called them Scribner decimal A rule, Scribner decimal B rule, and Scribner decimal C rule. The decimal C rule is the only one of these rules still widely used.

There are other log rules based on diagrams. The best known are the Spaulding rule, which was devised by N. W. Spaulding of San Francisco in 1868, and the Maine rule, which was devised by C. T. Holland in 1856. The Spaulding rule is used on the Pacific Coast of the United States; the Maine rule is used in northeastern United States. The Spaulding rule, which closely approximates the values of the Scribner rule, may be expressed by the following regression equation, where V is volume in board feet, D is scaling diameter in inches, and L is log length in feet.

$$V = (0.778D - 1.125D - 13.482)\frac{L}{16} \tag{9-3}$$

9-1.3 Mathematical Log Rules

In this method of constructing log rules one makes definite assumptions on saw kerf, taper, and milling procedures and prepares a formula that gives board-foot yield of logs in terms of their diameters and lengths. As will be seen, this is not a regression analysis.

The *Doyle log rule*—one of the most widely used, one of the oldest, and one of the most cursed log rules—was first published in 1825 by Edward Doyle. The rule states: "Deduct 4 inches from the diameter of the log, D, in inches, for slabbing, square one-quarter of the remainder, and multiply by the length of the log, L, in feet." As Herrick (1940) pointed out, when Doyle deducted 4 inches from the diameter of the log for slabbing, he was squaring the log. Then, he calculated the board-foot contents of the squared log, or cant, as follows.

$$\frac{(D - 4)(D - 4)L}{12}$$

To allow for saw kerf and shrinkage, he reduced the volume of the cant by 25 percent to obtain the final rule.

$$V = \frac{(D - 4)^2 L}{12} (1.00 - 0.25) = \left(\frac{D - 4}{4}\right)^2 L \tag{9-4}$$

For 16-foot logs the *Doyle rule of thumb* is

$$V = (D - 4)^2$$

This all points up an important reason this rule gained wide acceptance; it was genuinely simple, and it could be easily applied.

When the Doyle rule is applied to logs between 26 and 36 inches in diameter, it gives good results. When the rule is applied to large logs, it gives an underrun; when the rule is applied to small logs, it gives a high overrun. This comes about because the 4-inch slabbing allowance is inadequate for large logs and excessive for small logs.

The *International log rule*, one of the most accurate mathematical log rules, was developed by Judson F. Clark in 1900 when he was working for the Province of Ontario. It was published in 1906 (Clark, 1906). The derivation of the original rule is logical and simple. One first computes the board-foot contents of a 4-foot cylinder in terms of cylinder diameter in inches D, assuming the cylinder will produce lumber at the rate of 12 board feet per cubic foot.

$$\begin{array}{c} \text{Solid board-foot contents} \\ \text{of 4-foot cylinder} \end{array} = \frac{\pi D^2}{4(144)} (4)(12) = 0.262 D^2$$

To allow for saw kerf and shrinkage one assumes that, for each 1-inch board cut, $\frac{1}{8}$ inch will be lost in saw kerf and $\frac{1}{16}$ inch in shrinkage. Thus, the proportion lost from saw kerf and shrinkage is

$$\left(\frac{3/16}{1 + 3/16}\right) \qquad \text{or} \qquad 0.158$$

When reduced by 0.158, the volume of the 4-foot cylinder becomes

$$0.262 D^2 (1.000 - 0.158) = 0.22 D^2$$

Thus, the losses from saw kerf and shrinkage are proportional to the end area of the log (i.e., to D^2).

Clark determined that losses from slabs and edgings constitute a ring-shaped collar around the outside of the log and that they are proportional to the surface area, or the diameter D of the log. And from a careful analysis of the losses occurring during the conversion of sawlogs to lumber, Clark found that a plank 2.12 inches thick and D inches wide will give the correct deduction. (The thickness of the collar T is about 0.7 inches for all values of D. This can be determined from the following equation.

$$2.12 D = \pi \frac{D^2}{4} - \frac{\pi (D - 2T)^2}{4}$$

which leads to $T^2 - DT + 0.6748D = 0$.) Therefore, in terms of cylinder diameter in inches D the board-foot deduction is computed using equation 7-2.

$$\frac{2.12(D)(4)}{12} = 0.71D$$

Thus, the net board-foot volume V of a 4-foot cylinder is

$$V = 0.22 D^2 - 0.71 D \qquad (9\text{-}5)$$

After studying northeastern tree species, Clark decided to allow a taper of $\frac{1}{2}$ inch for each 4-foot section. With this assumption in mind, the basic formula was expanded to cover other log lengths.

$$V \ \ (8\text{-foot logs}) = 0.44 D^2 - 1.20 D - 0.30 \qquad (9\text{-}6)$$

$$V \ (12\text{-foot logs}) = 0.66 D^2 - 1.47 D - 0.79 \qquad (9\text{-}7)$$

$$V \ (16\text{-foot logs}) = 0.88 D^2 - 1.52 D - 1.36 \qquad (9\text{-}8)$$

$$V \ (20\text{-foot logs}) = 1.10 D^2 - 1.35 D - 1.90 \qquad (9\text{-}9)$$

Clark specified that lengths over 20 feet were to be scaled as two or more logs.

The volume of a log of intermediate length (e.g., 3, 10, 14, or 17 feet) can be calculated by application of the formula given by Grosenbaugh (1952a), or the following algorithm developed for use with programmable calculators (Beers, 1980).

1. Given log length L_i and scaling diameter D_i, calculate $L_i/4 = n.ff$; where n = integer part of the quotient and ff = fractional part of the quotient (.00, .25, .50, and .75).

2. Then, the volume of the ith log V_i is calculated
   a. For $4 \leqslant L_i \leqslant 20$, by

$$V_i = \sum_{j=1}^{n} \left[.22\left(D_i + \frac{n-j}{2}\right)^2 - .71\left(D_i + \frac{n-j}{2}\right) \right]$$

$$+ ff \left[.22\left(D_i + \frac{n}{2}\right)^2 - .71\left(D_i + \frac{n}{2}\right) \right]$$

   b. For $L_i < 4$, by

$$V_i = ff(.22D_i^2 - .71D_i)$$

   c. For $L_i > 20$, by dividing the log into shorter sections and following the above procedure.

The original International log rule may be modified to give estimates for saw kerfs other than $\frac{1}{8}$ inch. For example, for a kerf of $\frac{1}{4}$ inch (shrinkage of $\frac{1}{16}$ inch), the proportion lost is

$$\left(\frac{5/16}{1 + 5/16} \right) = 0.238$$

So the original rule may be converted to a $\frac{1}{4}$-inch rule by multiplying the values by the following factor.

$$\frac{1.000 - 0.238}{1.000 - 0.158} = 0.905$$

The factor to convert to a $\frac{7}{64}$-inch rule is 1.013; the factor to convert to a $\frac{3}{16}$-inch rule is 0.950.

When Clark (1906) published his log rule in table form, he rounded all values to the nearest multiple of 5 board feet. This is the form in which most International log rule tables appear today. The rule may also be presented as a decimal rule (U.S. Forest Service, 1977).

Other formula rules have been constructed, but they have never enjoyed the popularity of the Doyle and International log rules.

9-1.4 Combination Log Rules

This type of rule combines values from different log rules. Such a rule takes advantage of the best, or the worst, features of the rules used. For example, the *Doyle-Scribner rule*, a combination of the Doyle and Scribner rules, was prepared for use in defective and overmature timber. Since the Doyle rule gives an overrun for small logs, its values were used for diameters up through 28 inches. Since the Scribner rule gives an overrun for large logs, its values were used for diameters greater than 28 inches. Thus, the

Doyle-Scribner rule, which is still used by some private operators in the South, gives a consistently high overrun that is supposed to compensate for hidden defects.

The Scribner-Doyle rule, exactly opposite to the Doyle-Scribner rule, gives a consistently low overrun.

9-1.5 Comparison of Log Rules

Because different methods and assumptions are used in the construction of log rules, different rules give different results, none of which will necessarily agree with the mill tally for any given log (Table 9-1). This points out that the board-foot log scale, by any rule, is a unit of estimate, not a unit of measure.

9-2 CUBIC-VOLUME LOG RULES

The cubic volume of a log may be determined by any of the methods given in Section 8-1. However, for log scaling, gross cubic volumes are usually computed by Huber's or Smalian's formula. The formulas could be used directly, but it is more convenient to prepare tables. In the preparation of cubic meter tables, volumes are most often computed for 2-centimeter diameter classes and 0.2-meter length classes. In the preparation of cubic foot tables, volumes are most often computed for 1-inch diameter classes and 1-foot length classes. Diameter is often given as the midpoint diameter inside bark. (When the midpoint diameter cannot be measured, the average of the diameters inside bark at the two ends of the log is generally used for midpoint diameter.) In some cases, volume is given in terms of small-end diameter inside bark.

Table 9-1
Volume of 16-Foot Logs

Log Diameter (inches)	Mill Tally[1] (board feet)	Log Scale				
		Scribner	Maine	International $\frac{1}{4}$-Inch Rule (board feet)	Doyle	Spaulding
6		18	20	20	4	
10	75	50	68	65	36	
14	145	114	142	135	100	114
18	229	213	232	230	196	216
22	382	334	363	355	324	341
26	578	500	507	500	484	488
30	665	657	706	675	676	656
34	862	800	900	870	900	845
38	1037	1068	1135	1095	1156	1064

[1] Average yield of logs sawed in an Indiana band mill.

9-3 LOG SCALING

Scaling is the determination of the gross and net volumes of logs in board feet, cubic feet, cubic meters, or other units. The determination of gross scale consists of measuring log length and diameter and determining the volume by a log rule. The volume values are usually read from a scale stick, a flat stick that has volumes for different log diameters and lengths printed on its face.

9-3.1 Board-Foot Scaling

In board-foot scaling a maximum length of 40 feet is standard for the western regions of the United States; 16 feet is standard for the eastern regions. When logs exceed the maximum scaling length, they are scaled as two or more logs. If a log does not divide evenly, the butt section is assigned the longer length. The scaling diameter for the assumed point of separation can be estimated from the taper of the log. Although logs are most commonly cut and measured in even lengths (i.e., 8, 10, 12, 14, and 16 feet), they may be cut and measured, particularly with hardwood logs, in both odd and even lengths (i.e., 8, 9, 10, 11, and 12 feet). Logs must be cut longer than standard lumber lengths because it is impossible to buck logs squarely and because there is logging damage to log ends. This extra length, which will range from 3 to 6 inches, depending on the size of timber, products sawed, and logging methods, is called *trim allowance*.

The *National Forest Log Scaling Handbook* (U.S. Forest Service, 1977) gives these instructions on trim allowance for board-foot scaling.

> *Contract trim allowances are normally the permissive maximums. Regularly tape-measure enough lengths to insure proper observance of trim. Scale logs overrunning the trim allowance to the next 1-foot scaling measure in length unless otherwise instructed. For example, if 6 inches is the contract trim allowance for logs 8 to 20 feet in length, a log measuring 20 feet 10 inches is scaled as a 21; one measuring 24 feet 10 inches, as a 24; but one measuring 25 feet 2 inches, as a 25-foot log; 32 feet 0 inches, as a 31; or 32 feet 2 inches, as a 32; 41 feet 2 inches, as a 41.*

Most board-foot log rules call for diameter measurements inside bark, to the nearest inch, at the small end of the log. When a log is round, one measurement is enough. When a log is eccentric, as most logs are, the usual practice is to take a pair of measurements at right angles across the long and short axes of the log end and to average the results to obtain the scaling diameter.

To determine net scale one must deduct from gross scale the quantity of lumber, according to the log rule used, that will be lost due to defects. These deductions do not include material lost during manufacturing, or defects that affect the quality of the lumber. Instead, they include those defects that reduce the volume of lumber.

Detailed procedures have been worked out to estimate the volume of deductions in defective logs (see U.S. Forest Service, 1977). But the most logical and applicable system was proposed by Grosenbaugh (1952a). In this method the amount of material lost in defect is estimated by multiplying the gross scale by the proportion of the log affected. The system works, with minor modifications, regardless of the units in which the log is measured: board feet by any log rule, cubic feet, cubic meters, cords, pounds, kilograms, and so forth. The procedures for common defects can be summarized in the following five rules, where scaling diameter is defined as the average inside bark diameter at the small end, and measurements are in English units.

1. When defect affects entire section, the proportion P lost is

$$P = \frac{\text{Length of defective section}}{\text{Log length}}$$

EXAMPLE

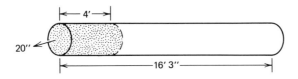

Gross scale, Scribner decimal C rule = 28
Cull = (4/16)(28) = 7   −7
Net scale   21

2. When defect affects wedge-shaped sector.

$$P = \left(\frac{\text{Length of defective section}}{\text{Log length}}\right)\left(\frac{\text{Central angle defect}}{360°}\right)$$

EXAMPLE

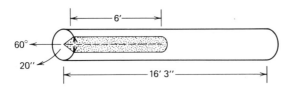

Gross scale, Scribner decimal C rule = 28
$\text{Cull} = \left(\dfrac{6}{16} \cdot \dfrac{60}{360}\right)(28) = 2$ −2
Net scale   26

3. When log sweeps (ignore sweep less than 2 inches).

$$P = \frac{\text{Maximum departure minus 2}}{\text{Scaling diameter}}$$

EXAMPLE

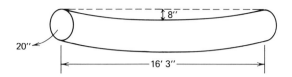

Gross scale, Scribner decimal C rule = 28

$$\text{Cull} = \left(\frac{8-2}{20}\right)(28) = 8 \qquad -8$$

Net scale $\overline{20}$

4. When log crooks.

$$P = \left(\frac{\text{Maximum deflection}}{\text{Scaling diameter}}\right)\left(\frac{\text{Length of deflecting section}}{\text{Log length}}\right)$$

EXAMPLE

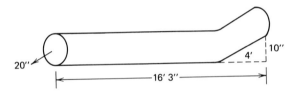

Gross scale, Scribner decimal C rule = 28

$$\text{Cull} = \left(\frac{10}{20}\cdot\frac{4}{16}\right)(28) = 3 \qquad -3$$

Net scale $\overline{25}$

5. When average cross section of interior defect is enclosable in an ellipse or circle.

$$P = \frac{(\text{Major diameter}+1)(\text{Minor diameter}+1)}{(\text{Scaling diameter}-1)^2}\left(\frac{\text{Defect length}}{\text{Log length}}\right)$$

(Defect in peripheral inch of log (slab collar) can be ignored.)

EXAMPLE

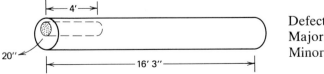

Defect Diameters:
Major = 9 inches
Minor = 7 inches

Gross scale, Scribner decimal C rule $= 28$

$$\text{Cull} = \frac{(9 + 1)(7 + 1)}{(20 - 1)^2}\left(\frac{4}{16}\right)(28) = 2 \qquad \begin{array}{r} -2 \\ \hline \end{array}$$

Net scale $\qquad\qquad\qquad\qquad\qquad\qquad\qquad\qquad\qquad \begin{array}{r} 26 \\ \hline \end{array}$

9-3.2 Cubic Volume Scaling

Cubic volume has its greatest utility when the volumes obtained give an accurate estimate of true volume. Thus, log measurement must be consistent. Then, log volumes in cubic units can be converted to the unit of measure appropriate to each manufacturing plant with less uncertainty than in convertting from board-feet log scale to board feet of lumber, or from board-feet log scale to square feet of veneer.

In cubic volume scaling, log diameters and log lengths are taken as explained in Section 9-2. In making deductions for defects the five rules given in Section 9-3.1 are applicable. If cubic feet are used, diameter is average small-end diameter in inches and length is in feet, and the rules are used without modifications. If cubic meters are used, diameter is average small-end diameter in centimeters and length is in meters, and the third and fifth rules are changed as follows.

3. When log sweeps (ignore sweep less than 5 centimeters).

$$P = \frac{\text{Maximum departure minus 5}}{\text{Scaling diameter}}$$

5. When average cross section of interior defect is enclosable in an ellipse or circle.

$$P = \frac{(\text{Major diameter} + 2)(\text{Minor diameter} + 2)}{(\text{Scaling diameter} - 2)^2}\left(\frac{\text{Defect length}}{\text{Log length}}\right)$$

9-3.3 Unmerchantable Logs

The definition of a cull, or unmerchantable, log is largely a local matter. Merchantability varies with species, economic conditions, and other factors. However, no matter what units are employed, specifications for a merchantable log should give the minimum length allowed, and minimum diameter allowed, and the minimum percent of sound material left after deductions are made for cull. For example, a cull log might be defined as any log less than 8 feet long, less than 6 inches in diameter, or less than 50 percent sound.

9-3.4 Sample Scaling

Under conditions where the scaling operation interferes with the movement of the logs, or where scaling costs are high, sample scaling should be considered. Sample scaling is generally feasible when: (1) logs are fairly homogeneous in species, volume, and value, (2) logs are concentrated in one place so they can be scaled efficiently, and (3) total number of logs is large.

Once one has decided to use sample scaling, one is faced with two basic questions: How many logs must be scaled to determine the total scale within limits of accuracy acceptable to both buyer and seller? How should the sample logs be selected?

The number of logs n to measure can be calculated from the formula applicable to a finite population.

$$n = \frac{CV^2 t^2 N}{Na^2 + CV^2 t^2} \qquad (9\text{-}10)$$

where

CV = coefficient of variation expressed as a percent
t = t-value corresponding to chosen probability
N = total number of logs in population
a = desired standard error of mean expressed as a percent of mean

An example will illustrate the use of equation 9-10. Let us assume CV is 50 percent for a population N of 10,000 logs, and the desired standard error is 3 percent (both CV and N are estimates). Then if we let t be 2, giving approximately 20 to 1 odds that a chance discrepancy between the estimated and true scale will not exceed 3 percent, we obtain

$$n = \frac{50^2(2^2)(10,000)}{10,000(3^2) + 50^2(2^2)} = 1000$$

A practical procedure to obtain the 1000-log sample would be to scale every tenth log—that is, take a systematic sample. Of course, to obtain total volume, every log must be counted since the total number of logs N used to calculate sample size n is an estimate.

Although random sampling is required if one desires to calculate valid sampling errors, it is not essential if the sole purpose of sampling is to obtain an unbiased estimate of the average volume per log and the total volume, for a given run of logs.

Johnson, Lowrie, and Gohlke (1971) describe how the 3P sample selection procedure can be applied to sample log scaling. This is a promising procedure that is probably more efficient in most situations than the conventional sample log scaling method described above. Indeed, even if the logs are not homogeneous in species, volume, and value, it can be used efficiently.

9-4 STACKED VOLUME

Stacked volume, which is discussed in Chapter 7, has traditionally been obtained for firewood, pulpwood, excelsior wood, charcoal wood, and other relatively low-value products that are assembled in stacks.

In scaling a stack of wood one first records the length—the average of measurements taken on both sides of the stack, to the nearest 0.1 foot. Then, stack height is obtained by averaging measurements taken at intervals of about 4 feet. The height, which is reduced about 1 inch per foot by some scalers to compensate for settling and shrinkage, is recorded to the nearest 0.1 foot. Finally, piece lengths are checked to see if they vary from the lengths specified in the sale or purchase contract (standard lengths for pulpwood cut in the United States are 4 feet, 5 feet, 5 feet 3 inches, and 8 feet 4 inches). If they do, the procedure given in the contract should be followed.

The volume in standard cords V_c of a stack of wood is calculated as follows.

$$V_c = \frac{L_s H_s L}{128} \tag{9-11}$$

where

L_s = stack length in feet
H_s = stack height in feet
L = stick length in feet

If stacks are piled on slopes, the length and height measurements should be taken at right angles to one another.

If the stacked volume is measured in cubic meters (1 cubic meter = 1 stere), stack length and height are measured in 2-centimeter classes, and piece lengths are checked as described above. Then, gross volume of stack in cubic meters V_m is

$$V_m = L \times H \times W \tag{9-12}$$

where

L = stack length in meters
H = stack height in meters
W = stick length in meters

Since the above procedure gives gross stacked volume, to obtain net volume deductions must be made for defective wood. The definitions of defects and the procedures of allowing for defects will vary from one organization to another. But in general deductions are made for *defective sticks* and *loose piling*.

Defective sticks include rotted sticks, burned sticks, undersized sticks, and peeled sticks with excessive bark adhering.

Loose piling may occur when knots have been improperly trimmed, when excessively crooked wood is present, and when sticks have been carelessly piled.

When making deductions for defective sticks, the scaler examines each stick in a pile and notes which sticks will not meet specifications. These sticks are then culled by deducting the cubic space they occupy from the gross cubic space occupied by the pile—either a stick is acceptable or it is not acceptable. Deductions for loose piling are made by estimating the cubic space that would be occupied by sticks that could be included in the loose pile and subtracting this volume from the gross cubic space occupied by the pile.

The term *rough wood* is used to designate wood with bark in contrast to the term *peeled wood*, which refers to wood with bark removed. It should be made clear in a sales contract whether wood is to be measured *rough* or *peeled*. If the sale price is based on rough wood volume, then if peeled wood must be measured, volume must be increased 10 to 20 percent, depending on bark thickness (Section 8-3).

9-4.1 Determination of Solid Cubic Contents of Stacked Wood

It is often necessary to know the solid cubic contents (standard cord, long cord, stere, etc.) of wood that is stacked on the ground or on trucks. Although average converting factors, such as those given in Chapter 7, are often used, better factors are generally required. These can be determined by the following methods.

1. Direct Measurement.   The cubic volume of individual sticks, or of groups of sticks, can be determined by displacement (Chapter 8). The cubic volume of individual sticks can also be computed by using Huber's, Smalian's, or Newton's formula. In any case, when the cubic space occupied by a pile is known, the ratio of solid cubic volume to total cubic volume can be calculated from the equation

$$f = \frac{\text{Solid cubic volume of pile}}{\text{Total cubic volume of pile}}$$

The factor f multiplied by the space occupied by a cord, or by an entire pile, will give the solid cubic volume of wood in the stack.

2. Photographic Methods.   The factor f can be estimated from photographs of the ends of sticks in a pile. One sets up, or holds, the camera at a distance from the pile that will give a scale of about 1:30 (Chapter 13). The optical axis of the lens should be perpendicular to the side of the pile. Normally, only a portion of a stack or truck load is included in a single photograph. After a photograph is developed, a templet consisting of systematically spaced pinholes is placed over the photograph and perforated with a needle at each pinhole. Then the photograph is placed on a light table for counting (Fig. 9-1). (A transparent dot grid with systematically spaced dots may be used, but it is more difficult to count dots than pinholes.) The factor f is computed as follows.

$$f = 1 - \frac{\text{(Total dots in air spaces)}}{\text{(Total dots in photograph)}}$$

Although one photograph of a truck load will usually give an adequate sample, several photographs of large stacks may be required.

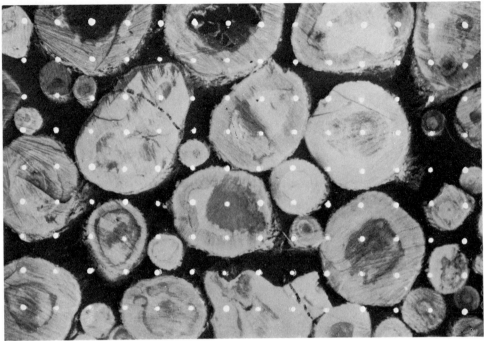

Fig. 9-1 *Perforated polaroid photograph of portion of a truck load of low-quality hardwoods. Solid wood contents of this sample is 79 cubic feet per standard cord (f = 0.62). A 20 percent photo sample of each load is sufficient to give ±2.4 percent accuracy at the 95 percent confidence level when a dot grid with 16 systematically spaced dots, or pinholes, per square inch is used (Garland, 1968).*

Cameras with film dimensions of $2\frac{1}{4}$ by $2\frac{1}{4}$ inches to 4 by 5 inches are recommended, because contact prints prepared from these films may be used without magnification to obtain accurate dot counts. If 35-mm film is used, enlargement or magnification is required to obtain accurate dot counts.

Kallio, Lothner, and Marden (1973) tested another logical photographic method to obtain the solid cubic contents of stacked wood. They placed an 8-foot scale rule against the stack so scale could be determined, and took the photographs, as explained above, to obtain 35-mm colored slides. Then they projected the slides on a rear projection screen at about half actual size and measured each stick directly on the screen. They found that solid cubic volume can be determined more accurately by this method than by dot counting. The method, however, is slower.

In general, it is not feasible to determine bark volume by the photographic method, because the line between the inner bark and the wood is not always clear.

3. Angle-Gauge Method. By "projecting" an angle of about 23 degrees parallel to the face of a stack from randomly selected points, the conversion factor *f* may be quickly and efficiently obtained. This method, which is a modification of horizontal point sampling, is discussed by Loetsch, Zöhrer, and Haller (1973).

9-5 ESTIMATION OF TREE VOLUMES

Tree volumes can be estimated from previously established relationships between certain tree dimensions and tree volume. Diameter, height, and form are the independent variables that are commonly used to determine the values of the dependent variable—tree volume. The final result is presented in formula or table form. The *volume formula* or *volume table*, then, gives the average contents of individual trees (in board feet, cubic feet, cubic meters, cords, or other units) in terms of one or more of the previously mentioned tree dimensions.

Local Volume Tables.   Local volume tables give tree volume in terms of diameter at breast height only. The term *local* is used because such tables are generally restricted to the local area for which the height-diameter relationship hidden in the table is relevant. Although local volume tables may be prepared from raw field data—that is, from volume and diameter measurements for a sample of trees—they are normally derived from standard volume tables. Table 9-2 shows a typical local volume table.

Standard Volume Tables.   Standard volume tables give volume in terms of diameter at breast height and merchantable or total height. Tables of this type may be prepared for individual species, or groups of species, and specific localities. The applicability of a standard volume table, however, depends on the form of the trees to which it is applied rather than on species or locality; for each diameter-height class the form of the trees to which the table is applied should agree with the form of the trees from which the table was prepared. Table 9-3 shows a typical standard volume table.

Form Class Volume Tables.   Form class volume tables give volume in terms of diameter at breast height, merchantable or total height, and some measure of form, such as Girard form class or absolute form quotient. Such tables come in sets, with one table for each form class. The format of each table is similar to that of a standard volume table. Note that if a single form class table is chosen as representative of a stand, volume determinations may be in error because it is unlikely that all trees will be of the same form class. Furthermore, since form class varies with tree size, species, and site, it is unlikely that variations in form class will be random. Thus, it is difficult to obtain an accurate average form class for a stand and is therefore undesirable to use a single form class table for any extensive area.

9-5.1 Descriptive Information to Accompany Volume Tables

A volume table should include descriptive information that will enable one to apply it correctly. This information includes:

Table 9-2
Example of a Local Volume Table for Yellow Poplar
(*Liriodendron tulipifera*) in Stark County, Ohio, Using International Rule
(¼-Inch Kerf)—Merchantable Stem to a Variable Top Diameter

dbh Outside Bark (inches)	Volume per Tree (board feet)	Merchantable Length (feet)	Basis in Trees (number)
10	30	19.5	4
11	50	23	5
12	70	26.5	13
13	95	30	9
14	125	33	9
15	155	36.5	1
16	190	40	5
17	235	43	7
18	285	45.5	6
19	345	48	4
20	405	51	2
21	480	53.5	1
22	555	56	2
23	635	58	3
24	720	60	1
25	800	62	—
26	885	64	1
27	975	65.5	2
28	1065	67	1
29	1155	69	5
30	1245	70	9
31	1340	71.5	7
32	1435	72.5	7
33	1535	73.5	1
34	1630	74.5	1
35	1725	75	—
36	1825	76	—

Trees climbed and measured by personnel of Work Projects Administration Offical Project 65-1-42-166—the Ohio Woodland Survey. Measurements taken at 16-foot log lengths above a 2.0-foot stump height. Scaled as 16-foot logs, and additional shorter top logs; top sections less than 8 feet in length scaled as fractions of an 8-foot log. Basis, 107 trees.

Table prepared, in 1939, by curving volume of merchantable length over dbh.

Aggregate difference: Table is 0.8% low. Average percentage deviation of basic data from table, 19.4%.

SOURCE: Diller and Kellog, 1940.

1. Species, or species group, to which the table is applicable, or the locality in which the table is applicable.

2. Definition of dependent variable, that is, volume, including unit in which volume is expressed.

Table 9-3

Example of a Standard Volume Table, Using Board-Foot Volume, International ¼-Inch Rule, for Red Oak (*Quercus rubra*) in Pennsylvania

dbh (inches)	Merchantable Height—Number of 16-Foot Logs											
	½	1	1½	2	2½	3	3½	4	4½	5	5½	6
8	8	18	28	37	47	57						
9	11	23	35	48	60	73						
10	13	29	44	59	75	90	105	121				
11	17	35	54	72	91	109	128	146				
12	20	42	64	86	108	130	153	175	197			
13	24	50	76	102	127	153	179	205	231			
14	28	58	88	118	148	178	208	238	268	298	328	
15	33	67	102	136	170	205	239	274	308	343	377	
16	37	77	116	155	194	233	273	312	351	390	429	469
17		87	131	175	219	264	308	352	396	441	485	529
18		97	147	197	246	296	345	395	445	494	544	594
19		109	164	219	275	330	385	440	496	551	606	662
20		121	182	243	304	366	427	488	549	611	672	733
21			201	268	336	403	471	538	606	673	741	809
22			221	295	369	443	517	591	665	739	813	888
23			241	322	403	484	565	646	727	808	889	970
24			263	351	439	527	616	704	792	880	968	1057
25			285	381	477	572	668	764	860	955	1051	1147
26			309	412	516	619	723	826	930	1033	1137	1240
27				445	556	668	780	891	1003	1114	1226	1338
28				478	598	718	839	959	1079	1199	1319	1439
29				513	642	771	900	1028	1157	1286	1415	1544
30				549	687	825	962	1101	1239	1376	1514	1652
31					734	881	1028	1175	1323	1470	1617	1764
32					782	939	1096	1253	1409	1566	1723	1880
33					832	999	1165	1332	1499	1666	1833	1999
34					883	1060	1237	1414	1591	1768	1945	2122
35					936	1124	1311	1499	1686	1874	2062	2249
36					990	1189	1387	1586	1784	1983	2181	2380

Stump height, one foot. Top diameter, 8.0 inches, inside bark.
Block indicates extent of basic data. Basis, 210 trees.
Sample trees scaled as 16-foot logs; top section measured to nearest foot.
Standard error of regression coefficient = 0.00261.
Proportion of variation accounted for by the regression = 0.974.
Tabular values derived from regression $V = -1.84 + 0.01914D^2H$.

SOURCE: Bartoo and Hutnik, 1962.

3. Definition of independent variables, including stump height and top diameter limit, if merchantable height is used.
4. Author.
5. Date of preparation.
6. Number of trees on which table is based.
7. Extent of basic data.
8. Method of determining volumes of individual trees.
9. Method of construction.
10. Appropriate measures of accuracy.

Table 9-2 and Table 9-3 include these items.

The first three items in the above list should always be given. The remaining items are of less interest and are sometimes omitted. When measures of accuracy are given, they should be understood to be measures of accuracy of the table when it is applied to the data used in its construction. Such measures give no assurance that a volume table will apply to other trees. Thus, when an accurate estimate is required, a table should be checked against the measured volumes of a representative sample of trees obtained from the stands to be estimated.

9-5.2 Checking Applicability of Volume Tables

In an applicability check, one should compare the volume of sample trees with the estimated volume from the volume table to be checked. Three conditions should be observed in selecting sample trees.

1. Sample trees for a given species, or species group, should be distributed through the timber to which the volume table will be applied.
2. No sizes, types, or growing conditions should be unduly represented in the sample.
3. If a sample of cut trees is used, this sample, if not representative of the timber, should be supplemented by a sample of standing trees. (Measurements on standing trees can be made by methods described in Chapter 2.)

Definite rules for measuring sample trees should be established. For example, the following rules are satisfactory for the eastern United States.

1. Diameters along the tree stem, inside and outside bark, should be taken at 8-foot intervals above a 1-foot stump, and at stump height, breast height, and merchantable height.
2. Diameter should be measured to nearest $\frac{1}{10}$ inch and bark thickness to nearest $\frac{1}{20}$ inch.
3. Knots, swellings, and other abnormalities should be avoided at points of measurement by taking measurements above or below them.

4. Total or merchantable heights should be measured to nearest foot. (Utilization standards for the timber in question should be considered in determining the upper limit of merchantable height.)

Table 9-4 illustrates how the comparison of measured and estimated volumes of sample trees should be made. For practical purposes, the aggregate difference of a test sample should not exceed $2CV/\sqrt{n}$, where CV is the coefficient of variation of the volume table being tested, and n is the number of trees used in the test. Since the coefficient of variation for the table tested in Table 9-4 is 15 percent, the table is applicable without correction because

$$\frac{2(15)}{\sqrt{62}} = 3.8\% > 1.0\%$$

If desired, checks may be made by diameter classes. And, of course, more complicated statistical tests, such as Chi square, might be used. The above procedure, however, is generally satisfactory.

When a table is judged to be inapplicable, one should adjust the table or obtain a better table. Practical methods of making adjustments are described by Gevorkiantz and Olsen (1955).

9-6 CONSTRUCTION OF VOLUME TABLES

The principles of volume table construction given by Cotta early in the nineteenth century are still valid.

Table 9-4
Comparison of Measured and Estimated Volumes of Sample of Red Oak

dbh Class (inches)	Sample Trees (number)	Measured Volume (board feet)	Estimated Volume[1] (board feet)	Aggregate Difference (percent)
13.0–15.9	14	2,010	2,045	−1.7
16.0–18.9	10	2,003	1,943	+3.1
19.0–21.9	9	3,041	3,106	−2.1
22.0–24.9	21	9,257	8,895	+4.1
25.0–27.9	4	2,084	2,223	−6.3
28.0–30.9	3	2,130	2,110	+0.9
31.0–33.9	0	—	—	—
34.0–36.9	1	870	860	+1.2
All classes	62	21,395	21,182	+1.0

[1] From volume table.

SOURCE: Gevorkiantz and Olsen, 1955.

Tree volume is dependent upon diameter, height, and form. When the correct volume of a tree has been determined, it is valid for all other trees of the same diameter, height, and form.

Since the time of Cotta, hundreds of volume tables have been constructed and used. Numerous methods have been used to construct the tables. But since 1946 there has been a trend, particularly for hardwoods, to reduce the number of volume tables used by adopting composite volume tables, tables applicable to average timber, regardless of species. Indeed, where the same standards of utilization are employed, differences in tree volumes among species are often of no practical consequence. Excellent examples of composite volume tables are Beers' (1973) for hardwoods in Indiana, and Gevorkiantz and Olsen's (1955) for timber in the Lake States. These tables have been extensively tested and have been found to replace individual species tables, especially for the estimation of volume on large tracts. Adjustment factors can be used for individual species that vary from the average.

Why have so many volume tables been constructed? Why has so much research gone into the development of volume tables? The answer is that foresters have been looking for methods that are simple, objective, and accurate. However, because trees are highly variable geometric solids, no single table, or set of tables, could possibly satisfy all of these conditions, regardless of the method of construction. Consequently, one by one the older methods of volume table construction have been abandoned. For example, the once popular harmonized-curve method (Chapman and Meyer, 1949), which requires large amounts of data to establish the relationships and considerable judgment to fit the curves, is rarely used today. The alignment-chart method, another subjective method, has been generally discarded. Other discarded methods have been described by Spurr (1952). Today interest has focused on the use of mathematical functions, or models, to prepare volume tables. There is no advantage for the majority of foresters in using any other method.

9-6.1 Mathematical Models for Construction of Volume Tables

The following equations can serve as mathematical models for the construction of volume tables or as bases to develop other models.

Local volume table:	$V = aD^b$	(9-13)
Standard volume table:	$V = bD^2H$	(9-14)
	$V = a + bD^2H$	(9-15)
	$V = aD^bH^c$	(9-16)
Form class volume table:	$V = aD^bH^cF^d$	(9-17)

where

$$
\begin{aligned}
V &= \text{volume in cubic units or board feet} \\
D &= \text{dbh} \\
H &= \text{total or merchantable height} \\
F &= \text{a measure of form (Girard form class or absolute form} \\
&\quad \text{quotient)} \\
a, b, c, d &= \text{constants}
\end{aligned}
$$

Equation 9-14 is called the *constant form factor volume equation;* equation 9-15 is known as the *combined variable volume equation.* However, both equations have combined variables. The nature of the equations becomes clear when they are rewritten: $V = bX$ and $V = a + bX$, where $X = D^2H$.

Both of these models are predicated on the assumption that, when volume is plotted over X (i.e., D^2H), the trend is linear. The constants a and b may be determined graphically, but the least squares solution is preferable. For equation 9-14 the line is forced through the origin; for equation 9-15 the Y intercept is computed. The theory and details of the calculations can be found in any standard text on regression.

Equation 9-16, which was proposed by Schumacher and Hall (1933), is given in its nonlinear form. Its logarithmic form, a form which has been utilized to fit nonlinear tree volume equations, is

$$
\log V = \log a + b \log D + c \log H \tag{9-18}
$$

A logarithmic equation is more compatible with the homogeneity of variance assumption for regression. On the other hand, a bias, called "logarithmic transposal discrepancy," is introduced in fitting the logarithmic equation and in recalculating the standard error in arithmetic units for comparison with nonlogarithmic equations. [This bias and its correction are discussed by Meyer (1953), Brownlee (1967), and Baskerville (1972).] By using nonlinear functions to estimate parameters, and by employing weighting methods to correct heterogeneous variance about the regression line, nonlinear tree volume equations may be developed that retain the statistical advantages and overcome the shortcomings of the logarithmic equation. In fact, with the availability of computers, there appears to be little justification for the use of logarithmic models, except to obtain initial estimates of the coefficients.

Moser and Beers (1969) give a method of utilizing equation 9-16 by nonlinear regression. Since their procedure is one of the most feasible methods of volume table construction, and since it is applicable to other nonlinear functions, it will be discussed in more detail than the other methods.

To fit sample data to equation 9-16, it is necessary to estimate values of the parameters that minimize

$$
\sum_{i=1}^{n} \varepsilon_i^2 = \sum_{i=1}^{n} (Y_i - \hat{a}D_i^{\hat{b}}H_i^{\hat{c}})^2 \tag{9-19}
$$

A rapid method of obtaining a convergent solution, which has been implemented in several standard statistical packages, is an algorithm developed by Marquardt (1963).

Moser and Beers used a variant of a SHARE program, which included a weighting option, to prepare a volume equation for northern red oak trees that gives merchantable cubic foot volume inside bark V.

$$V = 0.003173D^{1.988825}H^{0.981921} \tag{9-20}$$

The theory and details of weighted fitting of regression equations are part of standard regression theory. To solve the weighting problem, Moser and Beers grouped their data into D^2H classes and determined the volume variance for each class. Then, on plotting variance over D^2H they noted that variance was exponentially related to D^2H. Thus, the class variances, weighted by the number of observations in each D^2H class, were fitted to derive the following variance function.

$$S_i^2 = 0.62359e^{0.11082(D^2_iH_i/1000)} \tag{9-21}$$

Moser and Beers (1969) found that the coefficients obtained in fitting the red oak data to equation 9-16 are different from those obtained by the least squares fit of the logarithmic transformation, and that the weighted nonlinear equation gives a smaller index of fit (Furnival, 1961).

The coefficients for equation 9-13, a local volume table, may be calculated in a manner analogous to the calculation of the coefficients for equation 9-16. Furthermore, the utilization of equation 9-17, a form class volume table, offers no great computational obstacles. However, Behre (1935) and Smith, Ker, and Csizmazia (1961) concluded that no practical advantage is gained from the use of a measure of form in addition to dbh and height.

9-7 DERIVATION OF LOCAL VOLUME TABLE FROM STANDARD VOLUME TABLE

A local volume table may be derived from a standard volume table by "localizing" the heights by dbh classes. The procedure is simple.

1. Measure the heights and dbh's of a sample of trees representative of those to which the local volume table will be applied. Record dbh to nearest tenth of an inch, and height to the nearest even foot.
2. Prepare a curve of height over dbh by the freehand method.
3. Read average heights to nearest even foot from the curve for each dbh class (usually 2-inch classes).
4. Interpolate from the standard volume table the volume of the tree of average height for each dbh class, or if a standard volume equation is available, substitute the appropriate values in the equation and compute volume for each dbh class.

9-8 TREE VOLUME TARIF TABLES

The term *tarif*, which is Arabic in origin, means *tabulated information*. In continental Europe the term has been applied for years to volume table systems that provide, directly or indirectly, a convenient means of obtaining a local volume table for a given stand (Garay, 1961).

British tarifs (Hummel, 1955), which have been quite successful, stimulated the preparation of "Comprehensive Tree-Volume Tarif Tables" by Turnbull, Little, and Hoyer (1963). This clever system, which is summarized in Table 9-5, merits wider consideration in all types of inventories. It requires no curve fitting to obtain a local volume table; it provides a convenient method of converting from one unit of measure to another, or from one merchantable limit to another. To determine average annual volume increment per tree in any desired unit of volume and merchantable limit, one simply multiplies the average annual diameter increment in inches by the growth multiplier (GM).

9-9 VOLUME DISTRIBUTION AND CULL ESTIMATION IN TREES

A knowledge of volume distribution over the tree stem can be used to improve volume estimates and to aid in estimating volume losses from defects.

Table 9-6 and Fig. 9-2, which was derived from Table 9-6, illustrate two useful

Table 9-5
Specimen of Comprehensive Tree Volume Tarif Table

Height—dbh Access Table for Douglas Fir

dbh	Total Height (feet)				
	60	62	64	66	68
12.2	22.2	23.1	23.9	24.8	25.7
12.4	22.0	22.9	23.7	24.6	25.5
12.6	21.8	22.7	23.6	24.4	25.3
12.8	21.6	22.5	23.4	24.2	25.1
13.0	21.5	22.3	23.2	24.0	24.9
13.2	21.3	22.1	23.0	23.8	24.7
13.4	21.1	22.0	22.8	23.7	24.5
13.6	20.9	21.8	22.6	23.5	24.3
13.8	20.8	21.6	22.5	23.3	24.2
14.0	20.6	21.4	22.3	23.1	24.0
14.2	20.4	21.3	22.1	23.0	23.8
14.4	20.3	21.1	22.0	22.8	23.6
14.6	20.1	21.0	21.8	22.6	23.5

Instructions

1. Measure height and dbh of sample of trees representative of stand.
2. Look up tarif numbers of sample trees in appropriate *Height—dbh Access Table* and average them. For example:

Height	dbh	Tree Tarif No.
60	12.2	22.2
68	14.3	23.7
Etc.		
	Mean =	24.5

3. In tarif book find tarif table with mean tarif number.

Table 9-5 (continued)
Tarif Table No. 24.5

	Total Tree Volume						Volume to 6-Inch Top					
	Including Top and Stump (cubic feet)		Including Top Only (cubic feet)		Volume to 4-Inch Top (cubic feet)		(cubic feet)		Scribner (board feet)		International ¼-Inch Rule (board feet)	
dbh (inches)	Vol A	GM A	Vol B	GM B	Vol C	GM C	Vol D	GM D	Vol E	GM E	Vol F	GM F
2	0.3	0.2	0.2	0.2	·	·	·	·	·	·	·	·
3	0.7	0.7	0.6	0.6	·	·	·	·	·	·	·	·
4	1.5	1.0	1.4	1.0	·	·	·	·	·	·	·	·
5	2.6	1.4	2.5	1.3	1.4	1.5	·	·	·	·	·	·
6	4.1	1.7	4.0	1.6	3.0	1.8	·	·	·	·	·	·
7	5.9	2.0	5.7	2.0	4.9	2.1	1.9	2.2	6	7.2	9	10.7
8	8.1	2.4	7.8	2.3	7.1	2.4	4.3	2.8	14	9.8	21	13.9
9	10.5	2.7	10.2	2.6	9.6	2.7	7.2	3.1	25	11.9	36	16.2
10	13.3	3.0	12.9	2.9	12.3	3.0	10.5	3.4	38	13.7	53	18.0

dbh	V/BA Ratio		V/BA Ratio	% of Vol A	V/BA Ratio	% of Vol B	V/BA Ratio	% of Vol C	V/BA Ratio	B/CU Ratio	V/BA Ratio	B/CU Ratio
2	9.3		8.3	89.0	·	·	·	·	·	·	·	·
3	12.5		11.6	92.5	·	·	·	·	·	·	·	·
4	16.3		15.4	94.4	·	·	·	·	·	·	·	·
5	19.0		18.1	95.5	9.7	51.0	·	·	·	·	·	·
6	20.7		19.9	96.1	14.9	71.9	·	·	·	·	·	·
7	22.0		21.2	96.4	18.1	82.1	6.8	37.7	21.1	3.1	32.0	4.7
8	23.0		22.2	96.6	20.1	87.6	12.3	61.1	40.6	3.3	59.9	4.9
9	23.7		23.0	96.7	21.5	90.7	16.3	75.6	56.8	3.5	81.5	5.0
10	24.3		23.6	96.7	22.5	92.6	19.1	84.6	69.6	3.6	97.5	5.1

This tarif table gives volume in cubic feet for entire tree and volume in cubic and board feet to various merchantable limits. Volume/basal area ratios for horizontal point sampling and growth multipliers (GM) to determine growth are also given. The letters, A, B, C, etc., that follow Vol and GM are used for convenient identification of columns.

SOURCE: Turnbull and Hoyer, 1965.

Usable Length (16-foot logs)	Percent of Total Tree Volume in Each Log, by Position					
	1st	2nd	3rd	4th	5th	6th
1	100					
2	58	42				
3	42	33	25			
4	34	29	22	15		
5	29	25	21	15	10	
6	24	23	20	16	11	6

SOURCE: Mesavage and Girard, 1946.

methods of expressing volume distribution. Although the percentages vary slightly with tree diameter and unit of volume, they may be used without serious error for merchantable trees of all sizes that are measured in cubic or board-foot volume. Note that Fig. 9-2, which is basically for 16-foot logs, provides a satisfactory guide when heights are measured in 8- or 12-foot lengths.

For the experienced timber estimator who can estimate the length deductions for various indicators, such as conks, cankers, and injuries, graphs, such as Fig. 9-2, provide convenient guides. For example, for a tree that contains three 16-foot logs, it is judged that 4 feet of the butt must be removed, and that an 8-foot section between 32 and 40 feet on the merchantable stem must be cut out. The cull, then, will be 12 percent (12 − 0) for the defective butt, and 13 percent (88 − 75) for the defection section, or 25 percent for the tree.

Since many estimators do not have the experience to estimate cull in this manner, they need a more systematic method. The following three methods are suggested.

1. Use indicator cull percentages developed by multiple regression analysis. The dependent variable is cull in percent of gross volume; independent variables are age, dbh, conks, cankers, basal injuries, and trunk injuries. (Variables such as conks, cankers, and injuries are given a value of 1 if one or more are present; 0 if none are present.)

2. Use tables of average length deductions for indicators such as conks, cankers, injuries, crooks, forks, and so forth, and apply these along with tables giving hidden defect percentages by dbh classes. (Tables of average length deductions for various indicators are normally used in conjunction with graphs, such as Fig. 9-2.)

3. Use cull classes that give, on the basis of visible defects, the percentage of deductions to apply to gross volume. For example, cull classes might be set up in the following manner for Central States Hardwoods.

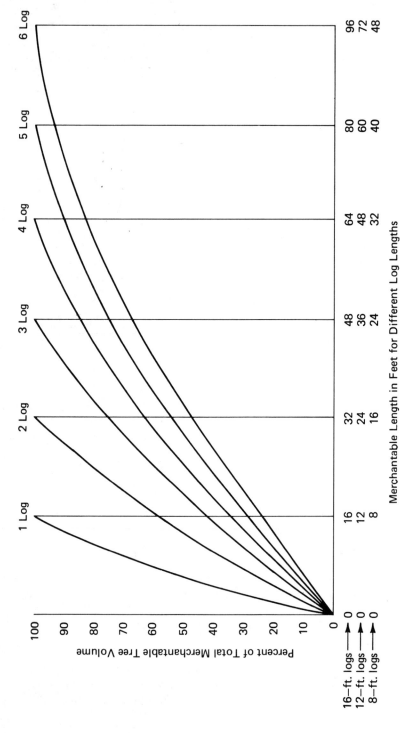

Merchantable Length in Feet for Different Log Lengths

Fig. 9-2 *Percentage of total merchantable volume at various heights for trees of different merchantable lengths (derived from Table 9-6).*

Cull Class	Defect in Percent of Gross Board-Foot Volume (Scribner)	Maximum Number of Defects Permitted
1	5	Two minor defects confined to upper two-thirds of merchantable stem
2	15	One major defect in merchantable stem
3	25	Three major defects in entire merchantable stem, or 2 major defects in lower one-third of merchantable stem
4	40	Four major defects in entire merchantable stem, or 3 major defects in lower one-third of merchantable stem
5	100	Will not meet Class 4 specifications

To apply this table, one must have specifications for minor and major defects, and a set of curves to estimate major defect equivalents for defects, such as butt rot, in the lower one-third of the merchantable stem (Miller and Beers, 1981).

Aho and Roth (1978) used the first two methods to estimate cull in white fir. Miller and Beers (1981) used the third method to estimate cull in Central States Hardwoods. The authors feel the third method is the easiest to develop and the easiest to modify.

Whatever method is used, a cull percent applicable to a species, or species group, in a stand is normally determined by estimating the cull of a sample of 40 to 50 trees of the species, or species group. Then,

$$\text{Cull percent} = \left(\frac{\text{Cull volume of sample}}{\text{Gross volume of sample}}\right)100 \qquad (9\text{-}22)$$

This percentage is applied to all trees of the species, or species group, in the stand.

10
Grading Forest Products

Grading is the process of grouping something that is produced by nature or made by industry into classes or grades that reflect characteristics of quality or worth. In forestry, grading systems find numerous applications: Christmas trees, nursery stock, poles, piling, ties, shingles, cooperage, lumber, veneer, logs, and standing timber are graded by qualities of interest to consumers; forest sites are ranked by timber-producing capacity; wildlands are rated by recreational potential. This chapter treats the grading of forest products; Chapter 17 covers site quality classifications; and Chapter 18 touches on the ranking of recreational areas.

10-1 QUANTIFICATION OF LOG AND TREE QUALITY

A unique characteristic of log and tree quality is that it may be measured as a continuous variable on an interval scale, although this is not commonly done, or as a discrete variable on an ordinal scale. Therefore, it is enlightening to speak of "quantification of log and tree quality" (Ware, 1964) rather than simply of "log and tree grades," because the phrase "quantification of quality" infers alternative methods of expressing quality. Quantification systems for log and tree quality must be intimately related to the quality of the products to be produced from any run of logs. Therefore, *quality* must be defined in terms of some derived product, preferably the product that gives maximum value for given logs. But there are problems since many trees are best suited for a mixture of derived products—lumber, veneer, pulp. Even a log best suited for lumber will produce several grades, no grade being produced in the same quantity. Furthermore, the proportion of grades will vary with mill practices, mill efficiency, and markets. The same is true for a veneer log. And not only will the value differ with grades of lumber or veneer, but also with species, time, region, and market conditions.

Unfortunately, uniform specifications are not accepted throughout industry for some derived products. For example, the diversity of specifications for hardwood veneer has retarded the development of quality quantifiers for hardwood veneer logs. Also, the quality specifications may be unstable; from time to time there may be changes in the derived-product quality specifications that necessitate changes in the log and tree quality quantifiers. Since such changes will continue in the face of technological advances and increasing competition from substitute materials, there is a need to develop quality quantification systems that are adaptable to change.

The selection of variables to predict quality is also a problem. A good system for quantifying quality is one for which the quality being quantified is highly correlated with log or tree characteristics that can be determined objectively. But quantifying quality in advance of the manufacturing process is difficult because many of the characteristics that affect the quality of the derived products cannot be seen in trees or logs. A list of some of the variables that can be used to predict quality will illustrate the problem: log or tree size, log or tree form, knots, stain and decay, wood density, compression wood, and spiral grain. Except for size, it is difficult to measure or numerate the variables in this list. More research is also needed to learn to assess the influence of these variables on quality and to find other variables, perhaps ones related to genetic and environmental factors, that are correlated with log and tree quality.

10-1.1 Systems of Quantifying Log and Tree Quality

To quantify log or tree quality we must select suitable independent variables to predict quality. Quality may be expressed in terms of (1) value in dollars, (2) a single number, called *quality index*, that expresses the value of the lumber in a log or tree as a percentage of the value of an equal volume of lumber of a base lumber grade, or (3) distribution of volume by quality classes of derived product.

One might think that through a multiple regression analysis equations could be developed to predict log or tree quality. Indeed, if suitable equations could be developed, forestry inventory sampling designs could be employed that would include double sampling for quality. Foresters, however, have been unable to develop suitable regression equations to quantify quality. Regression models that use independent variables such as log or tree size, knots, decay, and so on, account for only a marginal portion of the total variance of log or tree quality, however we express quality. But regression analyses are still useful; the order of importance of the independent variables can be determined, and the variables best correlated with quality can be selected for log or tree grading systems.

Essentially all systems used today for quantifying log quality place logs in broad classes or grades, the limits of which are defined by the surface characteristics of logs. A good example is the U.S. Forest Service's standard hardwood log grades (Vaughan, Wollin, McDonald, and Bulgrin, 1966). The overall work plan for developing such log grades is covered in a Forest Products Laboratory Report (1958). Although there are some differences in detail, softwood log grades have been developed along the same lines (Campbell, 1964).

Since trees are composites of logs, and since in the quantification of quality foresters have been preoccupied with log grades, foresters have used log grades in the following ways to develop tree grades.

1. Base tree grade on the grade of a 16-foot butt log, or of the best 12- or 14-foot section of the 16-foot butt log. (For example, say logs are separated into No. 1, No. 2, and No. 3 grades. Then Grade A, Grade B, and Grade

C trees might be trees with No. 1, No. 2, and No. 3 butt logs, respectively; Herrick and Jackson, 1957; Hanks, 1976.)

2. Use the grades of all the logs in the tree. (For example, a two-log tree with a No. 1 butt log and a No. 3 second log would be a grade 1-3 tree; Guttenberg and Reynolds, 1953.)

3. Base tree grade on the unweighted average of the grades of the logs in the tree. (For example, a tree with two No. 1 logs and one No. 2 log would have a tree grade of 1.33; Herrick, 1956.)

Method 1, which is most commonly used, is simple, but the variation in the grades of the upper logs, from one tree to another, reduces its precision in predicting actual tree quality below that obtained when individual logs are graded. A modification of method 1 (Schroeder, Campbell, and Rodenbach, 1968b) that considers conks, punk knots, and other evidence of rot anywhere on the tree stem improves the prediction of tree quality for southern pine.

A consideration of methods 2 and 3 reveals why they have been little used. They both require an excessive amount of time to determine tree grade, unless very simple log grading rules are used; furthermore, method 2 gives an excessive number of tree grades.

10-2   LOG GRADING SYSTEMS

Since a great variety of derived products come from the many species of logs cut, a great variety of log grades have been developed. These grades are constantly in the process of improvement and adjustment. Hence, it is inadvisable to give anything more than a sample of the grades here. For exact information readers are referred to the current edition of the rules for the species of interest.

Several handbooks, although printed some years ago, are extremely helpful guides to defect identification, an ability that is essential for grading logs. These are Lockard, Putnam, and Carpenter (1963) and Shigo and Larson (1969) on identification of grade defects in hardwoods; Campbell (1962) on identification of grade defects in southern pines; Jackson (1962) on identification of grade defects in ponderosa pine and sugar pine.

10-2.1   U.S. Forest Products Laboratory Hardwood Log Grading System

This widely known system illustrates the requirements and complexities of a good log grading system, whether it be for hardwoods or softwoods.

The Forest Products Laboratory has set up four broad log-use classes for hardwood logs: Factory Class, Construction Class, Local-Use Class, and Veneer-Use Class. These four use classes are as follows.

1. *Factory Class.* Logs falling in this class are divided into three grades: F1, F2, and F3. Grading is based on the assumption that logs will be cut into lumber that will be graded by the rules of the National Hardwod Lumber Association for standard lumber.

2. *Construction Class.* This class is not separated into grades. A log of this class is graded on the assumption that it will be sawed into ties, timbers, or other products that are to be used for structural and weight-bearing purposes.

3. *Local-Use Class.* This class is not separated into grades. The class is a catchall for logs that do not qualify for the other classes. The main requirement is that logs qualify within limits of merchantability, a factor largely determined locally.

4. *Veneer Class.* This class includes logs that have potential value for veneer. At present there is no standard system for grading veneer logs that is based on research. However, there are numerous local grade specifications, such as those published by the Northern Hardwood and Pine Manufacturers' Association and the Fine Hardwoods-American Walnut Association. But the diversity of specifications for hardwood veneer has retarded the development of standard veneer-log grades.

It should be understood that the value of a log in one use class furnishes little clue to the value of a log in another use class. Therefore, it would be difficult to select the class in which to place a log if all utilization possibilities were open. For most operators, however, the business situation determines the use class into which the logs should be put.

The standard specifications for the three grades of the Factory Class are summarized in Fig. 10-1. Since most hardwood factory lumber is graded on the basis of the percentage of the board surface that will yield clear areas of specified sizes, the size and number of clear cuttings, as determined by position of grade defects, are of prime importance in determining the grade of hardwood factory lumber logs. Other grading factors are position of log in tree, log diameter, log length, sweep, cull, and end defects.

10-2.2 Comparison of U.S. Forest Products Laboratory System with Other Hardwood Log Grading Systems

A good log grading system should provide:

1. A sound basis for establishing log selling prices and log purchase prices.

2. An up-to-date estimate of derived-product grade yields of logs.

3. A chance to check the quality production of a mill against established standards.

Grading Factors		Log Grades							
		F1			**F2**				**F3**
Position in tree		Butts only	Butts & uppers		Butts & uppers				Butts & uppers
Diameter, scaling, inches		[1]13–15	16–19	20+	[2]11	12+			8+
Length without trim, feet		10+			10+	8–9	10–11	12+	8+
Clear cuttings[3] on each 3 best faces	Lenth, min., feet	7	5	3	3	3	3	3	2
	Number, maximum	2	2	2	2	2	2	3	No limit
	Fraction of log length required in clear cutting[4]	5/6	5/6	5/6	2/3	3/4	2/3	2/3	1/2
Sweep and crook allowance (maximum) in percent gross volume	For logs with less than 1/4 of end in sound defects	15%			30%				50%
	For logs with more than 1/4 of end in sound defects	10%			20%				35%
Total scaling deduction including sweep and crook		[5]40%			[6]50%				50%
End defects:	See instructions, Forest Products Laboratory, 1966								

[1] Ash and basswood butts can be 12 inches if otherwise meeting requirements for small No. 1's.

[2] Ten-inch logs of all species can be No. 2 if otherwise meeting requirements for small No. 1's.

[3] A clear cutting is a portion of a face free of defects, extending the width of the face.

[4] See table 46.

[5] Otherwise No. 1 logs with 41-60% deductions can be No. 2.

[6] Otherwise No. 2 logs with 51-60% deductions can be No. 3.

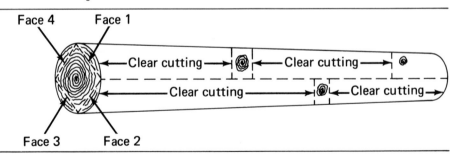

Fig. 10-1 *Forest Service standard specifications for hardwood factory lumber logs (Forest Products Laboratory, 1966).*

The Forest Products Laboratory grades accomplish all of these things to a reasonable extent. However, before effectively employing these grades, one must study them at length and then apply them for several weeks. Consequently, a number of simplified hardwood log grading schemes have been developed by wood-using industries, trade associations, and universities. These schemes, which can be employed with a minimum of study and a short period of application, are generally based on the Forest Products Laboratory System. Unfortunately, grade yield information is lacking or out of date for many of these "local" log grading systems. Consequently, they generally do not provide requirements 2 and 3, stated above, and, consequently, often do not provide a sound basis for log prices.

The Purdue System, which is typical of the systems that have grown out of the Forest Products Laboratory System, illustrates the nature of the simplified log grading schemes. This system does not recognize use classes, that is, it applies to any sawlog regardless of the end use of the material it contains. Compared to the Forest Products Laboratory grades for factory logs, the Purdue grades are based on fewer rules and no exceptions, which makes them extremely easy to apply (Fig. 10-2). The Purdue System does not predict log value as accurately as the Forest Products Laboratory System (Walters and Herrick, 1956), but it can be applied at about half the cost.

10-2.3   Softwood Log Grading Systems

There is no single system of softwood log grades that applies to as wide a range of species as does the Forest Products Laboratory System. The great variety of derived products that come from the many species of softwood logs has led to the development of numerous grade specifications. However, the majority of softwood logs fall into two use classes:

1. *Veneer Class.*   Includes high-value logs as well as some relatively low-value logs that can be utilized as veneer logs.
2. *Sawlog or Sawmill Class.*   Includes logs adapted to the production of yard and structural lumber.

Since most softwood veneer and most softwood lumber are graded on the assumption that the entire piece will be used, the number, location, and size of the grade defects determine the grade of the two classes of softwood logs. Other grading factors are log diameter, log length, sweep, and cull.

Grading bureaus have promulgated log grading rules for most of the important timber species of the western United States. Although grade yield information is lacking for a number of log grades used for certain species, the U.S. Forest Service is making efforts to remedy this situation for the more important species. In addition, the Forest Service has developed systems for grading southern pine logs (Schroeder, Campbell, and Rodenbach, 1968a).

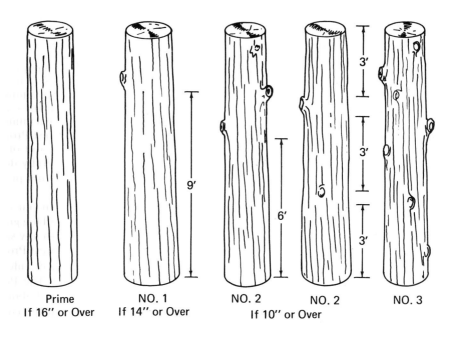

Prime	NO. 1	NO. 2	NO. 2	NO. 3
If 16" or Over	If 14" or Over			
		If 10" or Over		

GRADE SPECIFICATIONS

Prime — At least 90 per cent surface clear on three visible faces in one section (A face is any one-quarter of the log surface). Must be 16 inches or more in diameter inside bark at the small end.

No. 1 — At least 75 per cent surface clear on three visible faces in one section. Must be at least 14 inches in diameter inside bark at the small end.

No. 2 — At least 50 per cent surface clear on three visible faces in two sections, neither of which is less than three feet long. Must be at least 10 inches in diameter inside bark at the small end.

No. 3 — Will not meet No. 2 specifications.

Fig. 10-2 *Purdue specifications for hardwood sawlogs (Herrick, 1955).*

10-3   APPLICATION OF LOG AND TREE GRADES

To obtain volume by log grades, when one scales logs, one simply records the additional grade variable and tabulates volume by species, log grades, and log diameter, if diameter is important to determine value. Where logs are bought and sold by grade, one can use the current market prices for logs of recognized grades to determine value.

If suitable tree grades are available, quality estimation in forest inventories requires collecting the additional grade variable. By using point sampling, line sampling, or fixed-size plot sampling, one can easily devise tally forms to record the data, whether all or only a subsample of trees are graded.

Beers and Miller (1973) describe an efficient horizontal point-sampling system to obtain volume by species and log grades. However, if data are needed by log size as well as log grade, point or line sampling may be inefficient, unless used in conjunction with a fixed-size plot system. Then logs can be graded on fixed-size plots taken at every nth sampling point. Through this double sampling procedure the percent of the various log grades by diameters that can be recovered is determined and applied to the volume data obtained in the main inventory. Of course, fixed-size plots may be used to obtain all the required information. Then, in addition to collecting the usual data, the field crews might grade the logs in every merchantable tree on all plots, or on every fifth plot, or at some other interval, or on specially designated smaller plots that fall within the main plots. The higher the value of the timber, the greater the percentage of the trees that should be graded.

When log grades must be used to determine the quality of standing timber, simplified rules that include only characteristics that can be recognized in standing trees must be developed. This is necessary because most log grading systems require a more detailed inspection of the logs than is practical on standing trees. The log grading system shown in Fig. 10-2 is an example of a system that can be applied to standing timber.

11
Forest Inventory

A *forest inventory* is the procedure for obtaining information on the quantity and quality of the forest resource and many of the characteristics of the land area on which the trees are growing. Most forest inventories have been, and will continue to be, timber estimates. However, the need for information on recreation, water, wildlife, and other nonwood values has stimulated the development of integrated or multiresource inventories (see Rocky Mtn. Forest and Range Experiment Station, 1978; McClure, Cost, and Knight, 1979). The following discussion concentrates on inventories for timber values. When nontimber information is required, specialists in the pertinent fields should cooperate in planning and executing the inventory.

A complete forest inventory for timber evaluation provides the following information: estimates of area, description of topography, ownership patterns, accessibility, transportation facilities, estimates of timber quantity and quality, and estimates of growth and drain. When needed, additional information may be collected on the recreational resources, wildlife resources, and so forth. The emphasis placed on specific elements will differ with the purpose of the inventory (Husch, 1971). For example, if the purpose of an inventory is for the preparation of a harvesting plan, major emphasis should be put on a description of the topography, determination of accessibility and transportation facilities, and estimation of timber quantity; the other elements would be given little emphasis or would be eliminated. But, if the purpose of an inventory is for the preparation of a management plan, major emphasis would be put on estimation of timber quantity, growth, and drain.

Forest inventory information is obtained by measuring and assessing the trees and various characteristics of the land. The information may be obtained from measurements taken on the ground or on remote sensed imagery (aerial photographs, satellite imagery, etc.). When the measurements are taken for the entire forest, the inventory is a complete or 100 percent inventory. When the measurements are taken for a sample of the forest, it is a sampling inventory. The terms *cruise* in North America and *enumeration* in other English-speaking areas are frequently used instead of inventory.

11-1 QUANTITY RELATIONSHIPS IN FOREST INVENTORY

In executing a forest inventory it is impossible to measure directly quantities such as volume or weight of standing trees. Consequently, a relationship is established be-

150

tween directly measurable tree or stand characteristics (e.g., dbh, height) and the desired quantity. This may be done as follows.

1. Make detailed field measurements of trees or stands and compute the desired quantities from these measurements. For example, one might make detailed diameter measurements at intervals along the stems of standing trees and determine volumes by formulas or graphical methods (Section 8-1).

2. Estimate the desired quantities in trees or stands by utilizing relationships previously derived from other trees or stands. For example, one might measure tree or stand characteristics such as dbh, height, and form and determine the corresponding volume or weight from an equation or a table.

11-2 INVENTORY PLANNING

An important step in designing a forest inventory is the development of a comprehensive plan before initiating work. Such a plan ensures that all facets of the inventory, including the data to be collected, financial and logistical support, and compilation procedures, are thought through before the inventory begins. The student is referred to Husch (1971) for a discussion of inventory planning.

The following checklist includes all, or almost all, the items that should be considered in planning a forest inventory. The items do not always have the same importance; nor are all needed in all plans. If one is interested in additional resources—for example, wildlife or recreation—the necessary items can be added to the checklist.

1. Purpose of the inventory
2. Background information
   a. Past surveys, reports, maps, photographs, etc.
   b. Individual or organization supporting the inventory
   c. Funds available
3. Description of area
   a. Location
   b. Size
   c. Terrain, accessibility, transport facilities
   d. General character of forest
4. Information required in final report
   a. Tables and graphs
   b. Maps, mosaics, or other pictorial material
   c. Narrative report
5. Inventory design
   a. Estimation of area (from aerial photos, maps, or field measurements)
   b. Determination of timber quantity (e.g., volume tables, units of volume, etc.)
   c. Size and shape of fixed-size sampling units

d. Probability sampling
    i. Simple random sampling
    ii. Stratified random sampling
    iii. Multistage sampling
    iv. Double and regression sampling
    v. Sampling with varying probability
e. Nonrandom sampling
    i. Selective sampling
    ii. Systematic sampling
f. Setting precision for inventory
g. Sampling intensity to meet required precision
h. Times and costs for all phases of work

6. Procedures for photo interpretation
  a. Location and establishment of sampling units
  b. Determination of current stand information, including instructions on measurement of appropriate tree and stand characteristics coordinated with field work
  c. Determination of insect damage, forest cover types, forest fuel types, area, etc., coordinated with field work
  d. Personnel
  e. Instruments
  f. Recording of information
  g. Quality control
  h. Data conversion and editing

7. Procedures for field work
  a. Crew organization
  b. Logistical support and transportation
  c. Location and establishment of sampling units
  d. Determination of current stand information, including instructions on measurement of trees and sample units coordinated with photo interpretation
  e. Determination of growth, insect damage, mortality, forest cover types, forest fuel types, area, etc., coordinated with photo interpretation
  f. Instruments
  g. Recording of observations
  h. Quality control
  i. Data conversion and editing

8. Compilation and calculation procedures
  a. Instructions for reduction of photo and field measurements
    i. Conversion of photo or field measurements to desired expressions of quantity
    ii. Calculation of sampling errors
    iii. Specific methods to use (e.g., programmable calculators, computers, etc.)

 iv. Description of all phases from handling of raw data to final results, including programs

 9. Final report
  a. Outline
  b. Estimated time to prepare
  c. Personnel responsible for preparation
  d. Method of reproduction
  e. Number of copies
  f. Distribution

 10. Maintenance
  a. Storage and retrieval of data
  b. Plans for updating inventory

The decision to conduct an inventory depends on the need for information. An *inventory* is an information-gathering process that provides one of the bases for rational decisions. These decisions may be required for the purchase or sale of timber, for preparation of timber-harvesting plans, in forest management, or for obtaining a loan. If the purpose is clarified, one can foresee how the information will be used and thus know the relative emphasis to put on the elements of the inventory.

As a first step in planning an inventory, one should obtain the requisite background information and prepare a description of the area. One should then decide on the information required from the inventory and prepare outlines of tables that will appear in the final report. Table outlines should include titles, column headings, class limits, measurement units, and other categories needed to indicate the inventory results. These table outlines should be prepared before detailed planning begins because the inventory design will depend on the information required for the final report. Area information is usually shown by such categories as land-use class, forest-type or con- dition class, and ownership. Timber quantities are usually given in stand and stock tables. A *stand table* gives number of trees by species, dbh, and height classes. A *stock table* gives volumes (or weights) according to similar classifications. Stand and stock tables may be on a unit area basis (per acre or hectare) or for the total forest area and may be prepared for forest types or other classifications.

Since a forest is a living, changing complex, the inventory plan should consider the inclusion of estimates of growth and drain. For a discussion of tree and stand growth, see Chapters 15 and 16.

11-3 INVENTORY DESIGN

There must be wide latitude in designing an inventory to meet the variety of forest, topographic, economic, and transportation conditions that may be encountered. The required forest inventory information can be obtained by observations and measure- ments in the field and on aerial photographs (and through other remote sensing techniques). The most useful and practical approach is a procedure in which aerial photographs are used for forest classification or stratification, mapping, and area determination, while ground work is employed for detailed information about forest conditions, timber quantities, and timber qualities. One can design a forest inventory utilizing only field work but it is less efficient. If aerial photographs are available, they should be used. (Under some circumstances it is possible to design a forest inventory based entirely on photographic interpretations and measurements. However, this method will provide only rough approximations of timber species, quantity, quality, and sizes.)

The funds available and the cost of an inventory will strongly influence the design chosen. The main factors that will affect costs are the type of information required, the standard of precision chosen, the total size of the area to be surveyed, and the minimum size of the unit area for which estimates are required. General information is relatively inexpensive, but the cost increases as more details are required. The standard of precision chosen greatly influences costs; the greater the precision the higher the cost. Costs per unit area will decrease as the size of the inventory area increases. If independent estimates are required for subdivisions of a large forest area, this will also raise costs.

Descriptions of basic sampling designs applicable to forest inventory are given in Sections 12-5 to 12-9.

11-4 INVENTORY FIELD WORK

The size and organization of field crews will vary with sampling procedures, forest conditions, labor conditions, and tradition. For inventory work in temperate zone forests, crews of one or two workers are widely used. Small crews have proved satisfactory in these areas because roads are generally abundant (access by vehicle is good), forest travel is relatively easy (little brush cutting is required), and well-trained technicians are generally available (they require little supervision).

For inventory work in tropical zone forests, large crews are widely used because roads are generally sparse (access by vehicle is poor), forest travel is difficult (workers are required to cut vegetation), and few trained technicians are available (they require considerable supervision). Whatever the crew size or the assigned tasks, however, specific, clear instructions should be given to each crew member.

A basic rule of forest inventory is to prepare complete written instructions before field work starts. To minimize later changes, instructions should be tested before operations begin. Instructions should be clear enough and specific enough so that individual judgment by field crews on where and how to take measurements is eliminated. For example, field crews should not be permitted to subjectively choose the position of a sampling unit or to move it to a more accessible position or more "typical" stand. The aim should be to standardize all work to obtain uniform quality and the best possible reliability of the measurements, regardless of which individual does the work. The occurrence of mistakes or nonrandom errors must be held to a minimum.

It is best to settle on a standard set of instruments. The use of several kinds of instruments to make the same type of measurements should be avoided. To minimize systematic errors, the instruments should be periodically checked to see if they are in adjustment.

Decisions should be made on the precision required for each of the measurements. Thus, tree diameters (dbh) may be stipulated to the nearest centimeter, inch, tenth of an inch, etc., and heights to the nearest foot, meter, one-half log, and so on.

Inventory field operations must include a checking procedure. Whether plots, strips, or points are used, a certain percentage of the sampling units should be remeasured. The results of the remeasurement should be compared to the original measurement to see if the work meets the required standards. If it does not, remedial steps should be taken.

There is no standard form or system for recording field observations. The manner in which measurements are recorded depends in large part on the way that they will be processed. Here are four methods of recording field measurements.

1. The use of forms that are designed without concern for subsequent calculations. The needed information is extracted from the forms and computations done apart from them. This is an inefficient method.

2. The use of forms with space for subsequent calculations. Excellent for inventories for which limited information is desired.

3. The use of forms designed for efficient transfer of data to computer input devices (punch cards, computer terminals, etc.).

4. The use of methods that permit direct data entry into the computer (Porta-punch cards, mark-sensed cards, electronic data recorders, etc.).

11-5 CALCULATION AND COMPILATION

Plans for calculation and compilation of photo-interpretation data and field measurements should be made before field work begins. It is inefficient to postpone consideration of these things until data are obtained, because foreknowledge of the calculation and compilation procedures can influence collection of data. For example, if volume tables are to be used to determine tree volumes, there would be no point in measuring diameters to a tenth of a centimeter if the volume tables give volumes by full centimeter dbh classes.

The statistical formulas to be used for estimating means, totals, and standard errors should be selected early in the planning process. It is worthwhile to check the formulas with simulated data to work out the optimum sequence of computational steps.

12

Sampling in Forest Inventory

On a large forest, sampling can provide all the necessary information in less time and at a lower cost than a 100 percent inventory. Indeed, it is impractical on a large forest to measure all the trees. Since fewer measurements are needed, sampling may produce more reliable results than a complete tally because fewer, better-trained personnel can be used and better field supervision of work can be exercised. In addition, the idea of precision is in the forefront throughout a sampling inventory, whereas a complete tally may give the delusion of the acquisition of information without error.

This chapter and Chapter 14 are concerned with the application of sampling theory and techniques to forest resource evaluation. The reader is referred to Cochran (1977) for a basic treatment of sampling theory and technique.

12-1 SAMPLING TERMINOLOGY

In forest inventory work, sampling consists of making observations on portions of a population (the forest and its characteristics), to obtain estimates that are representative of the parent population. An observation is a recording of information, such as the volume of a fixed-area plot. The sampling units on which observations are made may be stands, compartments, administrative units, fixed-area plots, strips, or points. The aggregation of all possible sampling units constitutes the population. The group of sampling units chosen for measurement constitutes the sample. For purposes of selecting the sampling units, a frame can be prepared that can be a diagram or list of all the possible sampling units in the population. Some populations are finite; that is, they consist of a fixed number of sampling units such as all nonoverlapping 0.2-acre square plots on a 160-acre forest. Other populations, such as all possible points (as used in horizontal point sampling), are infinite; that is, they consist of an unlimited number of sampling units. In the statistical treatment of data it is important to know whether the population is infinite or finite. As a practical matter, however, very large finite populations can be treated as infinite.

The essential problem in sampling is to obtain a sample that is representative of the population. If the sample is representative of the population, then useful statements can be made about the characteristics of the population (volume per unit area, number of trees per unit area, etc.) on the basis of the characteristics observed in the sample observations. The characteristics of the population are referred to as *param-*

eters. The exact values of the parameters would be known if the entire population were measured. In most situations, however, complete enumerations are too time consuming and costly, and sampling is used to estimate the values of parameters. Such estimates are calculated from a sample and are called *statistics.* A statistic, therefore, is a summary value calculated from a sample to estimate a population parameter.

12-2 ERRORS IN FOREST INVENTORY

The precision of a forest inventory based on sampling is indicated by the size of the sampling error and excludes the effects of nonsampling errors. Accuracy of an inventory refers to the size of the total error and includes the effect of nonsampling errors. In forest inventory, as in any sampling procedure, we are primarily concerned with accuracy. We try to achieve accuracy by designing and executing an inventory for an acceptable precision and by eliminating or reducing nonsampling errors to a minimum.

Sampling errors result from the fact that the sample is only a portion of the population and may not produce estimates identical to the population parameter. The sampling error is expressed as the standard error of the mean. Nonsampling errors are errors not connected with the statistical problem of the selection of the sample and may therefore occur whether the entire population or a sample of the population is measured. Thus, nonsampling errors are always present; sampling errors are present only when sampling methods are employed.

Nonsampling errors arise from defects in the sampling frame, mistakes in the collection of data due to bias or negligence, and mistakes in the processing stage. When nonsampling errors are large, the total error will be reduced only slightly by taking a large sample to decrease the sampling error. Indeed, when the nonsampling errors are large, attention should be focused on reducing them before the sample size is increased to reduce the sampling errors.

12-3 CONFIDENCE LIMITS

Forest inventory estimates can be expressed as a range—the confidence interval, bounded by confidence limits—within which the population parameter is expected to occur at a given probability. The confidence interval is expressed by

$$CI = \bar{x} \pm ts_{\bar{x}} \tag{12-1}$$

The value of t for a chosen probability level can be found from Student's t distribution using $n - 1$ degrees of freedom, where n is the size of the sample.

If $\bar{x}$ is the mean volume per unit area and $s_{\bar{x}}$ is the standard error of the mean from a sampling inventory, then the confidence interval for the total timber volume estimate for a forest area a is

$$CI \text{ (total)} = a\bar{x} \pm ats_{\bar{x}} \tag{12-2}$$

(Note that the forest area is assumed to be known without sampling error.)

As an example, assume a 5000-acre forest area has been inventoried. Based on a sample of 144 fixed-area plots, the mean volume per acre was estimated as 11,500 board feet, with a standard deviation of 6000 board feet. The .95 confidence interval for the per acre volume would be calculated as follows.

$$s_{\bar{x}} = \frac{s}{\sqrt{n}} = \frac{6000}{\sqrt{144}} = 500$$

and

$$CI \text{ (per acre)} = 11,500 \pm 1.96(500) = 11,500 \pm 980$$
$$= 10,520 \text{ to } 12,480 \text{ board feet per acre}$$

Assuming there is no error in the area a, then the confidence interval and its limits for the total timber volume estimate is

$$CI \text{ (total)} = 5000(11,500) \pm 1.96 \,(5000)(500)$$
$$= 57,500,000 \pm 4,900,000$$
$$= 52,600,000 \text{ to } 62,400,000 \text{ board feet}$$

Equation 12-2 assumes that the total forest area a is known without error. This is seldom the case. Usually the forest area is estimated by some procedure, such as repeated planimeter or dot grid measurement (see Section 6-4), and the standard error computed. Then, considering both the sampling error of volume per unit area and the sampling error of the area estimate s_a

$$CI \text{ (total)} = a\bar{x} \pm t\sqrt{(as_{\bar{x}})^2 + (\bar{x}s_a)^2} \tag{12-3}$$

Assuming that the estimate of forest area in the example has a standard error of 75 acres, then

$$CI \text{ (total)} = (5000)(11,500)$$
$$\pm 1.96\sqrt{[(5000)(500)]^2 + [(11,500)(75)]^2}$$
$$= 57,500,000 \pm 5,183,415$$
$$= 52,316,585 \text{ to } 62,683,415 \text{ board feet}$$

The estimate of a forest area a may be determined as a proportion p of a total area using a procedure such as dot counts on aerial photos (see Section 13-5.1). In this case one assumes the total area A is known without error. However, the estimate of the proportion in forest p will have a standard error s_p. The estimate of the confidence interval for the total volume would then be

$$CI = a\bar{x} + t\sqrt{(as_{\bar{x}})^2 + (\bar{x}As_p)^2} \qquad (12\text{-}4)$$

where $a = Ap$.

Assuming that the forest area of 5000 acres in the previous example was determined from a dot count showing 62.5 percent in forest of a total land area of 8000 acres and that the estimate of $s_p = 0.007$, then

$$a = 8000 \, (0.625) = 5000$$

and

$$CI = 5000(11,500) \pm 1.96 \sqrt{[(5000)(500)]^2 + (11,500)(8000)(0.007)]^2}$$

$$= 57,500,000 \pm 5,059,965$$

$$= 52,440,035 \text{ to } 62,559,965 \text{ board feet}$$

The standard error of the mean is often substituted in the following formula to determine the percent error of an estimated mean $E\%$ at the desired probability level.

$$E\% = \frac{ts_{\bar{x}}}{\bar{x}} \, 100 \qquad (12\text{-}5)$$

This is also called the sampling error in percent and expresses the precision of the inventory.

For the example we have been discussing in this section, E (in percent) for the .95 probability level would be

$$E\% = \frac{(1.96)(500)}{11,500} \, 100 = 8.5\%$$

Instead of expressing a timber estimate by its mean and confidence interval, an analogous expression called the *reliable minimum estimate* (RME) can be used (Dawkins, 1957). The RME estimates the minimum quantity expected to the present with its probability.

$$RME = \bar{x} - ts_{\bar{x}} \qquad (12\text{-}6)$$

The value of t for a probability level is obtained only from one side of the distribution. In using a table of t values where the sign is ignored (i.e., a "two-tailed" table), the appropriate value would be obtained using the column for double the probability level required. Thus, the t value for a probability level of 0.05 would be read under the column headed .10, recognizing the appropriate degrees of freedom.

12-4 SAMPLING DESIGN

The sampling design is determined by the kind of sampling units used, the number of sampling units employed, and the manner of selecting the sampling units and distributing them over the forest area, as well as the procedures for taking measurements and analyzing the results. The specifications for each of these elements can be varied to yield the desired precision at a minimum specified cost.

Basic inventory designs generally fall into the following categories.

A. Probability sampling
    1. Simple random sampling
    2. Stratified random sampling
    3. Multistage sampling
    4. Multiphase sampling
    5. Sampling with varying probabilities
B. Nonrandom sampling
    1. Selective sampling
    2. Systematic sampling

In probability sampling, the probability of selecting any unit is assumed known prior to the actual sampling. The probability is greater than zero and may be equal for all units or it may vary. Frequently, in forest inventory work the probabilities are not known but are assumed to be equal. This is referred to as equal chance selection.

In nonrandom sampling the units that constitute the sample are not chosen by the laws of chance but systematically or by personal judgment.

This chapter will describe basic designs for forest inventory using simple random, stratified random, multistage, multiphase, selective, and systematic sampling. Sampling with varying probability is discussed in Chapter 14.

12-4.1 Precision Level and Intensity

The choice of the precision level for a forest inventory depends on the sampling error one is willing to accept in the estimates. Thus one might ask: What would occur if decisions on investment, forest management, and so forth, were based on estimates with sampling errors of $\pm 1\%$, $\pm 5\%$, or $\pm 10\%$? In most forest inventory work this is not done. Precision levels used are ones traditionally employed in similar inventories. The difficulty in attacking this problem comes from the lack of methods of quantifying the effects on decisions of inventory sampling errors of different sizes (Hamilton, 1978; Husch, 1980).

The precision level can be expressed in relative terms as a percent (equation 12-5) or as a standard error of the mean. In the following sections of this chapter the formulas for the determination of sample size utilize these expressions of acceptable precision.

The *intensity* of sampling indicates the percentage of the total area of a population that is included in the sample. Thus, if 200 0.2-acre plots are measured in a 1000-acre forest, the sampling intensity I is

$$I = \frac{(200)(0.2)}{1000}\,(100) = 4.0\%$$

12-4.2 Size and Shape of Fixed-Area Sampling Units

Fixed-area sampling units in a forest inventory are called plots or strips depending on their dimensions. The term *plot* is loosely applied to sampling units of small area and square, rectangular, circular, or triangular shape. A *strip* is a rectangular plot whose length is many times its width.

Unbiased estimates of timber volumes and other stand parameters can be obtained from any plot size or shape, although the precision and cost of the survey may vary significantly. Small sampling units are frequently more efficient than larger ones. On a fairly homogeneous forest, the precision for a given sampling intensity tends to be greater for small sampling units than for large ones because the number of independent sampling units is greater. However, the size of the most efficient unit is also influenced by the variability of the forest. When small sampling units are taken in heterogeneous forests, high coefficients of variation will be obtained. In such cases larger sampling units are more desirable.

The relative efficiency e of different-sized plots can be measured by

$$e = \frac{(s_{\bar{x}})_1^2 t_1}{(s_{\bar{x}})_2^2 t_2} \tag{12-7}$$

where

$(s_{\bar{x}})_1$ = standard error in percent for one plot size as the basis for comparison

$(s_{\bar{x}})_2$ = standard error in percent for the other plot size or shape to be compared

t_1 = cost or time for base plot size or shape

t_2 = cost or time for compared plot size or shape

The equation gives the efficiency of plot size or shape 2, relative to plot size or shape 1. If e is less than 1, then plot 1 is more efficient than plot 2. If e is greater than 1, plot 2 is more efficient than plot 1.

Procedures for investigating optimum plot sizes are given by Mesavage and Grosenbaugh (1956), Freese (1962), Tardif (1965), O'Regan and Arvanitis (1966), and Zeide (1980).

Zeide presented a procedure for determining optimal plot size in one-stage systematic or random sampling designs. He stated that optimal plot size is that which minimizes total time required for location and measurement to achieve a given precision. He then developed the following formula for calculating optimal plot size.

$$\text{Optimal plot size} = P_1\left(\frac{t}{m}\right)^2 \tag{12-8}$$

where

P_1 = size of plot used in preliminary sample to assess time and variation

t = average travel time between two neighboring plots of size P_1 (the distance between these plots should be for the number of plots that provides the desired precision)

m = average plot measurement time for plot size P_1

The formula indicates that the greater the distance between plots the larger they must be. When plot size is optimal, equal amounts of time will be spent on travel and on plot measurement.

In North America the most commonly used plot sizes for mature or near-mature timber are 0.20 to 0.25 acre. In Europe plot sizes vary between 0.01 and 0.05 hectare. In tropical areas, larger plots have proven more useful because of the greater heterogeneity of the forests. The guiding principle in the choice of plot size should be to have a plot large enough to include a representative number of trees, but small enough so that the time required for measurement is not excessive. For dense stands of small trees, plots should be relatively small; for widely spaced stands of large trees, plots should be relatively large.

Circular plots have been most widely used, since a single dimension, the radius, can be used to define the perimeter. The dimensions of commonly used sample plots are given in Appendix Table A-5.

Strips may be used instead of plots. For recording purposes, the continuous strip may be subdivided into smaller recording units. It is important to remember, however, that the entire strip is still considered the ultimate sampling unit and is the basis for the number of degrees of freedom in subsequent statistical computations. (If interrupted strips are used—e.g., alternate chains of strip tallied—the units of strip tallied should be treated like plots in subsequent calculations.)

An advantage of a continuous strip over plots is that a tally is taken for the entire strip traversed so that there is no unproductive walking time between sampling units. But continuous strips have a disadvantage: the number of sampling units, and thus the degrees of freedom, is small. Comparing strip sampling to plot sampling of equal intensity, the size of the sampling unit is larger and the number of sampling units smaller. The larger sampling unit results in a reduction in the variability, but this advantage is counteracted by the smaller number of sampling units. For this reason the sampling error of a strip sampling design is usually larger than for plot sampling, assuming the same sampling intensity.

12-5 SIMPLE RANDOM SAMPLING

Simple random sampling is the fundamental selection method. All other sampling procedures are modifications of simple random sampling that are designed to achieve greater economy or precision. Simple random sampling requires that there be an equal chance of selecting all possible combinations of the n sampling units from the population. The selection of each sampling unit must be free from deliberate choice and must be completely independent of the selection of all other units.

In simple random sampling, the entire forest is treated as a single population of N units. If fixed-area sampling units are used, the population of N has definable limits. If points are used, the population of N can be considered infinite. From the population, a sample of n sampling units with equal probability of selection is randomly chosen.

Simple, or unrestricted, random sampling in forest inventory yields an unbiased estimate of the population mean and the information to assess the sampling error. However, it has the following disadvantages.

1. Requirement of devising a system for randomly selecting the plots or points.
2. Difficulties in locating widely dispersed field positions of selected sampling units.
3. Time-consuming and expensive nonproductive traveling time between units.
4. Possibility of a clumpy distribution of sampling units may result in atypical estimates of the mean, standard deviation, and other measures.

Forest inventory using random sampling requires aerial photographs or a map to establish a frame from which to draw the sample. Figure 12-1 shows the simple case of a rectangular-shaped forest that can be subdivided into N sampling units of fixed area. After the number of sampling units n has been determined, they are chosen from the frame using any accepted procedure for random selection. The selection can be with or without replacement. If sampling with replacement is followed, since there is a possibility of the same sampling unit being selected more than once, the population can be considered infinite. For large finite populations, the calculation of mean and standard errors can be done as though dealing with an infinite population since the finite population correction factor $(N - n)/N$ approaches unity. Most sampling with fixed-area plots or strips is done without replacement. Using points, the population is infinite and the sampling is carried out with replacement.

The analysis of the resulting data from the sampling units is summarized below. Note that the formulas are applicable whether dealing with plots, strips, or points. (It should be noted that separate tallies for each sampling unit in a forest inventory must be recorded to calculate the sampling errors.)

Given:

n = number of sampling units measured
N = total number of sampling units in the forest
X_i = the quantity X measured on the ith sampling unit
$\bar{x}$ = the mean of X per sampling unit; an estimate of the population mean
s = standard deviation of the sample
$s_{\bar{x}}$ = standard error of the mean
$\hat{X}$ = estimated total of X for the population
c = coefficient of variation = $\dfrac{s}{\bar{x}}(100)$

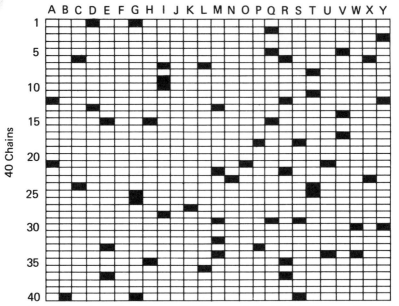

(a) Using plots. Forest area of 200 acres divided into 1000 sampling units of .2-acre (2 X 1 chain). Sixty plots randomly chosen.

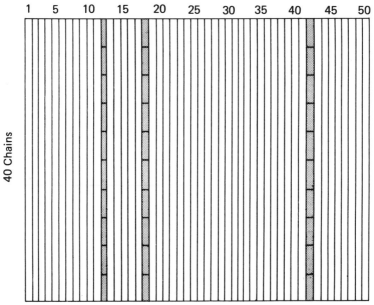

(b) Using strips. Forest area of 200 acres divided into 50 strips 1-chain wide and 40 chains long (4 acres). Three strips randomly selected with recording units of 4 X 1 chains.

Fig. 12-1 *Simple random sampling.*

E = allowable standard error in units of X

$E\%$ = allowable standard error as a percent of the mean

t = Student's "t" for desired probability level

Then,

$$\bar{x} = \frac{\sum_{i=1}^{n} X_i}{n} \tag{12-9}$$

$$s^2 = \frac{\sum_{i=1}^{n} (X_i - \bar{x})^2}{n - 1} \tag{12-10}$$

or

$$s^2 = \frac{n\sum_{i=1}^{n} X_i^2 - \left(\sum_{i=1}^{n} X_i\right)^2}{n(n - 1)} \tag{12-11}$$

$$s_{\bar{x}}^2 = \frac{s^2}{n}\left(\frac{N - n}{N}\right) \tag{12-12}$$

(For an infinite population, the finite correction factor would be omitted.)

The estimate of the population total is:

$$\hat{X} = N\bar{x} \tag{12-13}$$

The number of sampling units needed to yield an estimate of the mean with a specified allowable error and probability can be calculated from

$$n = \frac{Nt^2 s^2}{NE^2 + t^2 s^2} \tag{12-14}$$

For an infinite population

$$n = \frac{t^2 s^2}{E^2} \tag{12-15}$$

Expressing the standard deviation and allowable error in percentages,

$$n = \frac{Nt^2 c^2}{N(E\%)^2 + t^2 c^2} \tag{12-16}$$

For an infinite population

$$n = \frac{t^2 c^2}{(E\%)^2} \tag{12-17}$$

Note that in the equations, s and c refer to the standard deviation and coefficient of variation of a preliminary sample taken to give an indication of the variability of the population. E is for an arbitrarily chosen level and the t value depends on the required probability and degrees of freedom. The t value should be determined from Student's t distribution with $n - 1$ degrees of freedom, where n refers to the number of sampling units in the preliminary sample. Of course, to be correct, n should be the number of sampling units that is being sought. However, since this is unknown, the n of the preliminary sample can be used. Thus, the actual probability level for the indicated number of sampling units will change. If a preliminary sample has not been taken and if the expected sample size is large (over 30), the t values for an infinite number of degrees of freedom may be used.

If the amount of money for a survey is fixed, the number of sampling units must be determined within this restriction. If the total cost for a survey is given as c_t, then

$$c_t = c_o + nc_1 \tag{12-18}$$

where

c_o = overhead cost of a survey including planning, organization, analysis, and compilation
c_1 = cost per sampling unit
n = number of sampling units

The number of sampling units is indicated as

$$n = \frac{c_t - c_o}{c_1} \tag{12-19}$$

If preliminary information on the mean and variance of the population is known from experience or a preliminary sample, it is then possible to estimate the precision that can be obtained for a given cost.

12-6 RATIO AND REGRESSION ESTIMATION

Unfortunately, there exists some ambiguity in the term "regression" in the sampling context. It always pertains to the situation where two (or more) variables are under study, but it may be used to indicate either the case where the auxiliary variable mean or total is known without error (as opposed to "double sampling" when an error is involved) or the case where the model assumed has an intercept value (as opposed to "ratio" estimation where the intercept is zero). By careful use of the words *sampling* and *estimation*, the ambiguity can perhaps be minimized. Thus, we might consider the following classification using the notions presented by Freese (1962).

A. Regression sampling (true mean of X known)
    1. Regression estimation: $\bar{y}_R = \bar{y} + b(\mu_x - \bar{x})$, where $\bar{y}$, b, and $\bar{x}$ are estimated from the sample and μ_x is the known population mean.

2. Ratio estimation: $\bar{y}_R = \hat{R}\mu_x$, where $\hat{R}$ is an estimated ratio obtained from the sample by either $\hat{R} = \dfrac{\bar{y}}{\bar{x}}$ (ratio of means) or $\hat{R} = \dfrac{\Sigma\dfrac{y}{x}}{n}$ (mean of ratios), dependent on certain variance assumptions.

B. Double sampling (true mean of X unknown, and must be estimated from a sample)
   1. Regression estimation: $\bar{y}_{Rd} = \bar{y}_s + b(\bar{x}_l - \bar{x}_s)$, where $\bar{y}_s$, b, and $\bar{x}_s$ are determined from a small sample and $\bar{x}_l$ is determined from a large sample.
   2. Ratio estimation: $\bar{y}_{Rd} = \hat{R}\bar{x}_l$, where $\hat{R}$ is either a ratio of means or a mean of ratios obtained from the small sample and $\bar{x}_l$ is obtained from the large sample.

For complete details regarding sample sizes needed and formulas for the estimation of sampling errors of the various estimates in general situations, a sampling text such as Cochran (1977) should be consulted.

In this textbook we will discuss only certain applications of these techniques. Specifically, in the following paragraphs we will look at the use of ratio and regression estimation to correct irregular-sized sample plots, and in Section 12-9 we will consider the use of double sampling in a two-phase inventory such as that encountered when combining aerial photograph and ground plot estimations.

When the forest to be sampled, using fixed-area units, is irregular in form, the possibility exists of having fractional plots or irregular length strips. This is illustrated in Fig. 12-2. (The problem in relation to sampling points is discussed in Chapter 14.) Situations of this type arise frequently in actual application. Note that to meet the requirement of sampling units of equal size, it has become necessary (as in Fig. 12-2) to sketch imaginary boundaries to enclose the irregular boundaries of the forest. The area within each boundary has been divided into equal-sized sampling units. Some of the plots and strips fall entirely in the forest and others include both the forest area and a zone beyond. For this reason, it is necessary to measure two variables for each plot or strip; the area in the sampling unit which falls in the forest area and the quantity of timber on this area. The mean volume per unit area for the forest must then be calculated using the ratio or regression methods of estimation.

In the majority of cases involving plot sampling, the ratio or regression procedures have been bypassed in the interest of simplifying field work, saving time and reducing the later calculations. Thus, if a plot falls so that it straddles a forest or stratum boundary, instead of measuring both the area and quantities of that portion of the unit in the forest, a number of other questionable procedures have been used. These involve moving the plots so that they fall entirely within or without the forest or assigning them to the stratum in which the center point falls. To avoid the possibility of bias arising from these procedures, the ratio or regression methods may be used. However, because of the simplicity of application for both fixed-area plots and points, the mirage method (Section 14-5.6) is frequently justified.

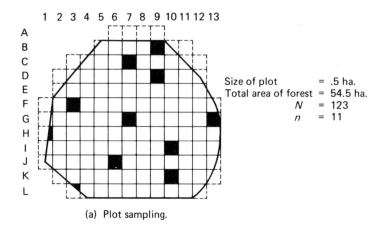

Size of plot = .5 ha.
Total area of forest = 54.5 ha.
N = 123
n = 11

(a) Plot sampling.

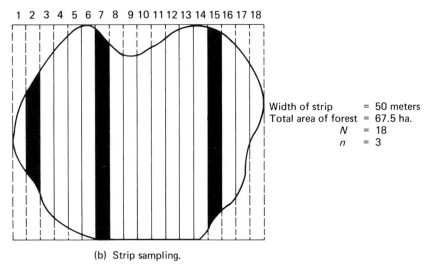

Width of strip = 50 meters
Total area of forest = 67.5 ha.
N = 18
n = 3

(b) Strip sampling.

Fig. 12-2 *Simple random sampling of irregularly shaped forest areas.*

When the ratio estimation procedure is employed, the sample should consist of 30 or more units so that the inherent bias of the method becomes neglible. In addition, the method is only truly applicable when the linear relationship between the two variables, timber quantity and area per sampling unit, passes through the origin. When the linear regression line does not pass through the origin (i.e., volume per unit can be zero when plot area is nonzero), the regression method of estimation is preferable.

12-6.1 Ratio Estimation

Given:

V_i = quantity measured on the ith sampling unit
X_i = area of the ith sampling unit
$\bar{v}$ = mean of the observed V_i on the selected sampling units
$\bar{x}$ = mean area of the sampling units selected
$\bar{r}$ = estimate of the mean quantity per unit area
$\hat{v}$ = estimate of the total quantity (e.g., volume) of the forest
X = known total area of the forest
N = number of sample units in the population
n = number of units in the sample

Then,

$$\bar{r} = \frac{\dfrac{\sum\limits_{i=1}^{n} V_i}{n}}{\dfrac{\sum\limits_{i=1}^{n} X_i}{n}} = \frac{\sum\limits_{i=1}^{n} V_i}{\sum\limits_{i=1}^{n} X_i} \tag{12-20}$$

$$\hat{v} = X\bar{r} \tag{12-21}$$

If the area is known without any sampling error, then the variance of the total estimate of quantity is

$$s_v^2 = X^2 s_{\bar{r}}^2 \tag{12-22}$$

If the total area of the forest is an estimate x and has a sampling error, then the variance is

$$s_v^2 = x^2 s_{\bar{r}}^2 + \bar{r}^2 s_x^2 \tag{12-23}$$

where

s_v^2 = variance of the total quantity
$s_{\bar{r}}^2$ = variance of the ratio $\bar{r}$
s_x^2 = variance of the total area estimate

$$s_{\bar{r}}^2 = \left(\frac{N-n}{N}\right)\frac{s^2}{n} \tag{12-24}$$

where

$$s^2 = \frac{\bar{r}^2}{n-1}\left(\frac{\sum\limits_{i=1}^{n} V_i^2}{\bar{v}^2} + \frac{\sum\limits_{i=1}^{n} X_i^2}{\bar{x}^2} - \frac{2\sum\limits_{i=1}^{n} V_i X_i}{\bar{v}\bar{x}}\right) \tag{12-25}$$

The number of sampling units for a given precision can be estimated by solving equation 12-24 for n and using a preliminary estimate of s^2. Student's t can be incorporated to vary the probability level.

12-6.2 Regression Estimation

Using this method, an estimate of the mean volume of the sampling units is adjusted by means of a regression coefficient. The regression coefficient indicates the average change in volume per unit change in area between the sampling units in the sample and the population. The total size of the forest area and the total number of sampling units in the population and their average size must be known.

Given

X_i, V_i, X, $\hat{v}$, $\bar{v}$, N, and n are as defined for ratio estimation
b = regression coefficient
$\bar{v}_{\text{reg}}$ = adjusted estimate of the mean volume per sampling unit in the population
$\overline{X}$ = mean area per sampling unit in the population
$\bar{x}_n$ = mean area of the sampling units selected

$$\bar{v}_{\text{reg}} = \bar{v} + b(\overline{X} - \bar{x}_n) \tag{12-26}$$

$$b = \frac{\displaystyle\sum_{i=1}^{n} X_i V_i - \frac{\displaystyle\sum_{i=1}^{n} X_i \sum_{i=1}^{n} V_i}{n}}{\displaystyle\sum_{i=1}^{n} X_i^2 - \frac{\left(\displaystyle\sum_{i=1}^{n} X_i\right)^2}{n}} \tag{12-27}$$

The estimate of the mean volume per unit area $\bar{r}_{\text{reg}}$ is

$$\bar{r}_{\text{reg}} = \frac{\bar{v}_{\text{reg}}}{\overline{X}} \tag{12-28}$$

The estimate of the variance of the regression $s^2_{\bar{v}\text{reg}}$ is

$$s^2_{\bar{v}\text{reg}} = \frac{s^2}{n}\left(\frac{N - n}{N}\right) \tag{12-29}$$

where

$$s^2 = \frac{1}{n-2} \left\{ \left[\sum_{i=1}^{n} V_i^2 - \frac{\left(\sum_{i=1}^{n} V_i \right)^2}{n} \right] - b \left[\sum_{i=1}^{n} V_i X_i - \frac{\sum_{i=1}^{n} V_i \sum_{i=1}^{n} X_i}{n} \right] \right\} \qquad (12\text{-}30)$$

The estimate of the variance of the volume per unit area s^2_{reg} is

$$s^2_{reg} = \frac{s^2_{\bar{v}_{reg}}}{\bar{X}^2} \qquad (12\text{-}31)$$

The estimate of the total volume of the population $\hat{v}$ is

$$\hat{v} = \bar{r}_{reg} X \qquad (12\text{-}32)$$

or

$$\hat{v} = \bar{v}_{reg} N \qquad (12\text{-}33)$$

and its variance is

$$s^2_v = X^2 s^2_{reg} \qquad (12\text{-}34)$$

12-7 STRATIFIED RANDOM SAMPLING

In many cases a heterogeneous forest may be broken down by stratification into subdivisions called *strata*.

In forest inventory work the purpose of stratification is to reduce the variation within the forest subdivisions and increase the precision of the population estimate.

Stratified random sampling in forest inventory has the following advantages over simple random sampling.

1. Separate estimates of the means and variances can be made for each of the forest subdivisions.

2. For a given sampling intensity, stratification often yields more precise estimates of the forest parameters than does a simple random sample of the same size. This will be achieved if the established strata result in a greater homogeneity of the sampling units within a stratum than for the population as a whole.

On the other hand, the disadvantages of stratification are that the size of each stratum must be known or at least a reasonable estimate be available, and that sampling units must be taken in each stratum if an estimate for that stratum is needed.

Stratification is achieved by subdividing the forest area into strata on the basis of some criteria such as topographical features, forest types, density classes, or volume, height, age, or site classes. If possible, the basis of stratification should be the same characteristic that will be estimated by the sampling procedure. Thus, if volume per unit area is the parameter to be estimated, it is desirable to stratify the forest area on the basis of volume classes. Aerial photographs are of tremendous assistance in stratification for forest inventory. In fact, it is safe to say that aside from mapping, the greatest use so far for aerial photographs in forest inventory has been precisely for this purpose.

The different strata into which a forest may be divided can be irregular in shape, of many sizes, and of varying importance. Stratification permits the sampling intensity and precision to be varied for the several strata.

An arbitrary form of stratification is often used in sampling large forest areas where there is little basis for some kind of natural subdivision. This often occurs in inventories of large, remote forest areas where maps or photographs are not available or where photo interpretation reveals little basis for stratification. In this case the forest can be broken into uniform-sized squares or rectangles even though it is known that the resulting blocks may not contain homogeneous subpopulations. But it is reasonable to assume greater homogeneity within a smaller block than in the entire forest.

Within each of the M strata into which the forest is divided, a number of sampling units are randomly chosen. Figure 12-3 shows a forest broken down into three strata of fixed-area sampling units, all of the same size.

The analysis of the data obtained from stratified random sampling is summarized below for the situation where the plots or strips are all of a uniform size, as illustrated by Fig. 12-3.

Given

M = number of strata in the population
n = total number of sampling units measured for all strata
n_j = total number of sampling units measured in the jth stratum
N = total number of sampling units in the population
N_j = total number of sampling units in the jth stratum
X_{ij} = quantity X measured on the ith sampling unit of the jth stratum
$\bar{x}_j$ = mean of X for the jth stratum
$\bar{x}$ = estimated mean of X for the population
P_j = proportion of the total forest area in the jth stratum $= \dfrac{N_j}{N}$
$\hat{X}$ = estimated total of X for the population
s_j^2 = variance of X for the jth stratum
$s_{\bar{x}}^2$ = estimated variance for the mean for the population
$s_{\hat{X}}^2$ = estimated variance of $\hat{X}$
E = allowable standard error in units of X

The estimate of the mean per stratum is

$$\bar{x}_j = \frac{\sum\limits_{i=1}^{n} X_{ij}}{n_j} \tag{12-35}$$

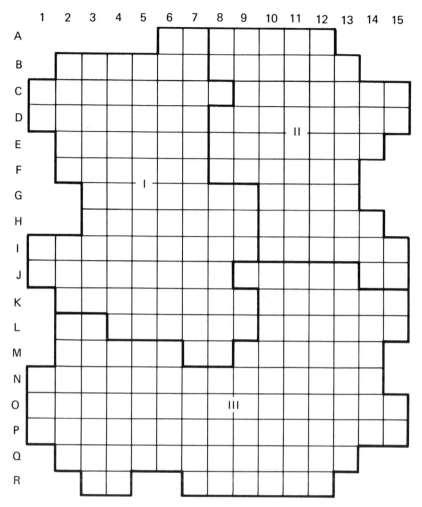

Fig. 12-3 *Forest area divided into three unequal-sized strata. All sampling units are 0.5 acre.*

The estimate of the mean for the population is

$$\bar{x} = \frac{\sum\limits_{i=1}^{n} N_j \bar{x}_j}{N} = \sum\limits_{j=1}^{M} P_j \bar{x}_j \qquad (12\text{-}36)$$

The estimate of the total for X for the entire population is

$$\hat{X} = \sum\limits_{j=1}^{M} N_j \bar{x}_j \qquad (12\text{-}37)$$

or

$$\hat{X} = N\bar{x} \qquad (12\text{-}38)$$

The variance for each stratum in the population s_j^2 is calculated as described for simple random sampling. The variance of the mean for the population is then calculated from

$$s_{\bar{x}}^2 = \frac{1}{N^2} \sum_{j=1}^{M} \left[\frac{N_j^2 s_j^2}{n_j} \left(\frac{N_j - n_j}{N_j} \right) \right] \qquad (12\text{-}39)$$

If the population consists of strata sufficiently large that the finite correction factor is insignificant or information is available on the relative sizes of the strata, then we can estimate the variance from

$$s_{\bar{x}}^2 = \sum_{j=1}^{M} P_j^2 \frac{s_j^2}{n_j} \qquad (12\text{-}40)$$

The standard error of the total estimate of X is then

$$s_X^2 = N^2 s_{\bar{x}}^2 \qquad (12\text{-}41)$$

The n_j sampling units actually chosen and measured per stratum should not be used to estimate the relative size of a stratum. Proportions in the strata should be estimated "a priori" to the actual sampling.

12-7.1 Estimation of Number of Sampling Units

To estimate the number of sampling units needed, it is necessary to have preliminary information on the variability of the strata in the population and to choose an allowable error and probability level. With this information, the intensity of sampling can be estimated. The total number of sampling units can then be allocated to the different strata either by proportional or optimum allocation.

In proportional allocation, the number of sampling units in a stratum, out of the total sample, is in proportion to the area of the stratum. In optimum allocation, the number of sampling units per stratum is proportional to the standard error of the stratum weighted by area. Optimum allocation will give the smallest standard error for a stratified population when a given total of sampling units is taken. If we wish to get the most precise estimate of the population mean for the expenditure of money, then optimum allocation should be used. The allocation can be done either if the costs of sampling units in all strata are equal or if they differ.

Using proportional allocation, advance knowledge of the variability in the several strata is desirable to determine the total sample size. Where no information on the variability of the individual strata is available, it is necessary to estimate the total sample size for the entire population as though simple random sampling was being employed.

Using proportional allocation, the determination of sample size n for a given precision when information on variability per strata is available is

$$n = \frac{Nt^2 \sum_{j=1}^{M} P_j s_j^2}{NE^2 + t^2 \sum_{j=1}^{M} P_j s_j^2} \qquad (12\text{-}42)$$

and

$$n_j = P_j n \qquad (12\text{-}43)$$

If the population can be considered infinite, then

$$n = \frac{t^2 \sum_{j=1}^{M} P_j s_j^2}{E^2} \qquad (12\text{-}44)$$

Using optimum allocation, the determination of sample size n with specified precision is shown in equation 12-45 for the simplest case when costs per sampling unit are the same in all strata. The sampling intensity is changed in each stratum, according to its variability, to achieve a given precision with the smallest possible number of sampling units.

$$n = \frac{Nt^2 \left(\sum_{j=1}^{M} P_j s_j \right)^2}{NE^2 + t^2 \sum_{j=1}^{M} P_j s_j^2} \qquad (12\text{-}45)$$

and

$$n_j = \frac{P_j s_j}{\sum_{j=1}^{M} P_j s_j} n \qquad (12\text{-}46)$$

If the population can be considered infinite, then

$$n = \frac{t^2 \left(\sum_{j=1}^{M} P_j s_j \right)^2}{E^2} \qquad (12\text{-}47)$$

For the determination of sample size by optimum allocation when the costs per sampling unit vary per stratum, or when the total cost of the inventory is fixed, the student is referred to Cochran (1977).

Frequently, the use of stratified random sampling may result in the boundary lines of strata passing through the uniform sampling units near strata boundaries, resulting in varying areas tallied. (This is analogous to the situation described in Section 12-6.) When this occurs, ratio estimation or regression procedures as described by Cochran (1977) and Loetsch and Haller (1964) should be used.

12-8 MULTISTAGE SAMPLING

In multistage sampling, a population consists of a list of sampling units (primary stage), each of which is made up of smaller units (second stage), which in turn could be made up of smaller units (third stage). A random sample would be chosen from the primary units. A random subsample of the secondary units would then be taken in each of the selected primary units, and the procedure would be continued to the desired stage. In general, this procedure is called multistage sampling. Two-stage sampling, the commonest application, which is discussed in this section, indicates the sampling stops at the secondary stage. For example, a forest to be inventoried might consist of numerous compartments that could be considered the primary units in a sampling design. Plots chosen in the selected compartments would then form the secondary units. Similarly, an inventory design using plots on randomly chosen lines or strips is a form of two-stage sampling. Another frequently used two-stage sampling design in forest inventory employs groups or clusters of plots or sampling points at randomly chosen locations.

Multistage sampling in forest inventory is not restricted to fixed-area sampling units but can also be employed with variable plot procedures. Thus, a series of primary locations could be randomly chosen in a forest and, at each location, a number of secondary points chosen for the selection of trees using the variable plot procedure. In all cases the group of secondary units selected within each of the primary units can be referred to as a *cluster*.

Multistage sampling has the principal advantage of concentrating the measurement work close to the locations of the chosen primary sampling units rather than spreading it over the entire forest area to be inventoried. This is advantageous when it is difficult and costly to locate and get to the ultimate sampling unit, while it is comparatively easy and cheap to select and reach the first-stage unit.

To permit the calculation of unbiased estimates of means and standard errors, random selection of sampling units at all stages should be used. It would also be possible to select primary units such as forest compartments with probability proportional to their sizes and then choose secondary units on a random basis.

Frequent use is made of systematic selection in two-stage sampling (see Section 12-10.1). A common design employed in forest inventory utilizes randomly chosen primary sampling units but then systematically selects the secondary units within them.

In two-stage sampling, m sampling units are selected from the M primary units of the population as a first stage. From each of the selected m units, n secondary units are then chosen from the population of N secondary units within each primary sampling unit. Note that if all M primary units are selected, the design is equivalent to stratified random sampling.

The most common cases encountered using two-stage sampling in forest inventory are:

1. All primaries of equal size containing equal numbers of secondaries of uniform size.
2. Primaries of unequal sizes containing varying numbers of secondary units of uniform size.
3. Primaries of unequal sizes containing varying numbers of secondary units of variable sizes.

Case 3 frequently occurs in forest inventory since uniform-sized secondary units falling on forest or strata boundaries will be divided. The analysis requires the use of ratio estimation. Only Case 1 will be discussed here. The student is referred to Cochran (1977) and Freese (1962) for analyses of cases when primaries are of unequal sizes. The formulas for estimates of the means, totals, and their variances are shown below.

M = total number of primary sampling units in the population
N = number of secondary units per primary, equal for all primaries
m = number of primary units in the sample
n = number of secondary units per primary in the sample, equal for all primaries
X_{ij} = quantity (e.g., volume) measured on the ith secondary unit in the jth primary
$\bar{x}_j$ = estimate of the mean of X for the jth primary unit
$\hat{X}_j$ = estimate of the total of X for the jth primary unit
$\hat{X}$ = estimate of the total of X for the entire forest
$\bar{x}$ = estimate of the mean of X for the entire forest
$s_{\bar{x}}^2$ = estimate of the variance of the mean for the entire forest
s_X^2 = estimate of the variance of the total for the entire forest

The estimate of the population mean is

$$\bar{x} = \frac{1}{mn} \sum_{j=1}^{m} \sum_{i=1}^{n} X_{ij} \tag{12-48}$$

The estimate of the variance of the mean is

$$s_{\bar{x}}^2 = \left(1 - \frac{m}{M}\right) \frac{s_B^2}{m} + \left(1 - \frac{mn}{MN}\right) \frac{s_W^2}{mn} \tag{12-49}$$

where

s_B^2 = estimate of the variance between the means of the secondary sampling units within primaries
s_W^2 = estimate of the variance within the groups of secondary sampling units

The values of s_B^2 and s_W^2 are found from

$$s_B^2 = \frac{\dfrac{\sum\limits_{j=1}^{m}\left(\sum\limits_{i=1}^{n}X_{ij}\right)^2}{n} - \dfrac{\left(\sum\limits_{j=1}^{m}\sum\limits_{i=1}^{n}X_{ij}\right)^2}{nm}}{m-1} \tag{12-50}$$

$$s_W^2 = \frac{\sum\limits_{j=1}^{m}\sum\limits_{i=1}^{n}X_{ij}^2 - \dfrac{\sum\limits_{j=1}^{m}\left(\sum\limits_{i=1}^{n}X_{ij}\right)^2}{n}}{m(n-1)} \tag{12-51}$$

When the number of primary units in the population is large, the formula for the variance of the mean can be reduced to

$$s_{\bar{x}}^2 = \frac{s_B^2}{m} + \frac{s_W^2}{mn} \tag{12-52}$$

Then,

$$\hat{X}_j = \frac{N}{n}\sum_{i=1}^{n}X_{ij} \tag{12-53}$$

$$\hat{X} = \frac{M}{m}\sum_{j=1}^{m}x_j \tag{12-54}$$

$$s_X^2 = M^2 N^2 s_{\bar{x}}^2 \tag{12-55}$$

Estimates of the numbers of sampling units at the two stages can be made as shown in equation 12-56.

$$n = \sqrt{\frac{c_1\, s_W^2}{c_2\, s_B^2}} \tag{12-56}$$

where

$$
\begin{aligned}
n\quad &= \text{optimum number of secondary units per primary}\\
c_1\quad &= \text{cost of establishing a primary unit}\\
c_2\quad &= \text{additional cost of establishing and measuring a second-}\\
&\quad\ \text{ary unit}\\
s_W^2 \text{ and } s_B^2 &= \text{estimates from a preliminary sample}
\end{aligned}
$$

Then,

$$m = \frac{t^2 \left(s_B^2 + \dfrac{s_W^2}{n} \right)}{E^2 + \dfrac{1}{M} \left(s_B^2 + \dfrac{s_W^2}{N} \right)} \qquad (12\text{-}57)$$

or, for an infinite population,

$$m = t^2 \frac{\left(s_B^2 + \dfrac{s_W^2}{n} \right)}{E^2} \qquad (12\text{-}58)$$

where

m = number of primary units required for an estimate of the mean with a given precision
E = a chosen allowable error in units of X
t = Student's t for a chosen probability level

We can estimate the number of primary units required to obtain optimum precision when we have a fixed total amount of money c available for the inventory. It is first necessary to estimate the optimum number of secondary units per primary as above. Then, the number of primary units that should be taken is

$$m = \frac{c}{c_1 + nc_2} \qquad (12\text{-}59)$$

12-9 DOUBLE SAMPLING

Double sampling is a form of multiphase sampling limited to two phases. In double sampling an estimate of the principal variable is obtained by utilizing its relationship to a supplementary variable. The method is useful when information on the principal variable is costly and difficult to obtain, whereas the supplementary and related variable can be more easily and cheaply observed. The aim of double sampling is to reduce the number of measurements of the costly, principal variable without sacrificing the precision of the estimate.

The general procedure in double sampling is that in a first phase a large random sample is taken of the secondary or auxiliary variable X that will yield a precise estimate of its population mean or total. In a second phase, a random subsample is taken from the previous sample and on these sampling units measurements are taken of the principal variable Y. Note that the first and second phases are mutually dependent since the measurements in the second phase are taken from a portion of the sampling units of the first phase. Thus, we have a small sample on which both the supplementary and principal variables, X and Y, have been measured. With these data a regression can be developed between the two variables that can be utilized with the large sample of the auxiliary variable to make an estimate of the mean and total for the principal variable.

The relationship of Y and X may have one of numerous forms. For illustrative purposes a simple linear relationship will be demonstrated. However, it is well to bear in mind that in many instances a curvilinear relationship may be required.

In double sampling the corrected estimate is obtained from a regression of the form

$$\bar{y}_{\text{reg}} = \bar{y}_s + b(\bar{x}_l - \bar{x}_s) \tag{12-60}$$

where

$$
\begin{aligned}
\bar{y}_{\text{reg}} &= \text{regression estimate of the mean of } Y \text{ (principal variable) from} \\
&\quad \text{double sampling} \\
\bar{y}_s &= \text{estimate of the mean of } Y \text{ from the second, small sample} \\
\bar{x}_l &= \text{estimate of the mean of } X \text{ (auxiliary variable) from the first,} \\
&\quad \text{large sample} \\
\bar{x}_s &= \text{estimate of the mean of } X \text{ from the second, small sample} \\
b &= \text{linear regression coefficient} \\
n &= \text{number of sampling units in the first, large sample} \\
m &= \text{number of sampling units in the second, small sample}
\end{aligned}
$$

The regression coefficient is calculated from

$$
b = \frac{\displaystyle\sum_{i=1}^{m} (X_i - \bar{x}_s)(Y_i - \bar{y}_s)}{\displaystyle\sum_{i=1}^{m}(X_i - \bar{x}_s)^2}
$$

$$
= \frac{\displaystyle\sum_{i=1}^{m} X_i Y_i - \dfrac{\displaystyle\sum_{i=1}^{m} X_i \displaystyle\sum_{i=1}^{m} Y_i}{m}}{\displaystyle\sum_{i=1}^{m} X_i^2 - \dfrac{\left(\displaystyle\sum_{i=1}^{m} X_i\right)^2}{m}} \tag{12-61}
$$

where

$$
\begin{aligned}
X_i &= \text{quantity } X \text{ measured on the } i\text{th sampling unit of the second,} \\
&\quad \text{small sample} \\
Y_i &= \text{quantity } Y \text{ measured on the } i\text{th sampling unit of the second,} \\
&\quad \text{small sample}
\end{aligned}
$$

The estimate of the variance of the regression estimate is then

$$s_{\bar{y}reg}^2 = \left[\frac{s_{y \cdot x}^2}{m} + \frac{s_{y \cdot x}^2(\bar{x}_l - \bar{x}_s)^2}{\sum\limits_{i=1}^{m} (X_i - \bar{x}_s)^2} \right] \left(1 - \frac{m}{n} \right) + \frac{s_y^2}{n} \left(1 - \frac{n}{N} \right) \qquad (12\text{-}62)$$

N represents the total number of sampling units in the first phase of double sampling (large sample). Note that, in equation 12-62, $(1 - n/N)$ is the finite population correction that is dropped if the population is considered infinite, or if n is small relative to N.

The values for insertion in the formula for the variance are calculated from

$$s_{y \cdot x}^2 = \frac{\sum\limits_{i=1}^{m}(Y_i - \bar{y}_s)^2 - b^2 \sum\limits_{i=1}^{m} (X_i - \bar{x}_s)^2}{m - 2} \qquad (12\text{-}63)$$

$$s_y^2 = \frac{\sum\limits_{i=1}^{m}(Y_i - \bar{y}_s)^2}{m - 1} = \frac{\sum\limits_{i=1}^{m} Y_i^2 - \dfrac{\left(\sum\limits_{i=1}^{m} Y_i \right)^2}{m}}{m - 1} \qquad (12\text{-}64)$$

An example of double sampling in forest inventory is the procedure using a combination of aerial photographic estimation and field plots. The first phase of sampling consists of estimating the volumes on aerial photographs of a large number of relatively inexpensive sampling units employing photo-interpretation and measurement techniques. Thus, the supplementary variable X could be volume per acre from interpretation of photo plots. In the second phase, a subsample of these plots is selected and visited in the field for direct determination of their volumes. The estimates of the volumes per acre from field plots would be the observation of Y. This subsample is much smaller since field plots are more expensive than photo plots. A regression is then prepared between field-plot volumes and photo-plot volumes, permitting a corrected volume estimate to be made from the large inexpensive sample of the first phase.

Double sampling can also be carried out using ratio estimates, remembering that the ratio estimate is a conditioned regression in which the relationship between the two variables X and Y is such that a zero value of X means a zero value of Y (Cochran, 1977).

12-10 NONRANDOM SAMPLING

Selective and systematic sampling are the two main applications of nonrandom sampling in forest inventory work. In selective sampling an observer selects a sample of plots that appears to be representative of the forest and measures them. Selective sampling may give good approximations of population parameters, but it is not recommended. Since human choice is often prejudiced by individual opinion, it may result in bias. In addition, selected samples will not yield a measure of reliability of the estimate because probability theory is based on the laws of chance. Selective sampling was used in the pioneer days of American forestry, but is now rarely used or justified.

12-10.1 Systematic Sampling

In systematic sampling, the sampling units are spaced at fixed intervals throughout the population. Systematic sampling designs, which are widely used, have a number of advantages. They provide reliable estimates of population means and totals by spreading the sample over the entire population. They are usually faster and cheaper to execute than designs based on probability sampling because the choice of sampling units is mechanical, eliminating the need for a random selection process. Travel between successive sampling units is easier since fixed directional bearings are followed and the resulting travel time is usually less than that required for locating randomly selected units. The size of the population need not be known since units are chosen at a fixed interval after an initial point has been selected. In addition, mapping can be carried out concurrently on the ground since the field party traverses the area in a systematic grid pattern. In forest inventory work the systematic distribution of sampling units can be used with fixed-area plots or strips, or points or lines in P.P.S. sampling (Chapter 14).

Since the units for a systematic sample are fixed at some regular interval, there will be a fixed set of possible samples. If a sampling interval k is chosen, there will be k possible samples. For the mean of a systematic sample to be unbiased, some form of random choice must be incorporated in the sampling process. The only randomization possible is the selection of one of the fixed sets of systematic samples. The set chosen will depend on the selection of the initial sampling unit in the population.

The initial sampling unit can be randomly chosen out of the entire population of units, or it may be randomly selected from the first k units in the population. In either case, once the first unit is chosen all following units will be selected at the interval k. Many forest inventories start a set of systematically distributed sampling units at some easily accessible, arbitrarily chosen point, assuming neglible bias in the resulting estimate of the mean.

12-10.2 Sampling Error for a Systematic Inventory

If the total population of sampling units in a forest were randomly distributed, exhibiting no pattern of variation, then a systematic sample would be equivalent to a random sample, and the random sampling formulas would be applicable for estimating the sampling error. In biological populations, such as a forest, the components are rarely, if ever, arranged completely independent of each other but, instead, show a systematic or periodic variation from place to place. If sampling units are systematically selected, then variation in the observed values may no longer be ascribable to randomness if the interval between sampling units happens to coincide with the pattern of population variation.

The larger the forest area inventoried, the greater the variation that can be expected and the more likelihood that a systematic sample will give a better estimate of the mean than a completely random sample. Even for a stratified population, a systematic sample will probably yield a better estimate of the mean if the strata are large and variable. As the homogeneity of the defined strata increases, the estimates from a random and systematic sample will tend to agree.

Fundamentally, the reason a systematic sample will not yield a valid estimate of the sampling error is that variance computations require a minimum of two randomly selected sampling units. A systematic sample (the entire set of units) consists of a single selection from the population. The sampling interval k divides the population in k clusters or sets of n sampling units and then only one of these clusters constitutes the sample.

Various methods to approximate the sampling error of a systematic sample have been devised. The systematic sample of equidistant sampling units can be considered as a simple or unstratified random sample and the sampling error computed as for a random sample. Osborne (1942) has shown that the sampling error computed in this way estimates the maximum sampling error, which may considerably overestimate the actual sampling error. Although a number of approximation procedures have been devised, there is no valid method for estimating the sampling error of a systematic sample.

Shiue (1960) proposed a method of systematic sampling that maintains the advantages of systematic sampling and also provides a reasonable means of estimating the sampling error. In this method several systematic samples are taken, with the initial sampling unit chosen randomly for each start. Using a line-plot procedure, the first sample of systematically located plots constitutes the first cluster. Another systematic sample would be the second cluster, and so on. Based on these clusters, estimates of the mean plot volume and its sampling error can be computed. To avoid a large t value and to maintain a small confidence interval for a given probability, at least five random starts should be used.

12-10.3   Systematic Strip Sampling

By using strips as the sampling unit, the systematic distribution is accomplished by first dividing the forest area into N strips of uniform size. Sampling units would then be taken at intervals of every kth strip to form the sample of n strips. Figure 12-4 shows a forest area divided into 50 strips of uniform length. The selection of n strips at a sampling interval of k strips can be carried out in two ways.

1. A random selection of a number from 1 to N can be made and the corresponding strip chosen as the initial sampling unit. Sampling units at the interval k are then taken in both directions from this initial strip. A practical way of carrying this out is to select some random number between 1 and

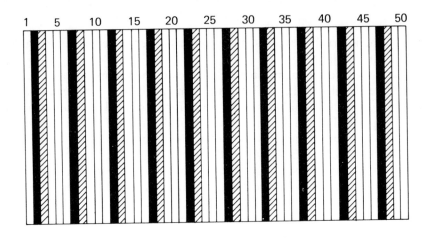

Fig. 12-4 *Forest area divided into strips for systematic strip inventory.*

N, divide by the interval k, and obtain the remainder of the division. This remainder will have the value between 1 and k. Then select the strip with this number as the first sampling unit. All subsequent strips at intervals of k are then selected.

2. Randomly select a number between 1 and k for the first strip. All subsequent strips are taken at intervals of k strips.

Both procedures will yield the same probable number of systematic samples. The first procedure will yield an unbiased estimate of the mean, whereas the second procedure may give a slightly biased result if the value of N is not an exact multiple of k. The first procedure should be used if possible. The second procedure, however, must be used if the size of the population is not known, as may occur in sampling a forest where no map is available.

Often a forest is irregular in shape rather than square or rectangular. If the area is then divided into strips of equal width, the strips will differ in length and consequently in area. Thus, the possible systematic samples that could be obtained by taking every kth strip may differ in size. Zenger (1964) has described a method that draws a systematic sample in such a way that the probability of selection of a strip is proportional to its length.

Applying strip sampling, the field party starts from a baseline, or one side of the tract, and runs a straight strip on a compass bearing across the tract stopping at the other side. The party then offsets the predetermined interval and runs back to the baseline or boundary. The procedure continues until all the strips in the sample have been measured. The timber occurring on the strips is tallied and represents the sample of the entire forest. Separate tally sheets may be kept for different forest types so that separate estimates may be made for them.

It is desirable to orient strips at right angles to the drainage pattern in order to increase the likelihood of having the strip transect all stand conditions. In North

America the common width of strip is one chain (66 feet) or less. The width of the strip and the interval between strips determine the intensity or percentage of the total area tallied. The intensity of a systematic strip design is expressed by

$$P = (W/D) \tag{12-65}$$

where

P = proportion of area covered by strips
W = width of strip in a given unit
D = distance between strips in same units as W

Equation 12-65 is used principally to determine the spacing of strips when the cruise intensity and strip width are known. For example, if a 10 percent cruise ($P = 0.10$) is to be conducted with 1-chain wide strips, the spacing of strips would be: $D = W/P = 1/0.10 = 10$ chains.

The intensity of a systematic design using 1-chain wide strips spaced at intervals of 10 chains is then 1/10 (100) = 10 percent. For a given intensity, narrow strips at closer spacing will give more uniform distribution and better coverage of the stand area than fewer strips that are wider but more distantly spaced, although the cost will be greater.

When all the strips are of equal size, the estimate of the population mean is often calculated from equation 12-9 and the sampling error from equation 12-12 under the assumption that the strips constitute a random sample.

If the sampling units are not of equal size, as would occur in a strip sample of an irregular-shaped forest area, the ratio method of estimation can be used. Thus, an estimate of the area of each sampling unit X_i and the observation, such as volume, on the units V_i are required to obtain the ratio estimate as described in Section 12-6.1.

12-10.4 Systematic Plot Sampling

If sampling units such as plots or points are used, the sample is systematic in two dimensions—that is, the sampling units are at the chosen interval k in the two directions normal to each other. The following discussion pertains to systematic sampling with fixed-area plots. A discussion of the distribution of sampling points is given in Chapter 14.

As with the strip method, there is a fixed number of possible samples that is determined by the sampling interval chosen. Figure 12-5, illustrating a systematic distribution of plots, shows a rectangular area made up of 60 vertical columns and 40 horizontal rows forming a population of 2400 fixed-area units. If we choose a sampling interval of $k = 10$ in each direction, there will be (10)(10) = 100 possible independent samples. The selection of sample plots at sampling interval of k can be carried out in a manner analogous to those described for the strip method. The only difference is that there are two dimensions instead of one.

1. A random selection of a number from 1 to the total number of columns is first made. The same procedure is followed for a random selection of

one of the rows. The two random numbers indicate the coordinates of the starting point for the sample grid.

2. Starting at one corner of the tract, a random selection is made from the first k by k grid. All subsequent sample plots are then taken at the consistent interval of k in both directions.

The more common case will be a forest of irregular shape, as shown in Fig. 12-6. The first method of selecting a sample requires establishing an imaginary boundary to completely enclose the area. The initial unit can then be chosen in a fashion similar to that used for Fig. 12-5. Since the number of columns and rows are not exact multiples of the sampling interval, the size of a sample will vary depending on the initial point chosen and the shape of the tract. In this case, 19 sample plots form the sample. By using the other method of selecting the sample, we can arbitrarily choose the lower left-hand corner of the tract and out of the first 10 by 10 squares randomly select one unit. All subsequent units will then be selected at the k by k interval (square-grid arrangement) resulting in $n = 18$.

When sampling units are spaced in a square-grid arrangement, the calculation of the desired statistics is often carried out as though the units were randomly chosen. Loetsch and Haller (1964) described a method for approximating the true standard error by taking the sum of the squared differences between successive sampling units of one of the two axes of the grid.

For practical reasons, the square-grid arrangement of sampling units is often dropped, resulting in the interval between lines being greater than between units. (If this modification is used with fixed-area plots, it is generally called line-plot sampling. However, some foresters also consider the square-grid arrangement of plots as line-plot sampling.) Numerous possible line-plot distributions can be devised, depending on the size of the plot, the distance between plots on a line, and the distance between lines. A line-plot design can be drawn up using the following relationships.

$$n = \frac{AP}{a} \tag{12-66}$$

$$P = \frac{a}{D_\ell D_p} \tag{12-67}$$

$$P = \frac{a}{D^2} \tag{12-68}$$

where

A = total stand area
P = proportion of area covered by plots
a = plot area in square units of D_ℓ and D_p
n = number of plots
D_ℓ = spacing of lines in a given unit
D_p = spacing of plots on lines in same units as D_ℓ
D = spacing of lines and of plots on lines for square-grid arrangement $(D_\ell = D_p)$

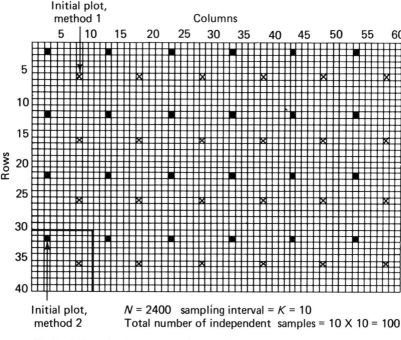

Initial plot, method 2

N = 2400   sampling interval = K = 10
Total number of independent samples = 10 X 10 = 100

Methods for selecting systematic sample:
  1) Random selection for column = 28

$\frac{28}{10}$ = 2 $\frac{8}{10}$. The remainder is 8, therefore start at column 8

Random selection for row = 16.
$\frac{16}{10}$ = 1 $\frac{6}{10}$. The remainder is 6, therefore start at row 6.

The 24 sampling units are indicated by x's.

  2) Choosing the lower lefthand 10 by 10 grid, one of the 100 sampling units is randomly selected. The 24 sampling units are indicated by shading.

Fig. 12-5 *Systematic distribution of equi-spaced plots for forest inventory.*

For a 5 percent cruise using 0.2-acre plots on lines spaced 10 chains apart, the spacing of plots on the lines is

$$D_p = \frac{a}{D_\ell P} = \frac{10(0.2)}{10(0.05)} = 4 \text{ chains}$$

(Note: The 10 in the numerator converts acres to square chains: 1 acre = 10 square chains.)

The design for the 10 by 4 chain arrangement is shown in Fig. 12-7.

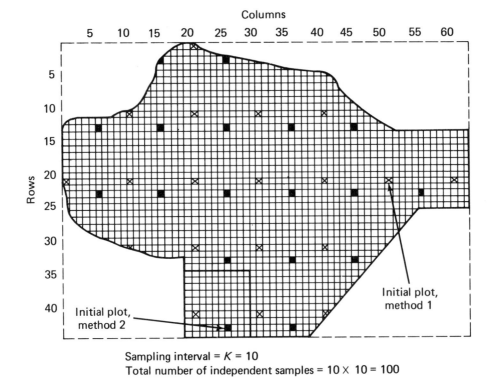

Sampling interval = K = 10
Total number of independent samples = 10 × 10 = 100

Methods for selecting systematic sample:
1) By selecting random numbers the initial plot is column 51, row 21. All subsequent plots are taken at intervals of 10 in both directions. The 19 sampling units are indicated by X's.

2) Choosing the lower left hand 10 by 10 grid, the sampling unit at column 26, row 43 is randomly selected. The 18 sampling units are indicated by shading.

Fig. 12-6 *Systematic distribution of equi-spaced plots for forest inventory of irregularly shaped area.*

For a 5 percent cruise using 0.2-acre plots set out in a square-grid arrangement, the spacing D would be

$$D = \sqrt{\frac{a}{P}} = \sqrt{\frac{10(0.2)}{0.05}} = 6.32 \text{ chains}$$

12-11 REPEATED SAMPLING IN FOREST INVENTORY

Repeated sampling (or sampling on successive occasions) in forest inventory has three objectives.

1. To estimate quantities and characteristics of the forest present at the first inventory.
2. To estimate quantities and characteristics of the forest present at the second inventory.
3. To estimate the changes in the forest during the intervening period.

The repetitive process can be continued and, on the occasion of all subsequent inventories, the previous inventory becomes the first and the new inventory becomes the second.

Repeated sampling can be carried out in any of the four ways illustrated in Fig. 12-8, as described in the next four sections. Of the four approaches, the third—successive sampling with partial replacement—is the most efficient.

12-11.1   New Sample Drawn at Each Inventory

As shown in Fig. 12-8a, sampling units at occasion 2 are different from those at occasion 1. The means, totals, and standard errors would be calculated separately as described in Section 12-5. The estimation of the change or growth would be the difference in the means for the two inventories. The estimate of the variance of this difference s_d^2 is

$$s_d^2 = \frac{s_x^2}{n_1} + \frac{s_y^2}{n_2} \tag{12-69}$$

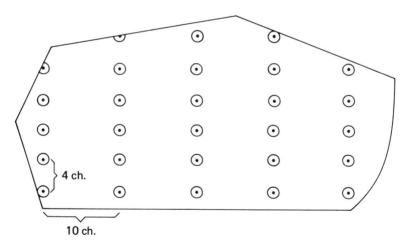

Five per cent line plot inventory using .2-acre plots spaced 4 chains apart on lines 10 chains apart.

Fig. 12-7   *Systematic line-plot inventory.*

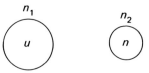

n_2 can be greater than, less than, or equal to n_1

$n_3 = 0$

(a) New sample drawn at each occasion.

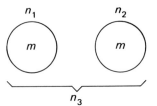

$n_1 = n_2 = n_3$

(b) Same sample remeasured on succeeding occasions.

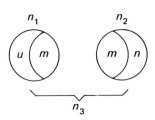

$n_3 < n_1$

$n_3 < n_2$

n_2 can be greater than, less than, or equal to n_1

$n_1 = u + m$

$n_2 = m + n$

$n_3 = m$

(c) Successive sampling with partial replacement.

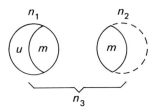

$n_1 = u + m$

$n_2 = m$

$n_3 = m$

$n_1 > n_2$

(d) Second sample a subsample of the first.

n_1 = sample at first inventory

n_2 = sample at second inventory

n_3 = m sampling units measured at both inventories (matched)

u = sampling units taken only at first inventory (unmatched)

n = sampling units taken only at second inventory (unmatched)

Fig. 12-8 *Types of repeated sampling.*

where

$$s_x^2 \quad = \text{variance at first inventory}$$
$$s_y^2 \quad = \text{variance at second inventory}$$
$$n_1 \text{ and } n_2 = \text{number of sampling units at first and second inventories}$$

12-11.2 Same Sample Remeasured on Succeeding Occasions

The sampling units taken at the first inventory are remeasured at the second and all succeeding inventories, as shown in Fig. 12-8b. This is the concept of permanent sample plots and the basis of the Continuous Forest Inventory (CFI) developed in North America. The estimates of the means, totals, and standard errors at each inventory would again be found as in the case of two separate inventories. Similarly, the differences between the means for each inventory would indicate the changes in the forest. However, since the same sampling units are taken on both occasions, the standard error of the difference would be calculated for paired plots as

$$s_d^2 = \frac{s_x^2 + s_y^2 - 2r_{xy}s_x s_y}{n} \tag{12-70}$$

The greater the correlation as expressed by r_{xy} between X and Y, measurements at the first and second inventories, the greater the reduction in the variance. The correlation can be expected to be large for short periods. However, it has been demonstrated by Hool and Beers (1964) that the correlation may remain surprisingly high, even for a 15-year remeasurement interval.

12-11.3 Successive Sampling with Partial Replacement

At the second inventory, a portion of the initial sampling units is remeasured and new ones are taken, as shown in Fig. 12-8c.

A detailed account of the development and application of this procedure to forest inventory is given by Ware and Cunia (1962), Cunia (1965), and Frayer (1966). Only a brief summary is presented here. The student is referred to the above references for an explanation of the statistical theory.

At the initial and second inventories, there are three kinds of sampling units (Fig. 12-8c):

$$u = \text{sampling units measured only at the first inventory}$$
$$m = \text{sampling units measured at the first and second inventories}$$
$$n = \text{sampling units measured only at the second inventory}$$

From observations on these sampling units, the quantities present at each inventory and the growth over the period can be calculated. Estimates at the initial inventory utilize data from all the u and m units measured. From the m units a relationship can be established between sampling unit volumes found at the initial and second inventories. A regression can then be determined and used to estimate the volumes of the u units at the second inventory and the volumes of the n units as they were at the initial inventory. As a result there are volume estimates for all units, some from direct measurement and others from a regression. The value of this procedure is that volume and growth estimates and their variances are obtained from all units, both temporary and permanent.

12-11.4 Sample at Second Inventory—Subsample of First

At the second inventory, a portion of the sampling units taken at the first inventory is remeasured, as shown in Fig. 12-8d.

The estimate of the mean at the first inventory uses the data from the u and m units. At the second inventory, measurements are only taken on the m units. From these m units a relationship is again established between the volumes found on units measured at both the first and second inventories. The mean volume at the second inventory is then determined using the data from the n_1 units in a regression based on this relationship.

The growth over the inventory period is expressed as the difference between the overall mean at the initial inventory and the regression estimate for the second inventory.

13
Aerial Photographs in Forest Inventory

Photographs may be taken from the air or from the ground. If an aerial photograph is exposed with the camera axis vertical, or nearly vertical, it is called a *vertical photograph*. If an aerial photograph is exposed with the camera axis intentionally tilted, it is called an *oblique photograph*. An oblique photograph in which the apparent horizon is shown is a *high oblique;* one in which the apparent horizon is not shown is a *low oblique*. If a photograph is taken from a fixed position on the ground, it is called a *terrestrial photograph*. Forest resource managers deal primarily with vertical aerial photographs. However, one should understand that both oblique aerial photographs and terrestrial photographs have important applications.

It is the aerial photograph that has given a new dimension to our study of planet earth. Humankind and nature shape and reshape the face of the earth, and through aerial photographs we can see for the first time in history the innumerable ramifications of our actions. Then too, aerial photographs speak the truth, if we are able to see it. Maps, which may be prepared from photographs, speak the mind of a society or a profession. Many people are likely to regard a highway map, a forest map, a city map, and so on, as revealed truth. But in each case reality is culturally determined. Maps, like language, select certain features and ignore others and, like language, maps are cultural expressions of elements significant to a profession or a society. Foresters, for instance, would emphasize commercial timber types in their cover maps, while wildlife scientists would emphasize habitat conditions for wildlife in their cover maps. There may be bias in the interpretation of aerial photographs, but the information is there, if one knows how to extract it.

In interpreting aerial photographs, resource managers must understand the basic mathematics of photographs and the principles of stereoscopic vision, and must supplement the objective findings thus obtained with deductive reasoning to answer such questions as: What timber type is this? What is the volume and area of this timber type? Where should we locate forest roads? Is this recreational area being overused? Where should we locate campgrounds? Where can we locate ski slopes? Where are the section lines? What land form is this?

Thus, in addition to understanding the basic mathematics of aerial photographs and the principles of stereoscopic vision, the skilled interpreter should have a background in forestry, geology, and other disciplines oriented toward the study of wildlands. The value of background and experience can hardly be overemphasized; the interpreter who does not recognize a white pine stand, an esker, campground degradation, forest sites, and so forth, when on the ground can hardly be expected to identify such things from aerial photographs. Indeed, resource managers with an understanding of the basic mathematics of aerial photographs and of the principles of stereoscopic vision can soon learn to interpret features in their area of expertise for a given forest, district, and so on, if they take the photographs to the field and compare the features on the ground with the images on the photographs.

13-1 SCALE FORMULAS FOR VERTICAL AERIAL PHOTOGRAPHS

There are only a few simple formulas that one needs to use to quickly determine important metrical information from vertical aerial photographs.

The scale, or representative fraction, of a vertical aerial photograph is the ratio of a distance on the photograph to the corresponding distance on the ground when object planes and image planes are parallel. Thus, from Fig. 13-1 we see that scale S is

$$S = \frac{ab}{AB} \tag{13-1}$$

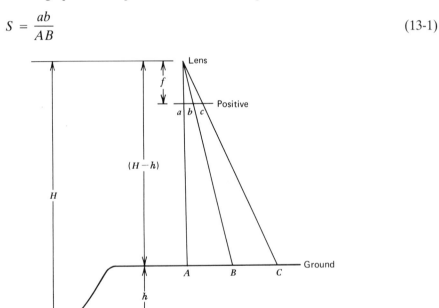

Fig. 13-1 *Diagram illustrating how ground objects are imaged in the positive plane for vertical aerial photographs. (a, b, and c are photo images of ground points A, B, and C. Thus, ab is a photo distance and AB is a ground distance. H = height of lens above mean sea level: h = height of terrain above mean sea level; (H − h) = height of lens above the ground (i.e., flying height above ground).*

It can also be seen from Fig. 13-1 that scale S is

$$S = \frac{f}{H - h} \tag{13-2}$$

It should be noted that scale S is expressed as a fraction with a numerator of 1. Thus, if a photograph has a scale of 1:12,000, it means that 1 unit on the photograph equals 12,000 units on the ground.

To facilitate calculations it is desirable to use the photo scale reciprocal (PSR = $1/S$) instead of scale. Then,

$$\text{PSR} = \frac{AB}{ab} \tag{13-3}$$

and

$$\text{PSR} = \frac{H - h}{f} \tag{13-4}$$

Note that in all of the above formulas distances must be in the same units.

From equations 13-3 and 13-4 we obtain the following important relationship from which valuable formulas can be derived.

$$\frac{AB}{ab} = \frac{H - h}{f} \tag{13-5}$$

Another useful relationship utilizes PSR and MSR (map scale reciprocal).

$$\frac{\text{PSR}}{\text{MSR}} = \frac{\text{Map distance}}{\text{Photo distance}} \tag{13-6}$$

A few examples will illustrate the use of the above equations and some scale conversions that are best computed from PSR.

EXAMPLE 1   The distance between two road intersections is 3350 feet on the ground and 4.22 inches on a photo. What is the PSR of the photo? (Use equation 13-3.)

$$\text{PSR} = \frac{(3350)12}{4.22} = 9526$$

EXAMPLE 2   Find the PSR of a photograph taken with a 152.36-mm focal length camera at an elevation of 1981 meters above mean sea level over terrain that is 300 meters above mean sea level. (Use equation 13-4.)

$$\text{PSR} = \frac{(1981 - 300)1000}{152.36} = 11,033$$

EXAMPLE 3   A photographic crew has a camera with an 8.25-inch focal length lens. At what altitude (in feet) above the ground must they fly to produce prints with a scale of 1/20,000 (i.e., PSR = 20,000)? (Use equation 13-4 and solve for $H - h$.)

$$H - h = \frac{8.25(20,000)}{12} = 13,750 \text{ feet}$$

EXAMPLE 4 Suppose that the smallest image that can be consistently distinguished on aerial photographs has a diameter of 0.002 inches. If you fly some photograph with an 8.25-inch focal length camera at an altitude of 11,000 feet above the ground, what would be the ground distance of the smallest tree crown you could distinguish? (Use equation 13-5 and solve for ground distance AB.)

$$AB = \frac{0.002(11,000)}{8.25} = 2.67 \text{ feet}$$

EXAMPLE 5 Assume you desire the PSR of a photograph depicting some of the same area covered by a quadrangle map. You measure the distance between two road intersections across the center of the photo and find it to be 6.25 inches. The corresponding distance on the map is 4.36 inches. If the MSR is 24,000, what is the PSR? (Use equation 13-6 and solve for PSR.)

$$PSR = \frac{24,000(4.36)}{6.25} = 16,742$$

EXAMPLE 6 Compute the following scale conversions for a 9 by 9 inch vertical aerial photograph with scale of 1/20,000 (i.e., PSR = 20,000): feet per inch, chains per inch, miles per inch, acres per square inch.

$$\text{Feet per inch} = \frac{PSR}{12} = \frac{20,000}{12} = 1,666.67$$

$$\text{Chains per inch} = \frac{PSR}{12(66)} = \frac{20,000}{12(66)} = 25.25$$

$$\text{Miles per inch} = \frac{PSR}{12(5280)} = \frac{20,000}{12(5280)} = 0.3157$$

$$\text{Acres per square inch} = \left(\frac{PSR}{12}\right)^2 \frac{1}{43,560} = \left(\frac{20,000}{12}\right)^2 \frac{1}{43,560} = 63.77$$

When one desires the reverse of any of these equations (e.g., inches per mile instead of miles per inch), one computes the reciprocal of the appropriate value. For example, if miles per inch = 0.3157, inches per mile = 1/0.3157 = 3.17.

13-2 SCALE VARIATIONS ON VERTICAL AERIAL PHOTOGRAPHS

Equation 13-2, $S = f/(H - h)$, indicates that the scale of a vertical aerial photograph is a function of f, H, and h. Thus, we can make the following conclusions.

1. At a given height above the ground $H - h$ the scale S varies directly as the camera focal length f.

2. For a given value of f, the scale varies inversely as the height above the ground $H - h$.

3. For given values of f and H, scale will vary directly from place to place on individual photographs as h varies from place to place on the ground.

On an accurate planimetric map, all features are depicted at their correct positions in a horizontal plane. The observer has a vertical view of every detail shown. This standard is rarely met by a vertical aerial photograph because of optical or photographic deficiencies, tilting of the camera lens axis at the instant of exposure, and variations in relief, as discussed above. Because of these things, the scale of a vertical aerial photograph will usually vary from place to place, or to put it another way, images may be displaced from their map positions. From the standpoint of the resource manager, optical and photographic deficiencies are not significant, and tilt amounting to less than three to four degrees can be ignored. (Most photographs taken by reliable aerial survey companies will have less than three to four degrees of tilt.)

The most significant source of scale variation, or image displacement, on vertical photographs is relief. All objects that extend above or below a specified ground datum plane register at different scales and so have their photographic images displaced from their true map positions. Figure 13-2 illustrates a depression and a height of land with an assumed datum plane at an intermediate level. Ground point A, for which the correct map position on the photograph is a, registers on the photograph at a'. Ground point B, for which the correct map position is b, registers at b'. Thus, it can be seen that objects below the datum plane register at too small a scale and are therefore displaced inward from their correct map positions. Objects above the datum plane register at too large a scale and are therefore displaced outward from their correct map positions. These displacements are radial to the plumb point, or nadir (nadir and principal point PP are the same in vertical aerial photographs).[1] Therefore, the true distance of points cannot be scaled on a vertical photograph if they represent topographic features above or below a reference plane. Moreover, if the points are displaced in opposite directions, such as are A and B in Fig. 13-2, the azimuth of a line joining the points on the photograph will not be the true direction, as can be inferred from the photo depicted in the upper left of Fig. 13-2.

The formula to determine the displacement of an image from its true map position is easily derived. In Fig. 13-2, the ground distance from the nadir to point B is R,

---

[1] The *principal point* is defined as the intersection of lines drawn between opposite fiducial marks.

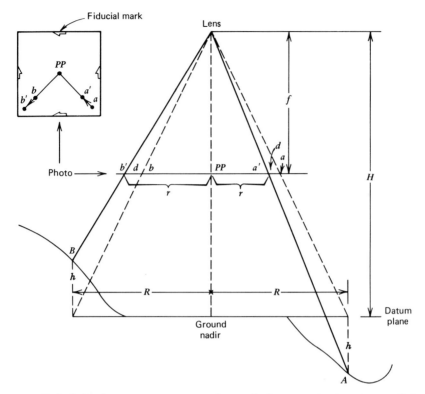

Fig. 13-2 *Relief displacement in a vertical aerial photograph. (In the small figure in the upper left, a and b are correct image positions, and a' and b' are displaced image positions.)*

the photo distance from the nadir, or *PP*, to the image b' of point B is r, and the displacement of the image of point B from its true map position b is d. From equation 13-5 it can be seen that for point B

$$\frac{R}{r} = \frac{H - h}{f} \qquad \text{and} \qquad \frac{R}{r - d} = \frac{H}{f}$$

Thus,

$$r = \frac{fR}{(H - h)} \qquad \text{and} \qquad r - d = \frac{fR}{H}$$

Subtracting the last two equations, we obtain

$$
\begin{array}{r}
r = fR/(H - h) \\
- [r - d = fR/H] \\
\hline
d = fR/(H - h) - fR/H
\end{array}
$$

and also

$$d = \frac{fRh}{(H - h)H}$$

but since

$$\frac{fR}{(H - h)} = r$$

$$d = \frac{rh}{H} \qquad (13\text{-}7)$$

In using equation 13-7, the symbols should be clearly understood.

d = displacement from true map position in inches
r = radial distance in inches between the displaced image and the principal point
h = difference in elevation in feet between the displaced image and the datum plane
H = altitude in feet of camera lens above the datum plane

The foregoing procedure will also yield equation 13-7 for point A, which is below the datum plane. Note that when h is for a point above the datum plane, h is plus; when h is for a point below the datum plane, h is minus. Thus when d is plus it indicates that the object is displaced outward; when d is minus it indicates that the object is displaced inward.

EXAMPLE Assume a photograph was flown at an elevation of 13,750 feet above the datum plane. If a point that is 650 feet below the datum plane is imaged 4.00 inches from the principal point, how much and in what direction would images at this point be displaced from their true map position?

$$d = \frac{4.00(-650)}{13,750} = -0.19 \text{ inches (inward)}$$

Equation 13-7 may also be used to calculate the height of objects such as buildings, smokestacks, and so on, for which displacement can be accurately measured. The formula is simply solved for h. Equation 13-7 is not suitable, however, for estimating tree heights on aerial photographs because it is almost impossible to make accurate displacement measurements of trees.

13-3   STEREOSCOPIC VISION

The range of binocular vision for unaided eyes is about 200 meters. This distance can be extended somewhat by the use of binoculars. However, stereoscopic photography makes it possible for us to extend our perception of solidity or relief far beyond the limits of binocular vision. It may be said that it furnishes us with a sixth sense. What binocular instrument could give the three-dimensional effect from altitudes of 10,000 feet? From altitudes of 50,000 feet? From space?

The visual world we enter when we view vertical aerial photographs stereoscopically is a world in which trees, hills, buildings, and so on, appear stretched (or occasionally compressed) vertically from their true scale heights. When the model is stretched, as it usually is in vertical photographs, the gradient of hills and roofs is increased, spheres appear egg-shaped, cubes look like rectangular solids with square bases, short trees appear tall, and tall trees appear very tall. The converse is true when the model is compressed. This effect is called *vertical exaggeration*.

In addition to vertical exaggeration there may be modifications in the directions of lines extending depthwise that may produce apparent changes in the form of hills, slopes, and valleys, and cause buildings and trees to appear to lean. These distortions do not affect the magnitude of vertical exaggeration. They may, however, affect interpretations.

Since distortions result, in large part, from the incorrect positioning of the eyes or the stereoscope relative to the photographs, and incorrect orientation of the photographs, they may be greatly reduced by following a simple procedure when viewing 9 by 9 inch vertical aerial photographs with a lens stereoscope. In addition, eye strain is minimized when the photographs are aligned in this fashion; here is the procedure (refer to Fig. 13-3).

1. Locate the principal point *PP* of each photograph. To do this, align opposite sets of fiducial marks with a straightedge, draw short lines at the center with a pencil to form a cross, and make a fine hole at the intersection with a needle. Rub chalk dust into the hole so it can be seen.

2. Locate the conjugate image of each principal point on the adjacent photograph. This point, the conjugate principal point *CPP*, may be located by inspection or stereoscopically.

3. Locate the flight line, or photo baseline, on each photograph by drawing a fine ink or pencil line between *PP* and *CPP*.

4. Orient the photographs so that the shadows fall toward you and the light source is at the upper left. Now tape down the outside edge of the left photograph of the pair. Place the right photograph over the left photograph, superimpose the flight lines, and position the right photograph so its *CPP* is about 2.2 inches from the *PP* of the left photo. Tape down the outside edge of the right photograph of the pair.

5. Set the stereoscope so the spacing of the lenses equals the interpupillary distance. (The *interpupillary distance* equals the distance between the right, or left, borders of the two pupils. The distance, which is 65 millimeters for the average individual, can be measured by looking into a mirror with a ruler before the eyes.)

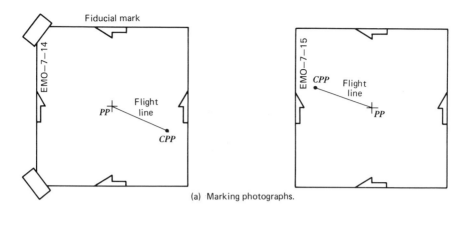

(a) Marking photographs.

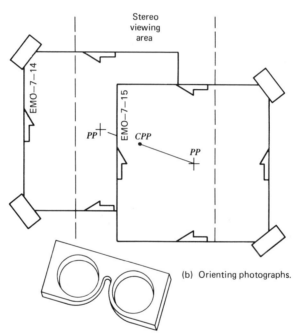

(b) Orienting photographs.

Fig. 13-3 *Aligning 9 by 9 inch vertical aerial photographs for viewing with a lens stereoscope.*

6. For viewing any place within the stereo viewing area, place the stereoscope with its long axis parallel to the flight line, and locate the stereoscope lenses as near as possible, in the same relative positions, to the object to be viewed as were the two positions of the camera lens at the instant of exposure. (Experiment on buildings. Slide the stereoscope from left to right, and up and down, and note that buildings will appear to lean one

way or another. By doing this one can learn to judge where to locate the lenses for best results.)

7. To view the right side of the stereo viewing area, flip the photographs so the left photograph overlaps the right photograph.

13-3.1 Vertical Exaggeration

Vertical exaggeration E is defined as the relative increase or decrease of the vertical scale relative to the horizontal scale. It can be evaluated (Miller, 1973) by the simple formula

$$E = 6.25 \frac{P}{f} \tag{13-8}$$

where P is the average distance from PP to CPP on the photographs of the stereopair, and f is the camera focal length. (Note that P and f are in the same units.)

For 9 by 9 inch vertical aerial photographs taken with 60 percent overlap, P will be 3.6 inches, and E will be as follows for various values of f.

f (in inches)	E
6.0	3.8
8.25	2.7
12.0	1.9
24.0	0.9
36.0	0.6

For the 8.25-inch lens this means that in the stereoscopic model seen under the stereoscope trees and buildings will be 2.7 times taller than in a scale model, and slopes will be 2.7 times steeper than in a scale model.

We can interpret the space scene more intelligently if we know how much it is stretched or compressed. Indeed, it has been observed (Miller, 1973) that when E is less than 1, stereoscopic viewing is not better than nonstereoscopic viewing for interpreting vegetation. And other things being equal, as E increases, the accuracy of height determinations with floating dot devices increases (Section 13-5).

13-4 OBTAINING AERIAL PHOTOGRAPHY

Resource managers often rely on prints of existent photography taken for public agencies: Agricultural Stabilization and Conservation Service, U.S. Forest Service, state highway departments, state departments of natural resources, among others. However, forest industries often have aerial surveys made of their lands; ownerships adjacent to their lands may be included in the coverage. Such surveys are thus a possible source of photography.

Most organizations that contract someone to do photography for wildland management prefer a scale of about 1/16,000. (In the United States a scale of 1/15,840—i.e., 4 inches = 1 mile—is often used because this scale is convenient to use on areas surveyed by the U.S. Public Lands Surveys.) This economical scale is somewhat of a compromise; it is good for the preparation of forest ownership maps and cover maps and is acceptable for use in forest inventory and forest management work. However, some organizations contract for photographs at a scale of 1/12,000, or even as large as 1/9600, for intensive inventory and forest management. (There is little advantage in using larger scales.) The standard scale of most of the photography held by the Agricultural Stabilization and Conservation Service, which holds the largest share of recent aerial photography of the United States, is 1/20,000. Satisfactory work can be done with this scale in regional surveys and in broad cover-type classification, but the scale is too small for good results in applications that apply to the day-to-day tasks of the resource manager.

Black and white panchromatic film is still the most widely used film. However, black and white infrared film is preferred by resource managers in some regions because many tonal variations between trees and other vegetation can be detected, and the shadows, which register in black, enable one to identify some tree species by shadow patterns. In regions where the terrain is rough and where slopes away from the sun are in shadow, much information is lost, because shadows register in black. Conventional color film and infrared color film, although more expensive than black and white film, are no longer excessively expensive when one considers the information that can be obtained from them.

It should be emphasized that regardless of the source of the imagery a resource manager uses the same techniques and skills to obtain information and make interpretations on vertical aerial photographs.

13-5   AERIAL PHOTO MENSURATION

Of the information desired on trees and stands, only certain characteristics can be obtained from vertical aerial photographs. These are: tree height, ground slope, aspect, crown counts, crown diameter, crown closure, forest types or species groups, and stand area. Other information, such as stem diameter, basal area, tree volume, stand volume, site quality, and so on, is obtained through correlation with these characteristics. One can determine characteristics with a few simple scales, the methodology discussed in this chapter, and a little ingenuity. Here are some practical suggestions.

- Determine *tree heights* from ocular estimates while viewing the photographs stereoscopically. Use some known heights as a frame of reference. Stereograms that include trees of known heights are helpful. Any of several stereometers—that is, a stereoscope with special attachments for measuring parallax—may be used to determine tree heights from stereo pairs.

The simpler stereometers, which are often called height finders, are available from suppliers of forestry and engineering equipment. Their theory and use are thoroughly discussed in photogrammetry texts (e.g., Moffitt, 1967; Wolf, 1974). We have found that the majority of people who use stereometers intermittently do not develop, or do not have, the requisite stereo acuity or patience to obtain consistent height measurements. Consequently, unless one uses a stereometer regularly and has the requisite stereo acuity, one is well advised to obtain tree heights by estimation, or to employ a skilled observer to take the stereometer measurements.

- Determine *ground slope* (in percent) by first determining vertical exaggeration E for the photographs used (Section 13-3.1). Then, while viewing the photographs stereoscopically, ocularly estimate the apparent slope percents of the slopes of interest. [Appropriate stereo models with slopes of known values may be used (Miller, 1973); however, without such aids one can ocularly classify apparent slope in 10-percent classes.] Finally, divide the apparent slope percent by E to obtain ground slope percent.

- Determine *aspect*, the direction toward which a slope faces, from cardinal directions determined for the photographs.

- Make *crown counts* within circular sample plots drawn to scale on the photographs or on transparencies placed on the photograph. (For example, radii of circular plots on photographs with PSR of 15,840 would be 0.0399 inches for $\frac{1}{5}$ acre, 0.0631 for $\frac{1}{2}$ acre, 0.0892 for 1 acre, and 0.1995 for 5 acres.) The circular plot should be large enough to include 20 to 30 visible crowns. Since only trees that are visible from directly above and that are large enough to be distinguished can be seen (see Example 4, Section 13-1), accurate counts are difficult to obtain and counts often show poor correlations with stand volume.

- Determine photo distance of *crown diameter* with a micrometer-measuring device or a crown diameter scale (Fig. 13-4). Convert photo distance to ground distance (equation 13-3). Average crown diameter is normally determined within circular sample plots. Photo measurements of the crown diameter of given trees are generally less than measurements made on the ground.

- Determine *crown closure*, the percent of the area of a stand covered by tree crowns (normally consider only the overstory or main stand), by ocular estimation or with the aid of a crown density scale (Fig. 13-5).

- Locate *forest type* boundaries on the photographs. From experience and ground checking, determine types within each boundary. One may also outline areas of different stand conditions (i.e., size and density) within each type. Section 13-5.2 explains how to code cover classifications.

- Determine *stand area* with a dot grid or a planimeter (Chapter 6).

Crown diameter scales, micrometer scales, crown closure scales, and dot grids may be obtained from forestry supply companies, some of the U.S. Forest Service's Forest

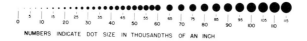

CROWN DIAMETER SCALE

CENTRAL STATES FOREST EXPERIMENT STATION

NUMBERS INDICATE DOT SIZE IN THOUSANDTHS OF AN INCH

Fig. 13-4 *Dot-type crown diameter scale for measuring crown diameters by comparison with photo images. Crown diameter scales are usually printed on transparent film.*

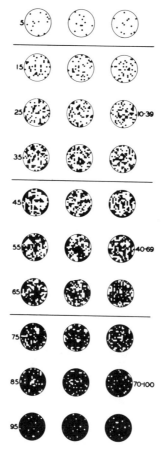

Fig. 13-5 *Crown density scale for estimating crown closure percent by comparison with photo images. Crown density scales are normally prepared so that the circles represent a specified plot size for a given scale (e.g., 1 acre at a scale of 1/10,000), and they are usually printed on glossy, single-weight paper.*

and Range Experiment Stations, and some forestry schools and departments. However, they can be made if one uses a little ingenuity.

13-5.1 Area by Dot Sampling

Dot sampling, one of the most commonly used methods of determining area on vertical aerial photographs, is done with a dot grid. A *dot grid* is simply a grid consisting of lines forming squares or rectangles (preferably squares) that has a dot placed in the center of each square or rectangle. All the lines, except those forming 1-inch squares, are usually deleted. Dot grids are printed as positive transparencies.

In placing a transparent dot grid over a photograph to obtain the area of a stand class, the grid should be positioned at random. Then, when the total area sampled is known, the area of a given stand class may be found from the relationship

$$\frac{a}{A} = \frac{x}{n} \qquad \text{and} \qquad a = \frac{x}{n}(A) = p(A) \tag{13-9}$$

where

a = area in stand class
A = total area sampled
x = number of dots counted in stand class
n = number of dots counted in total area sampled
$p = \dfrac{x}{n}$ = estimated proportion of total area in stand class

Equation 13-9 assumes that the total area A is known without error. Often, however, A is unknown or of no interest. Consequently, the most common method of determining the area of any class by dot grid is to compute the ground area represented by each dot at the appropriate photo scale and to multiply this factor by the number of dots counted in the class. For example, if one used a dot grid with 64 dots per square inch on a vertical aerial photograph with PSR of 15,840, and counted 77 dots in an Aspen-Birch type, then

$$\text{Acres per square inch} = \left(\frac{15,840}{12}\right)^2 \frac{1}{43,560} = 40.00$$

$$\text{Acres per dot} = \frac{40.00}{64} = 0.625$$

$$\text{Area of Aspen-Birch type} = 77(0.625) = 48.125 \text{ acres}$$

The precision of an estimated proportion may be computed by the following formula when dots are counted over the entire area.

$$E = t\sqrt{\frac{p(1-p)}{n}} \tag{13-10}$$

where

E = error of estimated proportion, expressed as a proportion of total area, for desired probability level

t = Student's t for desired probability level

p = estimated proportion of total area in class

n = number of dots counted in total area sampled

For example, assume on a photograph with PSR of 12,000 you use a 64 dot per square inch grid and count 5000 dots in the entire forest and 1500 dots in an Aspen-Birch type. Then, the sampling error of the estimated proportion of Aspen-Birch (p = 1500/5000 = 0.30) at the .05 probability level would be

$$E = 1.96 \sqrt{\frac{0.30(1 - 0.30)}{5000}}$$

$$= 0.0127$$

If the total area of the forest is 115,000 acres, the sampling error in acres at the .05 probability level would be

$$.0127(115,000) = \pm 1461 \text{ acres}$$

Equation 13-10 is rarely used because dot counts are seldom made over an entire forest. It is, however, useful to determine the number of dots required. The equation is simply solved for n to obtain

$$n = p(1 - p)\left(\frac{t}{E}\right)^2 \tag{13-11}$$

As an example, assume area breakdowns are needed for 12 square miles photographed at a scale of 1:15,840. Rough estimates of the proportions of each classification are: urban and built-up land, 0.12; agricultural land, 0.55; rangeland, 0.10; forestland, 0.20; and water, 0.03. If the classification of main interest is forestland, sampling intensity should be calculated with a p value of 0.20. Thus, to determine the proportion of forestland with a probability of .95 and a sampling error E of 0.02, it is necessary to count a total of

$$n = 0.20(1 - 0.20)\left(\frac{1.96}{0.02}\right)^2 = 1537 \text{ dots}$$

At the scale of 1:15,840, there are 40 acres per square inch. Consequently, the print area represented by 12 square miles is 12(640)/40 or 192 square inches. And the number of dots needed per square inch to obtain the desired intensity is 1537/192 or 8.0. Here it would be expedient to use a grid containing 9 dots per square inch located in a square grid arrangement.

Equation 13-11 requires a prior knowledge of p, the value being sought. However, since p always lies between 0 and 1, it can be shown that the statistic $p(1 - p)$ reaches a maximum value of 0.25 when p = 0.50. Thus, equation 13-11 is often revised to read

$$n = 0.25\left(\frac{t}{E}\right)^2 \tag{13-12}$$

Then, for the example being considered,

$$n = 0.25\left(\frac{1.96}{0.02}\right)^2 = 2401 \text{ dots}$$

This would call for a dot grid with 2401/192, or 12.5 dots per square inch. In this case it would be sensible to select a dot grid with 9 dots per square inch, because equation 13-12 will generally call for more dots than needed for the desired sampling error.

How does one proceed if one desires to determine the sampling error E for a given timber type on a forest for which one does not know, or has no interest in, the total area? In this case one should make several determinations of the area of the timber type. The standard deviation of these determinations is the standard error of the mean area estimate. A multiplication by the appropriate Student's t will provide the desired sampling error E.

13-5.2 Forest Mapping

The wildland manager needs to know the location of different cover types, their size and density, and whether they are in large unbroken units or in small scattered units. Such information is of great value to foresters, wildlife managers, and recreation planners, and is best presented in forest maps.

It is prohibitively expensive to prepare forest maps for extensive areas from ground surveys alone. Good quality aerial photographs are indispensable. Generally, a scale of 1/15,840 or larger is preferred.

Cover map classifications will vary from one organization to another, but four variables form the basis for many classification systems: species or species group, size class, stand density, and site (sometimes omitted). The components of the classification for northern forests might be coded as follows.

Species groups: HP = hemlock-pine; PS = pine-spruce; AB = aspen-white birch; MBB = sugar maple-beech-yellow birch (and/or basswood); MAB = sugar maple-aspen-white birch; and so forth.

Size classes: 1 = saplings   (Stands of trees not qualifying under one of the following size classifications, but with tree growth occupying at least 10 percent of the growing space.)

2 = poles   (Stands of trees averaging less than 250 cubic feet of gross volume per acre but with more than 30 trees per acre that are 5 inches or more in dbh.)

	3 = small saw-timber	(Stands of trees averaging 250 to 1500 cubic feet of gross volume per acre.)
	4 = large saw-timber	(Stands of trees averaging over 1500 cubic feet of gross volume per acre.)
Density classes:	1 = poor stocking (0–30% crown closure)	
	2 = medium stocking (31–60% crown closure)	
	3 = good stocking (61–100% crown closure)	
Site classes:	D = dry; F = fresh; W = wet	

Species groups and crown closure are determined as explained in Section 13-5.

To determine size classes, volumes for each classification are estimated on a predetermined number of plots of appropriate size (e.g., 1 acre, 0.5 hectare).

Plots are best located in a square-grid arrangement. Each plot center should be pin-pricked through a templet and numbered. With experience, one can learn to determine size classes from ocular estimates. However, it is more accurate to use a stand aerial volume table.

Stand aerial volume tables might be prepared with three independent variables: average photo height, crown closure percent, and average crown diameter; or with two independent variables: average photo height and crown closure percent. (The dependent variable is volume per acre or per hectare, or weight per acre or per hectare.) Since there is little or no correlation evident between volume and average crown diameter, that variable is often omitted. A composite stand aerial volume table is shown in Table 13-1.

Since each plot covers 1 acre, the mean per-acre volume for a given classification is obtained by averaging the volumes assigned to the plots in that classification. If one wishes to determine the total volume of a classification, one can multiply the area of the classification by volume per acre. Refinements of photo cruising can be achieved by applying some of the principles discussed in Chapter 12.

The simple site classes used in our example take into account topography and soil, two important determinants of site quality. In the West, one might refine the classification by considering elevation, aspect, and slope percent, as well as soil. But at best, site classes determined on aerial photographs are imprecise. The three classes given here—dry, fresh, and wet—are determined on the photographs by the relative topographic location of the site and the character of the vegetation (cover type is related to soil type).

If one determined the components of a given classification to be hemlock-pine small sawtimber, of medium stocking, on a fresh site, this could be coded: HP-3-2-F. Or if one had a two-storied stand of sugar maple-aspen-white birch small sawtimber, of medium stocking, over pine-spruce poles of good stocking, on a dry site, this could be coded

$$\frac{D}{3/2 \text{ MAB/PS } 2/3}$$

The understory of a two-storied classification must usually be determined on the ground. In all cases, a sample of the classifications should be ground checked.

Table 13-1
Composite Stand Aerial Volume Table for Northern Minnesota
(in gross cubic feet per acre)[1,2]

Average Total Height (feet)	Crown Closure (percent)									
	5	15	25	35	45	55	65	75	85	95
30	40	120	200	280	360	440	520	600	680	760
35	80	190	300	410	520	630	740	850	960	1070
40	180	310	440	570	700	830	960	1090	1220	1350
45	460	590	720	850	980	1110	1240	1370	1500	1630
50	740	870	1000	1130	1260	1390	1520	1650	1780	1910
55	1020	1150	1280	1410	1540	1670	1800	1930	2060	2190
60	1300	1430	1560	1690	1820	1950	2080	2210	2340	2470
65	1580	1710	1840	1970	2100	2230	2360	2490	2620	2750
70	1860	1990	2120	2250	2380	2510	2640	2770	2900	3030
75	2140	2270	2400	2530	2660	2790	2920	3050	3180	3310
80	2420	2550	2680	2810	2940	3070	3200	3330	3460	3590
85	2700	2830	2960	3090	3220	3350	3480	3610	3740	3870
90	2980	3110	3240	3370	3500	3630	3760	3890	4020	4150
95	3260	3390	3520	3650	3780	3910	4040	4170	4300	4430
100	3540	3670	3800	3930	4060	4190	4320	4450	4580	4710

[1] Based on fifty 1-acre plots in Carlton County, Minnesota. Heavy lines indicate limits of basic data.

[2] Gross volumes are inside bark and include all trees 5.0 inches dbh and larger from stump to a variable top diameter not less than 4.0 inches inside bark.

SOURCE: Avery and Meyer, 1959.

13-6 TERRESTRIAL STEREOSCOPIC PHOTOGRAPHY

Double-lens cameras possessing two matched lenses that are separated about 2.5 inches are convenient for taking stereoscopic photographs. However, single-lens cameras are preferred because they permit variation of lens separation. With a single-lens camera one takes two exposures of the same view from different positions. When these two photographs are properly mounted, we have a stereogram; and when this stereogram is viewed with the proper aids so that each eye sees only the appropriate image, the brain interprets the two different views as one three-dimensional scene. A single-lens camera cannot, of course, be used to obtain stereoscopic photographs of moving objects.

Stereograms are of great value in furnishing documentary material. This is particularly true for the wildland manager who desires to illustrate such things as ground cover, density of stands, methods of brush disposal, decay in trees, taxonomic features of plants, and site deterioration. Ordinary pictures made to illustrate such things are often disappointing. With them the ground cover on one area looks much the same as on another, one hardwood forest looks like the next, and undergrowth and brush piles blend together to give a confused mass of sticks and vegetation. In a stereogram the distinguishing characteristics of each scene are clearly illustrated. And if the stereo photos are taken from permanent camera stations, they provide permanent three-dimensional records that can be compared over time. Sequential stereograms of this type are particularly valuable for monitoring campground deterioration (Section 18-7).

In making a stereogram all that is required is to take one picture, make a simple lateral displacement of the camera, and make another exposure. Here are the basic rules.

1. The lens should be approximately the same distance from the subject in both views.
2. The optical axis of the camera should be parallel for both views.
3. The camera must never be tilted sideways for either exposure, and the lens must be in the same horizontal plane for both shots. It is permissible to tilt the camera forward or backward the same amount in both views.
4. Determine the camera settings as for any picture, but whenever possible use stops between f/8 and f/22 so adequate depth of field will be obtained.
5. The stops and lighting should be the same for both exposures.

The amount of lateral displacement may be determined from the following formula.

$$L = \frac{FN}{(F - N)50} \tag{13-13}$$

where

L = lens separation (in same units as F and N)
F = distance of farthest image of interest
N = distance of nearest image of interest

This equation gives the most vivid relief possible without fatigue in viewing the stereogram.

A simple wooden tray with the undersurface bushed for a tripod screw makes a satisfactory displacing device. The tray enables one to displace the camera laterally along a line parallel to the plane of the film. Its width is that of the camera plus 12 inches, thus allowing a movement of 12 inches. To determine the amount of displacement, a scale is placed on the back of the tray. Such a device can be adapted to any camera, but is especially adaptable to cameras with flat bases. Although 35-mm cameras provide satisfactory photographs for preparing stereograms, cameras with larger film dimensions ($2\frac{1}{4}$ by $2\frac{1}{4}$ to 4 by 5) are better, because contact prints prepared from these films are excellent for stereo viewing. Enlargements should be made when 35-mm film is used.

An ordinary plane table serves about as well as the above contrivance and allows for greater lateral displacement. Lines drawn on the board will aid in making the proper lateral displacement.

When photographs are taken, it is desirable to include targets of known dimensions (e.g., range poles) in the scene at known distances to provide scale, and to expose the film with the camera axis horizontal. Then PSR may be determined from equation 13-3 for each plane perpendicular to the camera axis that includes a target. Scales for intermediate planes may be determined by interpolation. For example, if the PSR at 50 feet is 100 and the PSR at 100 feet is 200, the PSR at 75 feet is 150. The intermediate distances may be estimated while viewing the photographs stereoscopically. This procedure provides an efficient method of measuring trail conditions (Rinehart, Hardy, and Rosenau, 1978), the diameter of tree stems and branches, and so on, from ground photographs.

13-6.1 Preparation of Stereograms

Trimming the prints is not difficult, but for best results proceed as follows.

1. Take either print and trim it so that the margins are parallel to vertical lines such as tree trunks, and so that only the portions that give value to the picture are included.

2. Take the other print and trim the bottom so as to cut the same points as on the first print.

3. Trim off the right side of this print so that any point at the base of the picture that is in the nearest foreground is the same distance from the edge as in the first print.

4. Trim the other two sides so that the dimensions check with the first print.

In mounting, the view taken on the right is mounted on the right; the view taken on the left is mounted on the left. Before the trimmed prints are glued on 5 by 7 inch, or larger, cards with rubber cement, Duco cement, or other waterproof glue, they must be aligned. To do this, draw a light pencil line parallel to the base of the mounting card and midway between the top and bottom; line up the bases of the two prints so they are equidistant from this centerline; and then move the prints laterally until they fuse and can be viewed comfortably with a stereoscope. The spacing of any set of conjugate images on the prints will normally range from 1.9 to 2.2 inches for different individuals. This procedure may also be used to prepare stereograms of selected areas on vertical aerial photographs.

An excellent stereoscope for viewing stereograms is the folding pocket stereoscope.

14
Inventory Using Sampling with Varying Probabilities

In the discussion of stratified sampling (Chapter 12), the advantages of sampling more intensively in the larger strata were evident. In stratified sampling the techniques of proportional and optimum allocation of sample units are used to assure that those strata comprising large portions of the population are also represented to the same extent in the sample. Samplers use another general technique to assure adequate representation of the larger components of the poplation; rather than adjust the number of sample units to locate in each group, the probability of selecting that group (or individual) is modified so that the larger or more valuable population components are given an increased chance of being selected for the sample.

Pioneer work on the subject of sampling with varying probabilities or, as it is called in classical sampling literature, "sampling with probability proportional to size" (p.p.s sampling) was done by Hansen and Hurwitz (1943) and later thoroughly described by Hansen, Hurwitz, and Madow (1953). Foresters made use of this general concept without realizing it when introduced to "angle-count cruising" by Bitterlich (1948). This was not recognized, however, as a type of p.p.s. sampling until pointed out by Grosenbaugh (1955).

In recent years several useful variations of the variable-probabilty-sampling notion have come to the fore. In a discussion of the advantages of sampling with unequal (i.e., varying) probabilities, Grosenbaugh (1967) classifies the major types of such sampling as:

1. *List sampling* (probability of sampling a previously listed item is made proportional to a listed quantity associated with the item)
2. *3P sampling* (probability proportional to prediction)
3. *p.p.s. sampling* (probability proportional to size)

This chapter follows Grosenbaugh's classification, although for clarity several points should be recognized for these three general sampling schemes, all of which in the broad sense are "sampling with probability proportional to size."

First, list sampling, as usually applied, requires a complete enumeration of *groups* of individuals; therefore, it is typically considered cluster sampling. Thus, in traditional sampling texts it will often be found under that category.

Second, 3P sampling, a term peculiar to forestry literature, is a type of sampling similar to list sampling. It differs in that a complete enumeration of the population of individuals (or groups) is not necessary prior to drawing the sample, although each individual (or group) is eventually examined for some attribute.

Third, p.p.s. sampling, a term which appears in traditional sampling literature as a broad type of sampling, usually infers in the forestry context some type of point or line sampling that employs variable-sized plots. It will be so used in this chapter; it is a type of sampling where trees (or other appropriate individuals) are assessable and selected for the sample by means of an instrument designed for that purpose. The term *polyareal plot sampling* is sometimes used to describe, generically, this type of sampling (see Section 14-3.2).

14-1 LIST SAMPLING (WITH VARYING PROBABILITIES)

List sampling with varying probabilities can be applied where the listed items are individual elements having different sizes, or where the listed items are clusters having different numbers of elements, or some other appropriate expression of size. In the usual forestry context we can think of individual trees as the elements, and compartments or stands as clusters. This type of variable probability sampling applied to individual listed trees is usually impractical, because of the necessity of listing each element prior to sampling. Therefore, we will discuss the technique for *compartments* (clusters of trees) having varying, but known, areas.

The sampling method is carried out first by listing the compartments in any order, along with their measure of size, say area. A sample size is decided on, and sampling is performed in such a way as to give the larger compartments a greater chance of being selected. The selected compartments are then visited and measurements are taken on the variable of interest. The analysis of the data, subsequently shown, yields unbiased estimates of means, totals, and their variances, as long as sampling is performed with replacement (i.e., once chosen, a compartment is "put back in" the population and can be drawn again). In application, sampling is often without replacement; in this case bias (often considered negligible) in the variance estimate will be incurred if equation 14-2 is used.

To illustrate the procedure, consider a forest made up of compartments having areas designated by X_i. We want to choose n compartments with selection probability proportional to X_i and measure the variable Y_i on each of the chosen compartments. We must first list the compartments as shown in Table 14-1, obtaining a column of cumulative areas in the process. It is also helpful to show a column of associated numbers that indicates the set of consecutive integers from the one above the previous cumulative total to, and including, the cumulative total of the compartment in question.

If we decide that $n = 5$, then 5 random numbers are appropriately drawn from the range of integers 1 through 419, the total area of the compartments. A compartment

Table 14-1
List of Compartments and Individual and Cumulative Areas for Use in List Sampling with Varying Probabilities

Compartment	Area in Acres (X_i)	Cumulative Total of X_i	Associated Numbers
1	25	25	1–25
2	10	35	26–35
3	30	65	36–65
4	28	93	66–93
5	15	108	94–108
6	30	138	109–138
7	38	176	139–176
8	51	227	177–227
9	40	267	228–267
10	12	279	268–279
11	60	339	280–339
12	80	419	340–419
Total:	419		

is chosen as part of the sample if the random number falls within the interval indicated in the column of associated numbers. Thus, the following might be chosen.

Random Numbers Drawn	Compartments Chosen
153	7
052	3
414	12
283	11
177	8

After the compartments are visited and measurements are taken regarding the variable of interest Y_i, the estimate of the mean value of Y per unit area is

$$\bar{r} = \frac{1}{n} \sum_{i=1}^{n} \frac{Y_i}{X_i}$$

$$= \frac{1}{n} \sum_{i=1}^{n} r_i$$

(14-1)

where

Y_i = the quantity measured on the ith compartment, e.g., total compartment volume

X_i = size of the ith compartment, e.g., compartment area

n = number of compartments chosen

$\bar{r}$ = mean value of Y per unit value of X, e.g., mean volume per acre (hectare)

r_i = ratio of Y_i to X_i

The variance of $\bar{r}$ can be estimated from

$$s_{\bar{r}}^2 = \frac{\displaystyle\sum_{i=1}^{n} r_i^2 - \frac{\left(\displaystyle\sum_{i=1}^{n} r_i\right)^2}{n}}{n(n-1)} \tag{14-2}$$

The estimate of the total Y for the population, then, is

$$\hat{y} = \bar{r}X \tag{14-3}$$

with estimated variance

$$s_{\hat{y}}^2 = s_{\bar{r}}^2 X^2 \tag{14-4}$$

where

$\hat{y}$ = estimate of the total Y
X = total of X_i for the entire population

Instead of a complete tally within each compartment as implied above, a more practical procedure would be to take a subsample of secondary sampling units within each of the chosen compartments. This practice implies a type of two-stage sampling (see Chapter 12), with secondary sampling units chosen from primary units of unequal size. Although formulas for estimating means are straightforward, those estimating the variance are complex. Reference should be made to sources such as Cochran (1977) for details.

List sampling with varying probabilities has been used in other forestry applications. For example, Scott (1979) describes its use for mid-cycle updating of permanent inventory plots. The initial plot volumes are used as the listed item from which a sample of n plots is chosen to be remeasured to obtain an estimate of current volume.

14-2 3P SAMPLING

The necessity of listing the units prior to sampling acts as a severe deterrent to list sampling in many forestry applications, especially those where individual trees are the potentially listed items. Grosenbaugh (1963b), making use of a principle similar to that proposed by Lahiri (1951) to overcome the prior listing requirement, proposed a type of sampling that utilizes the p.p.s. concept, but the element of size considered in his original application was the timber cruiser's on-the-spot estimate of tree volume. The name used by Grosenbaugh for this technique is "3P sampling," which should be interpreted as "sampling with probability proportional to prediction."

The purpose of our present treatment of 3P sampling is to make the reader aware of the general concept and application of the method. For details, reference should be made to the works of Grosenbaugh (especially 1963b, 1964, 1965, 1967, and 1979).

This type of sampling has been applied to timber sales where each tree in the population (all the marked trees) is assessed for a "crude" prediction of volume or value, and a subsample of these trees is selected for more detailed measurements. For this purpose it appears to be very efficient.

Before listing the steps in a simplified application of 3P sampling, it is perhaps best to dwell on the basic sampling concept which leads to the name "probability proportional to prediction." The analogy given by Mesavage (1965) is convenient for this purpose.

> . . . [S]uppose we have 20 cards numbered one to twenty. If we stipulate that the predicted volume of a sample tree must be equal to or greater than the number on a card subsequently drawn at random, a tree with a prediction of 1 would have only 1 chance in 20 to qualify [as a sample tree to be carefully measured for volume], whereas one with a predicted volume of 15 would have 15 chances in 20. The probability of selection is thus seen to be proportional to prediction. . . .

The steps in conducting a 3P sampling for the purpose of estimating the total volume of timber marked for sale might be as follows

1. Designate a sample size n, the number of trees to be carefully measured for volume. This can be done using the simple random sampling formula (Chapter 12), assuming an infinite population and noting that the coefficient of variation figure used should be based on the ratios of actual to estimated tree volume (defined in step number 8). This value is typically somewhat smaller than the usual coefficient of variation based on tree volume, which partially explains the usual high efficiency of 3P sampling. Alternatively, for sample size determination, one can make use of a crude guide such as that suggested by Mesavage (1965) for a large timber sale—for trained cruisers 100 or so trees are usually sufficient for an accuracy of 1.5 percent and, for inexperienced cruisers, approximately 200 trees are needed to achieve the same accuracy. It is worth noting here that it is the consistency (precision) of the cruiser's estimates that leads to high accuracy using 3P, and individual or volume table bias is of little consequence. Thus, the experienced cruiser, though possibly biased, is probably less erratic in estimates than the beginner and, therefore, likely to have a more efficient sample.

2. Estimate the sum of volumes for the N trees making up the sale. Thus,

$$\hat{x} = \text{estimated} \sum_{i=1}^{N} X_i$$

where

> X_i = cruiser's *estimate* of tree volume (this can also be an entry from a volume table utilizing the cruiser's estimate or measurement of tree diameter and possibly height)

Note that X, the actual sum of estimated volumes, is *known* only after the inventory is completed.

3. Designate the maximum individual tree volume expected as K. Thus,

$$K = \text{maximum } X_i$$

K, then, is used as the upper limit of the set of integers running from 1 through K, which will act as the means by which each tree will be checked for qualification as a sample tree to be measured in detail.

4. Adjust the set of integers to ensure that you obtain close to the sample size desired. That is, define

$$n' = \frac{\hat{x}}{K + Z} \tag{14-5}$$

where

n' = expected sample size
$\hat{x}$ = estimate of the total volume of all trees
Z = number of "rejection symbols" to be randomly mixed with the set of integers 1 through K

Thus, if we decide on $n' = 200$, and if $\hat{x} = 240,000$ board feet and maximum tree volume $K = 1000$ board feet, Z must be 200; otherwise, we would likely obtain 240 sample trees rather than the desired 200. Obviously, equation 14-5 can be solved for Z to streamline the calculations. In designing a bias-free 3P inventory, it is also worth following the guidelines suggested by Grosenbaugh (1963b) and cited by Johnson (1972), as summarized here.

    a. $n'K$ must be less than $\hat{x}$.

    b. $(Z/K)^2$ must be greater than $(4/n') - (4/N)$, where N equals the anticipated total number of trees in the timber sale.

5. Visit each of the N trees comprising the sale. At each tree follow this procedure.

    a. Estimate directly or indirectly, using a volume table, the tree volume (or value) X_i.

    b. Record the estimate.

    c. Draw a number (or symbol) at random from the set of integers 1 through K having the Z interspersed rejection symbols. A special device invented and described by Mesavage (1967) facilitates this operation.

Various programmable hand calculators have also been adapted to generate the random number or rejection symbol in this procedure (Estola, 1979; Beers, 1979).

6. If the volume estimate X_i is *greater than or equal to* the random integer, measure the tree for accurate volume determination. This volume is then recorded as Y_i, the actual volume of the ith tree. In Grosenbaugh's (1965) 3P program (THRP), the volume calculations (based on measurements using the Barr and Stroud dendrometer) as well as steps 8 and 9 are done by computer subsequent to the field work.

7. If the volume estimate X_i is *less than* the random integer, or if instead of a number a rejection symbol is drawn, nothing more is required from the tree and the crew moves on to the next marked or to-be-marked tree.

8. After completion of the inventory the total volume of the N marked trees can be estimated from the formula

$$\hat{y} = X\left(\frac{\sum\limits_{i=1}^{n}\frac{Y_i}{X_i}}{n}\right) = X\left(\frac{\Sigma R_i}{n}\right) \tag{14-6}$$

where

$$X = \sum_{i=1}^{N}X_i \quad \text{and} \quad R_i = \frac{Y_i}{X_i}$$

and where n equals the number of sample trees on which Y_i has been measured.

In words, then, the estimated total marked volume is equal to the sum of the estimates of tree volume obtained from the complete population, times the average ratio of actual to estimated volume that is obtained from the n sample trees.

At the completion of the inventory one should have approximately the sample size n' originally prescribed, but minor variations are possible because of vagaries associated with random numbers in the selection procedure.

9. The variance of the estimate $\hat{y}$ can be estimated, although it is pointed out by Ware (1967) that no exact expression for the true variance exists. The following approximation cited in Grosenbaugh's early work on 3P sampling has most often been applied, perhaps because as shown it is the same as that used for simple random sampling.

$$s_{\hat{y}}^2 = \frac{\sum\limits_{i=1}^{n}\left(\frac{Y_iX}{X_i} - \hat{y}\right)^2}{n(n-1)} \tag{14-7}$$

$$= \frac{X^2}{n}s_R^2 = X^2s_{\overline{R}}^2 \tag{14-8}$$

where

$s_R^2 = $ variance of the ratios $\left(\dfrac{Y_i}{X_i}\right)$

$\quad = \dfrac{\Sigma(R_i - \overline{R})^2}{n-1}$

$s_{\overline{R}}^2 = $ square of the standard error of the ratios

$\quad = \dfrac{s_R^2}{n}$

Further discussion of formulas for approximating the variance is provided by Grosenbaugh (1976).

It is apparent that 3P sampling has been adopted by many innovative inventory practitioners, and this trend will probably continue. For example, 3P applications have been described for traditional log scaling (Johnson, Lowrie, and Gohlke, 1971) and for sample weight scaling involving multiproduct tree length logging (Chehock and Walker, 1975). Other applications, such as for mid-cycle updating of forest inventories, and the combined use of point sampling and 3P sampling have been suggested in the literature; however, for these purposes the superiority of 3P sampling over the more traditional types of sampling has not been clearly demonstrated.

14-3   POINT AND LINE SAMPLING PRELIMINARIES

Beginning with the work of Bitterlich (1947) and continued by the work of Keen (1950) and Grosenbaugh (1952b, 1958), the concept known variously as "angle-count cruising," "plotless cruising," "point sampling," and "variable plot cruising" has earned a valuable place in the forester's tool kit. Probably no single forestry technique has been described so often, as indicated by the excellent bibliographies by Thomson and Deitschman (1959) and Labau (1967). Numerous articles have focused on the topic of why point sampling works, and it is evident that there is no "one way" to develop the workings of the technique. One of the more traditional discussions of Bitterlich's angle-count cruising (or, more definitively, "horizontal point sampling") would proceed as follows.

14-3.1   Fundamental Concept of Horizontal Point Sampling

In the application of horizontal point sampling, a series of sampling points is chosen, much as one would select plot centers for fixed-size plots. The observer occupies each sampling point, sights with an angle gauge (an instrument designed to "project" a horizontal angle of some arbitary size) at breast height on every tree visible from the point, and tallies all trees that are greater than the projected angle of the gauge. Figure 14-1 illustrates the procedure; the circles represent the cross sections of trees at breast height, and the lines indicate the angle projected from the sampling point. Any variables associated with the selected trees may be measured, just as in the case of fixed-size plots, but the unique feature of horizontal point sampling is that no tree measurements are needed to obtain an unbiased estimate of basal area per acre (hectare) from that sampling point. The number of trees counted that are larger than the projected angle, multiplied by a constant factor, dependent only on the size of the angle, yields the basal area per acre (hectare) estimate. Thus, it will be shown that each qualifying tree (i.e., each tree larger than the projected angle), regardless of its dbh, represents the same basal area per acre.

In order to show this, it is helpful to refer to Fig. 14-2, where two trees are depicted at locations where the gauge angle is precisely tangent to the breast height cross section, and to keep the following in mind.

1. The angle gauge is projecting a fixed horizontal angle, θ.
2. At any sampling point a series of concentric circular plots is established (conceptually), a different plot radius being associated with every different tree diameter.
3. The radius of each concentric plot is determined by each different tree diameter and is not influenced by the actual spatial location of the tree; therefore, for the purpose of development, all trees can be considered in the "borderline" condition as shown in Fig. 14-2.
4. At the borderline condition, the ratio of tree diameter to plot radius is a constant. Thus, for a given gauge angle θ, tree diameter in inches D, and plot radius in feet R, we can define this "gauge constant" k to be

$$k = \frac{D}{12R} = 2 \sin \frac{\theta}{2} \tag{14-9}$$

(Note that the only function of the 12 in this formula is to maintain the same units in numerator and denominator.)

The basal area per acre represented by each qualifying tree F can now be shown to be not dependent on tree diameter.

$$F = (\text{Basal area for the tree})(\text{Factor to convert to an acre basis})$$

$$= BA_i \left(\frac{\text{Area of one acre}}{\text{Plot area for the } i\text{th tree}} \right)$$

$$= \left(\frac{\pi D_i^2}{4(144)} \right) \left(\frac{43560}{\pi R_i^2} \right) \tag{14-10}$$

$$= 10{,}890 \left(\frac{D_i}{12R_i} \right)^2$$

$$= 10{,}890k^2$$

In the usual metric units D is in centimeters; R is in meters. Therefore,

$$k = \frac{D}{100R} = 2 \sin \frac{\theta}{2} \tag{14-11}$$

and the basal area in square meters per hectare F is

$$F = \left(\frac{\pi D_i^2}{4(10{,}000)} \right) \left(\frac{10{,}000}{\pi R_i^2} \right) = \frac{D_i^2}{4R_i^2} = 2500k^2 \tag{14-12}$$

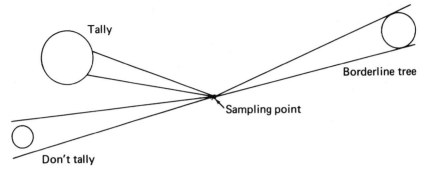

Fig. 14-1 *Selection of trees in point sampling.*

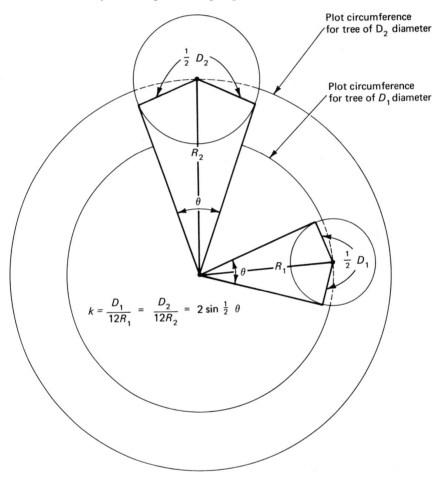

$$k = \frac{D_1}{12R_1} = \frac{D_2}{12R_2} = 2 \sin \tfrac{1}{2}\theta$$

Fig. 14-2 *Gauge constant* $k = \dfrac{D}{12R}$ *where D is tree diameter in inches and R is plot radius in feet.*

It should be clear now that we can choose a gauge constant k such that $10,890k^2$ is some convenient number. Therefore, in practice we will first choose a specific value of F, and so fix the ratio of $(D_i/12R_i)$ that, as a consequence, implies a different plot radius for every different tree diameter. Now, all trees of the same diameter that are located less than their "plot-radius distance" from a given sampling point will be "on the plot," and their basal areas can be converted to an acre basis. For example, if a gauge angle is used for which $k = 0.0303$, then for each tree greater than the projected angle

$$F = 10,890k^2$$

$$= 10,890(0.0303)^2$$

$$= 10.0 \text{ square feet per acre}$$

Although we might now develop horizontal point sampling in detail, it is more illuminating to approach the subject from a basic viewpoint so that other useful forms of p.p.s. sampling closely allied to horizontal point sampling can be understood and used. To do this, we will develop the logic from fixed-area plots of one size through concentric fixed-area plots of several sizes to point and line sampling, where there are many sizes of "plots" used at each sample location.

14-3.2 A Classification Scheme

To explain the operation of p.p.s. sampling in forest inventory, it is convenient to refer to one-size fixed-area plot sampling as *monareal* plot sampling, two-size concentric fixed-area plot sampling as *biareal*, three-size plot sampling as *triareal*, and, therefore, point and line sampling in general as *polyareal* plot sampling.

In monareal plot sampling the same size circular or rectangular (square or strip) plot is used for all trees. Then, all trees are measured on a number of monareal plots (such as $\frac{1}{5}$-acre circular plots or one-chain wide strips) that comprise a known portion of the area inventoried. A class of trees, in particular, diameter and height classes, can expect to be sampled in proportion to the frequency of trees in the given class. Thus, in many inventories, small diameter and small height classes will be sampled more intensively than large diameter and large height classes, because, in most stands, there are more small trees than large trees.

We may, of course, modify monareal plot sampling so that we will get fewer small trees—but still an adequate sample—and a more desirable sample of large trees by using two sizes of concentric circular plots, or two widths of strips. We could, for example, use $\frac{1}{10}$-acre circular plots for trees under 14 inches in diameter, and $\frac{1}{5}$-acre circular plots for trees 14 inches and over. If we desired to use strips, we could, in an analogous fashion, use two widths of strips, or on the other hand we could let the plot size be dependent on total or merchantable tree height rather than tree diameter. For example, we could use $\frac{1}{10}$-acre circular plots for trees under 30 feet in total height, and $\frac{1}{5}$-acre circular plots for trees 30 feet and over, or, if we preferred rectangular plots, we could use one-half chain wide strips for trees under 30 feet in total height, and one-chain wide strips for trees 30 feet and over.

In triareal plot sampling, for which we should obtain a more desirable sample of trees in three broad classes, we might, for example, use a $\frac{1}{10}$-acre circular plot for trees under 12 inches in diameter, a $\frac{1}{4}$-acre circular plot for trees 12 to 16 inches in diameter, and a $\frac{1}{5}$-acre circular plot for trees 16 inches and over in diameter. And, of course, all the alternatives of biareal plot sampling are possible for triareal plot sampling.

When this line of thought is taken to its logical conclusion, it is seen that a different size of circular or rectangular plot can be specified for each different tree diameter or tree height, resulting in polyareal plot sampling. But, until Bitterlich (1947) first wrote about the use of the horizontal angle gauge to estimate stand basal area per unit of land area, serious consideration was not given to the application of this concept. Indeed, few foresters felt it was practical even to use biareal or triareal plot sampling. Few foresters saw the implications and practicality of polyareal plot sampling until Grosenbaugh (1952b, 1955, 1958) pointed out that an angle gauge could be used to select sample trees, with probability proportional to some element of size, and postulated the theory of point sampling to obtain unbiased estimates of frequency, volume, growth, value, height, and so forth per acre from measurements of such p.p.s. sample trees.

Both the work of Bitterlich and that of Grosenbaugh were landmarks. They both displayed flashes of originality seldom seen. The original concept of angle-count cruising is unquestionably traced back to Bitterlich; however, the works of Grosenbaugh, although they emphasized horizontal point sampling, covered the entire theory of p.p.s. sampling as the basis of what we have called *polyareal plot sampling*. Hirata (1955) was the first to employ the vertical angle gauge in vertical point sampling to estimate mean squared height, and Strand (1957) the first to describe horizontal and vertical line sampling.

Line sampling theory, with which most foresters are unfamiliar, is but an extension or analogy of point sampling theory. Thus, it is desirable to discuss the theory of all polyareal plot systems at one time; that is, to use Grosenbaugh's specific terminology, to discuss the following systems together: horizontal point sampling, horizontal line sampling, vertical point sampling, and vertical line sampling.

To get a clear picture of all the polyareal plot sampling systems, it is well first to review monareal plot sampling. This can best be done by a study of Figs. 14-3a and 14-3b. In Fig. 14-3a, a monareal circular plot is drawn around a plot center. It is obvious which trees are to be measured. Then, in Fig. 14-3b, monareal circular plots are drawn with each tree as a center. In this case we are saying that each tree has a plot associated with it and that the plots are all of the same size. But no matter whether we think in terms of plots centered on points or plots centered on trees, we select the same trees for measurement. One may think of rectangular plots in the same manner.

The selection of trees using horizontal point sampling is illustrated in Figs. 14-4a and 14-4b. These two figures demonstrate that (1) plot size varies with tree size, (2) either the point-centered or tree-centered concept will result in the same trees being selected, and (3) a tree's associated plot size is determined by a borderline tree of its

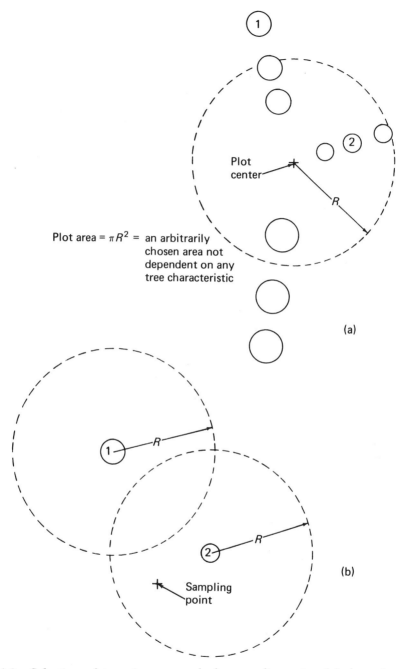

Fig. 14-3 *Selection of trees in monareal plot sampling using (a) the point-centered concept, and (b) the tree-centered concept. Tree number 2 qualifies; tree number 1 does not.*

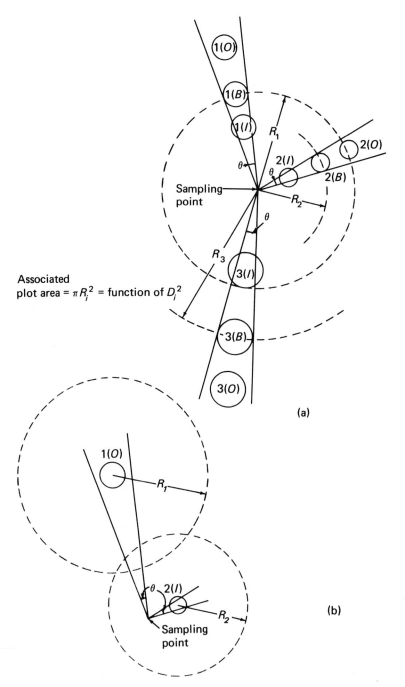

Fig. 14-4 *Selection of trees in horizontal point sampling using (a) the point-centered concept, and (b) the tree-centered concept. Tree number 2 (I) qualifies, tree number 1 (O) does not.*

size, although a tree may be *in* (*I*), *borderline* (*B*), or *out* (*O*), depending on its distance from the sampling point.

The remaining forms of polyareal plot sampling are illustrated in Figs. 14-5, 14-6, and 14-7, and, although only the point- or line-centered concept is shown, one should realize that the tree-centered approach is equally appropriate. Furthermore, it should be evident that in the two horizontal systems tree selection is accomplished by projecting a small horizontal angle, and in the two vertical systems tree selection is accomplished by projecting a large vertical angle.

In the next section we discuss the basic theory of all polyareal systems. To understand this discussion, it is essential to refer to Figs. 14-4 to 14-7 and to remember that

1. In horizontal point sampling the plot associated with any given tree is circular and its area (or radius squared) is a linear function of tree diameter squared.

2. In horizontal line sampling the plot associated with any given tree is rectangular and its area (or width) is a linear function of tree diameter.

3. In vertical point sampling the plot associated with any given tree is circular and its area (or radius squared) is a linear function of tree height squared.

4. In vertical line sampling the plot associated with any given tree is rectangular and its area (or width) is a linear function of tree height.

14-3.3 General Theory of Polyareal Plot Sampling

Assuming it is decided to employ one of the p.p.s. systems in a specific forest inventory, how are the necessary formulas and, ultimately, the numerical values developed for field application and data summarization?

Before the steps leading to these formulas are listed, we will describe a system of terminology and symbolism that will greatly facilitate the development. To introduce the proposed terminology, an example from monareal plot sampling will be used.

It is clear that each tree tallied on a $\frac{1}{5}$-acre plot represents $1\frac{1}{5}$ or 5 trees per acre. Furthermore, a 12-inch, one-log tree with a volume of 40 board feet represents $40(1\frac{1}{5}) = 200$ board feet per acre, or $12 (1\frac{1}{5}) = 60$ inches of diameter per acre. Thus, the ratio 1/(plot area in acres) is a per acre conversion factor and can be called the tree factor since it specifies the number of trees per acre represented by each tree on the plot. For any given plot size the tree factor is constant and easy to employ. Thus, it never has been necessary to label the per acre contribution of volume (e.g., 200 board feet) or the per acre contribution of diameter (e.g., 60 inches). However, by analogy these can be called the volume factor and the diameter factor, respectively, since they represent the number of board feet per acre and the number of inches of diameter per acre represented by each tree on the plot.

On the other hand, when dealing with polyareal plot sampling, the per acre conversion factor (tree factor) is not constant but is dependent on tree size; therefore, it will be found helpful to use specific factor names and to assign symbols and develop

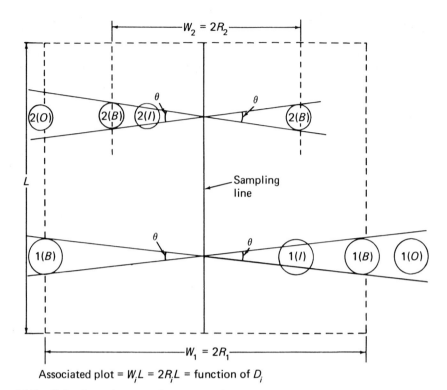

Associated plot = $W_iL = 2R_iL$ = function of D_i

Fig. 14-5 *Selection of trees in horizontal line sampling using the line-centered concept.*

formulas for them. When applied to polyareal plot sampling, the "factor terminology" then leads to the following definitions.

Tree factor. The number of trees per acre (hectare) represented by each tree tallied.

Basal area factor. The number of units of basal area per acre (hectare) represented by each tree tallied.

Volume factor. The number of units of volume per acre (hectare) represented by each tree tallied.

Height factor. The number of units of height per acre (hectare) represented by each tree tallied; and so on.

A table of "standard symbolism" is given in Table 14-2. the system was developed with a conscious effort to employ terms that have been commonly used. For example, the letter *F* was used to symbolize the horizontal factors and the letter *Z* to symbolize the vertical factors. Furthermore, the following "rules" were employed in the development of this symbolism.

1. Use the term *factor* to imply a certain number of units per acre (hectare) represented by each tree tallied for any tree characteristic.

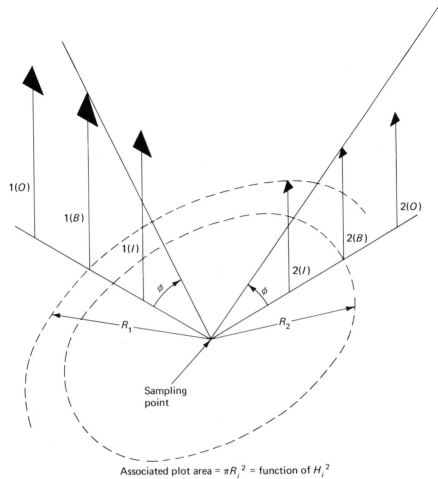

$$\text{Associated plot area} = \pi R_i^2 = \text{function of } H_i^2$$

Fig. 14-6 *Selection of trees in vertical point sampling using the point-centered concept.*

2. Use capital letters for point sampling factors, and lowercase letters for line sampling factors.

3. Use a lowercase subscript to indicate the tree characteristic involved (volume, diameter, circumference, etc.) if any tree dimensions are needed to obtain the actual value of the factor.

4. Use no subscript when no tree dimensions are needed to obtain the actual value of the factor. Thus, the four unsubscripted factors "characterize" the four types of polyareal sampling. For example, F = basal area per acre (hectare), represented by each tree tallied, characterizes horizontal point sampling.

Returning now to the basic question: For a given system, how does one develop

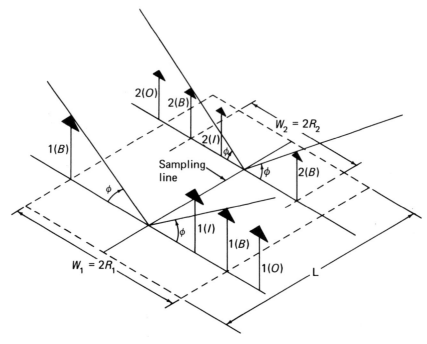

Associated plot area = $W_i L = 2R_i L$ = function of H_i

Fig. 14-7 *Selection of trees in vertical line sampling using the line-centered concept.*

formulas so that each different "factor" can be expressed in terms of tree character-istics?

Regardless of the type of polyareal plot sampling under consideration, there are only three steps needed to develop any factor.

1. Obtain the associated plot area in terms of the appropriate tree charac-teristic (i.e., diameter for horizontal sampling, height for vertical sampling).

EXAMPLE In horizontal point sampling, any tree of diameter D_i and plot radius R_i will have the following plot area in square feet.

$$\text{plot area} = \pi R_i^2$$

$$= \pi \left(\frac{D_i}{12k} \right)^2 \tag{14-13}$$

where

k = gauge constant = $(D/12R)$ = ratio of tree diameter in inches to plot radius in inches (Recall that to achieve equivalent units, the 12 in the denominator is necessary since plot radius is nor-mally expressed in feet.)

Table 14-2
Useful Symbolism for the Application of Polyareal Plot Sampling[1]

Item To Be Symbolized	Type of Polyareal Sampling				Units Represented by Each Tree Tallied	
	Horizontal		Vertical			
	Point	Line	Point	Line	English System	Metric System
Tree factor[2]	F_t	f_t	Z_t	z_t	Number trees per acre	Number trees per hectare
Basal area factor	F	f_b	Z_b	z_b	Square feet per acre	Square meters per hectare
Diameter factor	F_d	f	Z_d	z_d	Inches or feet per acre	Centimeters or meters per hectare
Quadratic height factor	F_{qh}	f_{qh}	Z	z_{qh}	Feet squared per acre	Meters squared per hectare
Height factor	F_h	f_h	Z_h	z	Feet per acre	Meters per hectare
Volume factor	F_v	f_v	Z_v	z_v	Board feet, cubic feet, cords, or tons per acre	Cubic meters or kilograms per hectare
Circumference factor	F_c	f_c	Z_c	z_c	Inches or feet per acre	Centimeters or meters per hectare
Bole surface area factor	F_s	f_s	Z_s	z_s	Square feet per acre	Square meters per hectare
Gauge constant	k	k	q	q		
Gauge angle	θ	θ	ϕ	ϕ		
Distance multiplier	HDM	HDM	VDM	VDM		
"Plot" radius	R		R		Feet	Meters
"Plot" length		L		L	Chains	Number of 20-meter lengths
"Plot" width		$W = 2R$		$W = 2R$	Feet	Meters
Expansion constant	E					

[1] Horizontal point = probability proportional to diameter squared.
   Horizontal line = probability proportional to diameter.
    Vertical point = probability proportional to height squared.
     Vertical line = probability proportional to height.
[2] Tree factor   = the number of trees per acre (hectare) represented by each tree tallied.
Basal area factor = the number of square feet per acre (square meters per hectare) represented by each tree tallied, etc.

    2. Obtain the "tree factor" (i.e., the per acre conversion factor, or number of trees per acre represented by each tree tallied) by dividing the associated plot area into the square-foot area of 1 acre (square-meter area of 1 hectare in the metric system).

EXAMPLE   In horizontal point sampling, when we express area in square feet, the tree factor will be

$$\text{Tree factor} = F_{t_i} = \frac{43{,}560}{\text{Associated plot area}}$$

$$= \frac{43{,}560}{\pi\left(\dfrac{D_i}{12k}\right)^2} \tag{14-14}$$

$$= \frac{43{,}560(144k^2)}{\pi D_i^2}$$

and dividing both numerator and denominator by (4×144), we get

$$F_{t_i} = \frac{(43{,}560/4)k^2}{\pi D_i^2/(4 \times 144)} = \frac{10{,}890k^2}{0.005454D_i^2}$$

and, finally, since the horizontal point basal area factor F equals $10{,}890k^2$, and $0.005454D_i^2$ equals tree basal area BA_i,

$$F_{t_i} = \frac{F}{BA_i} \tag{14-15}$$

3. Obtain any other factor for the system under consideration by multiplying the appropriate tree characteristic by the tree factor and simplifying.

EXAMPLE   In horizontal point sampling,

$$\text{Basal area factor} = F = (\text{Tree } BA)(\text{Tree factor})$$

$$F = BA_i\left(\frac{10{,}890k^2}{BA_i}\right)$$

$$= 10{,}890k^2 \tag{14-16}$$

$$\text{Volume factor} = F_{v_i} = (\text{Tree volume})(\text{Tree factor})$$

$$F_{v_i} = V_i\left(\frac{F}{BA_i}\right) \tag{14-17}$$

$$\text{Diameter factor} = F_{d_i} = (\text{Tree diameter})(\text{Tree factor})$$

$$F_{d_i} = D_i\left(\frac{F}{BA_i}\right)$$

$$= D_i\left(\frac{F}{0.005454D_i^2}\right)$$

$$= \frac{F}{0.005454D_i} \tag{14-18}$$

and since the constant $(F/0.005454)$ occurs quite often, we will call it the expansion constant E. Then,

$$F_{d_i} = \frac{E}{D_i} \tag{14-19}$$

Circumference factor $= F_c =$ (Tree circumference)(Tree factor)

$$
\begin{aligned}
F_{c_i} &= C_i\left(\frac{F}{BA_i}\right) \\
&= \pi D_i\left(\frac{F}{0.005454D_i^2}\right) \\
&= \frac{576F}{D_i} \tag{14-20}
\end{aligned}
$$

This simple procedure will lead to all the "factor" formulas given in Table 14-3a for polyareal plot sampling.

If metric units are used, the three steps to develop factor formulas can be summarized.

1. Associated plot area in square meters:

Plot area $= \pi R_i^2$

$$= \pi\left(\frac{D_i}{100k}\right)^2 \tag{14-21}$$

where

$$k = \frac{D}{100R} = \text{Ratio of tree diameter in centimeters to plot radius in centimeters.}$$

2. Obtain the tree factor:

$$F_{t_i} = \frac{10{,}000}{\pi\left(\dfrac{D_i}{100k}\right)^2} = \frac{10{,}000k^2}{\pi\left(\dfrac{D_i}{100}\right)^2} \tag{14-22}$$

and dividing numerator and denominator by 4,

$$F_{t_i} = \frac{2500k^2}{0.00007854D_i^2}$$

Therefore, since $F = 2500k^2$ and $0.00007854D_i^2$ equals tree basal area in square meters, the tree factor can alternatively be expressed as

$$F_{t_i} = \frac{F}{BA_i} \tag{14-23}$$

Table 14-3(a)
Equations for Obtaining Common Factors and Constants[1]—U.S. Units

Item	Type of Polyareal Plot Sampling			
	Horizontal		Vertical	
	Point	Line	Point	Line
Gauge constant	$k = \dfrac{D}{12R} = 2\sin\dfrac{\theta}{2}$		$q = \dfrac{H}{R} = \tan\phi$	
Plot radius or half-width (feet)	$R = \dfrac{D}{12k} = \dfrac{33\sqrt{10D}}{12\sqrt{F}}$	$R = \dfrac{D}{12k} = \dfrac{330D}{f}$	$R = \dfrac{H}{q} = \dfrac{66\sqrt{10H}}{\sqrt{\pi Z}}$	$R = \dfrac{H}{q} = \dfrac{330H}{Z}$
Associated plot area (square feet)	$Area = \pi R^2$ $= \pi\left(\dfrac{D}{12k}\right)^2$	$Area = 66L(2R)$ $= \dfrac{11LD}{k}$	$Area = \pi R^2$ $= \pi H^2\cot^2\phi$	$Area = 66L(2R)$ $= 132LH\cot\phi$
Tree factor (trees per acre)	$F_{ti} = \dfrac{43{,}560}{Area}$ $= \dfrac{10{,}890k^2}{0.005454D_i^2}$ $= \dfrac{F}{BA_i} = \dfrac{E}{D_i^2}$	$f_{ti} = \dfrac{43{,}560}{Area}$ $= \dfrac{3960k}{D_i}*$ $= \dfrac{f}{D_i}$	$Z_{ti} = \dfrac{43{,}560}{Area}$ $= \dfrac{43{,}560q^2}{\pi H_i^2}$ $= \dfrac{Z}{H_i^2}$	$z_{ti} = \dfrac{43{,}560}{Area}$ $= \dfrac{330q}{H_i}*$ $= \dfrac{z}{H_i}$
Basal area factor (square feet per acre)	$F = BA_i(F_{ti})$ $= 10{,}890k^2$	$f_{bi} = BA_i(f_{ti})$ $= 6.875\pi kD_i$	$Z_{bi} = BA_i(Z_{ti})$	$z_{bi} = BA_i(z_{ti})$
Diameter factor (inches per acre)	$F_{di} = D_i(F_{ti})$ $= \dfrac{E}{D_i}$	$f = D_i(f_{ti})$ $= 3960k$	$Z_{di} = D_i(Z_{ti})$	$z_{di} = D_i(z_{ti})$

Quadratic height factor (feet squared per acre)	$F_{qhi} = H_i^2(F_{ti})$	$f_{qhi} = H_i^2(f_{ti})$	$Z = H_i^2(Z_{ti})$ $= \dfrac{43{,}560}{\pi}q^2$	$z_{qhi} = H_i^2(z_{ti})$ $= H_i330q$
Height factor (feet per acre)	$F_{hi} = H_i(F_{ti})$	$f_{hi} = H_i(f_{ti})$	$Z_{hi} = H_i(Z_{ti})$ $= \dfrac{43{,}560}{\pi H_i}q^2$	$z = H_i(z_{ti})$ $= 330q$
Volume factor (cubic or weight units per acre)	$F_{vi} = V_i(F_{ti})$	$f_{vi} = V_i(f_{ti})$	$Z_{vi} = V_i(Z_{ti})$	$z_{vi} = V_i(z_{ti})$
Circumference factor (inches per acre)	$F_{ci} = C_i(F_{ti})$ $= \dfrac{576F}{D_i}$	$f_c = C_i(f_{ti})$ $= 3960\pi k$	$Z_{ci} = C_i(Z_{ti})$	$z_c = C_i(z_{ti})$
Bole surface area factor (square units per acre)		$f_{si} = bC_iH_i(f_{ti})$ $= bH_i3960\pi k$		$z_s = bC_iH_i(z_{ti})$ $= bD_i330q$
Expansion constant	$E = \dfrac{F}{0.005454}$ $= F(183.352)$			
Distance multiplier	$HDM = \dfrac{33\sqrt{10}}{12\sqrt{F}} = \dfrac{1}{12k}$	$HDM = \dfrac{330}{f} = \dfrac{1}{12k}$	$VDM = \dfrac{66\sqrt{10}}{\sqrt{\pi Z}} = \dfrac{1}{q}$	$VDM = \dfrac{300}{z} = \dfrac{1}{q}$

1 D_i = diameter in inches.
BA_i = basal area in square feet.
H_i = height in feet.
C_i = circumference in inches.
V_i = volume in any desired unit for the ith tree.

* f_i and z_i calculated on the assumption that L = 1 chain.

Table 14-3(b)
Equations for Obtaining Common Factors and Constants[1]—Metric Units

Item	Type of Polyareal Plot Sampling			
	Horizontal		Vertical	
	Point	Line	Point	Line
Gauge constant	$k = \dfrac{D}{100R} = 2\sin\dfrac{\theta}{2}$		$q = \dfrac{H}{R} = \tan\phi$	
Plot radius or half-widths (meters)	$R = \dfrac{D}{100k} = \dfrac{D}{2\sqrt{F}}$	$R = \dfrac{D}{100k} = \dfrac{250D}{f}$	$R = \dfrac{H}{q} = \dfrac{100H}{\sqrt{\pi Z}}$	$R = \dfrac{H}{q} = \dfrac{250H}{z}$
Associated plot area (square meters)	$Area = \pi R^2$ $= \pi\left(\dfrac{D}{100k}\right)^2$	$Area = 20L(2R)$ $= \dfrac{LD}{2.5k}$	$Area = \pi R^2$ $= \pi H^2 \cot^2\phi$	$Area = 20L(2R)$ $= 40LH\cot\phi$
Tree factor (trees per hectare)	$F_{ti} = \dfrac{10,000}{Area}$ $= \dfrac{2500k^2}{0.00007854D_i^2}$ $= \dfrac{F}{BA_i} = \dfrac{E}{D_i^2}$	$f_{ti} = \dfrac{10,000}{Area}$ $= \dfrac{25,000k}{D_i}\ *$ $= \dfrac{f}{D_i}$	$Z_{ti} = \dfrac{10,000}{Area}$ $= \dfrac{10,000q^2}{\pi H_i^2}$ $= \dfrac{Z}{H_i^2}$	$z_{ti} = \dfrac{10,000}{Area}$ $= \dfrac{250q}{H_i}\ *$ $= \dfrac{z}{H_i}$
Basal area factor (square meters per hectare)	$F = BA_i(F_{ti})$ $= 2500k^2$	$f_{bi} = BA_i(f_{ti})$ $= 0.625\pi k D_i$	$Z_{bi} = BA_i(Z_{ti})$	$z_{bi} = BA_i(z_{ti})$
Diameter factor (centimeters per hectare)	$F_{di} = D_i(F_{ti})$ $= \dfrac{E}{D_i}$	$f = D_i(f_{ti})$ $= 25,000k$	$Z_{di} = D_i(Z_{ti})$	$z_{di} = D_i(z_{ti})$

Factor				
Quadratic height factor (meters squared per hectare)	$F_{qhi} = H_i^2(F_{ti})$	$f_{qhi} = H_i^2(f_{ti})$	$Z = H_i^2(Z_{ti}) = \dfrac{10{,}000}{\pi} q^2$	$z_{qhi} = H_i^2(z_{ti}) = H_i 250q$
Height factor (meters per hectare)	$F_{hi} = H_i(F_{ti})$	$f_{hi} = H_i(f_{ti})$	$Z_{hi} = H_i(Z_{ti}) = \dfrac{Z}{H_i}$	$z_{hi} = H_i(z_{ti}) = 250q$
Volume factor (cubic or weight units per hectare)	$F_{vi} = V_i(F_{ti})$	$f_{vi} = V_i(f_{ti})$	$Z_{vi} = V_i(Z_{ti})$	$z_{vi} = V_i(z_{ti})$
Circumference factor (centimeters per hectare)	$F_{ci} = C_i(F_{ti}) = \dfrac{40{,}000F}{D_i}$	$f_c = C_i(f_{ti}) = 25{,}000\pi k$	$Z_{ci} = C_i(Z_{ti})$	$z_{ci} = C_i(z_{ti})$
Surface area factor (square units per hectare)		$f_{si} = bC_iH_i(f_{ti}) = bH_i 25{,}000\pi k$		$z_{si} = bC_iH_i(z_{ti}) = bD_i 250q$
Expansion constant	$E = \dfrac{F}{0.00007854} = F(12{,}732.37)$			
Distance multiplier	$HDM = \dfrac{1}{2\sqrt{F}} = \dfrac{1}{100k}$	$HDM = \dfrac{250}{f} = \dfrac{1}{100k}$	$VDM = \dfrac{100}{\sqrt{\pi Z}} = \dfrac{1}{q}$	$VDM = \dfrac{250}{z} = \dfrac{1}{q}$

[1] D_i = diameter in centimeters.
BA_i = basal area in square meters.
H_i = height in meters.
C_i = circumference in centimeters.
V_i = volume in any desired unit for the ith tree.
* f_i and z_i calculated on the assumption that L = number of 20-meter lengths.

3. Obtain other factors; for example,

Basal area factor:

$$F = BA_i \left(\frac{2500k^2}{BA_i} \right) = 2500k^2 \tag{14-24}$$

Volume factor:

$$F_{v_i} = V_i \left(\frac{F}{BA_i} \right) \tag{14-25}$$

Diameter factor:

$$F_{d_i} = D_i \left(\frac{F}{BA_i} \right)$$

$$= D_i \left(\frac{F}{0.00007854D_i^2} \right)$$

$$= \frac{F}{0.00007854D_i} \tag{14-26}$$

$$= \frac{E}{D_i} \tag{14-27}$$

where

$$E = \frac{F}{0.00007854}$$

Table 14-3b summarizes the factor formulas for the metric system that can be derived using the above procedure.

Until the time when the metric system is completely adopted by the forestry profession, certain useful conversions between factors expressed for each system will be required. For example, as Bruce (1979) points out, for horizontal point sampling, the relation between U.S. and metric basal area factors is exactly $F_{U.S.} = 4.356F_{metric}$; therefore, metric equivalents can easily be found from the inverse

$$F_{metric} = 0.2295684F_{U.S.}$$

Similar relations between the systems can be found in Table 14-4. Grosenbaugh (1973), using different symbolism, presents a similar table including the relation between unit area conversion factors.

The great utility of the factor formulas shown in Table 14-3a and Table 14-3b is that they indicate the use of simple field procedures to estimate certain stand parameters: in fact, it is frequently possible to eliminate the need to measure certain tree dimensions with no loss in accuracy.

A few examples will illustrate the possibilities.[1]

[1] Similar examples can be developed using the metric system.

Table 14-4
Basic Relationships for Polyareal Plot Sampling

Type of Sampling	Characteristic Factor Name	Symbol	Equation Relating Characteristic Factor and Gauge Constant[1] Using U.S. Units	Using Metric Units	U.S. to Metric Conversion Formula
Horizontal point	Basal area	F	$F = 10{,}890k^2$	$F = 2500k^2$	F in m^2/ha $= 0.2295684$ ($F_{\text{in ft}^2/\text{acre}}$)
Horizontal line	Diameter	f	$f = 3690k = 12\sqrt{10F}$	$f = 25{,}000k = 500\sqrt{F}$	f in cm/ha $= 6.313131$ ($f_{\text{in inches/acre}}$)
Vertical point	Quadratic height	Z	$Z = \dfrac{43{,}560}{\pi}q^2$	$Z = \dfrac{10{,}000}{\pi}q^2$	Z in m^2/ha $= 0.2295684$ ($Z_{\text{in ft}^2/\text{acre}}$)
Vertical line	Height	z	$z = 330q = \dfrac{\sqrt{10\pi Z}}{2}$	$z = 250q = 2.5\sqrt{\pi Z}$	z in m/ha $= 0.7575758$ ($z_{\text{in ft/acre}}$)

[1] For horizontal sampling $k = 2 \sin \dfrac{\theta}{2}$ and for vertical sampling $q = \tan \phi$, where θ and ϕ represent, respectively, the horizontal and vertical gauge angles in degrees.

EXAMPLE A To estimate basal area per acre, even though individual tree basal area depends on tree diameter, if *horizontal point sampling* is used, basal area per acre is a constant for each tree which qualifies. Specifically, since $F = 10890k^2$, if $k = 1/33 = 0.030303$, $F = (10890/1089) = 10$ square feet per acre.

EXAMPLE B To estimate the sum of circumferences per acre, even though individual tree circumference depends on tree diameter, if *horizontal line sampling* is used, the sum of circumferences per acre is a constant for each tree which qualifies. Specifically

$$f_c = C_i(f_{ti})$$

$$= \pi D_i \left(\frac{3960k}{D_i} \right) \tag{14-28}$$

$$= 3960\pi k$$

Thus, if $k = 1/33$, $f_c = 3960\pi(1/33) = 377$ inches of circumference per acre.

EXAMPLE C (This example illustrates a very powerful general approach for estimation of stand volume.) It will be noted in Table 14-3 that the horizontal point volume factor is

$$F_{v_i} = V_i(F_{ti})$$

which may also be written

$$F_{v_i} = V_i \frac{F}{BA_i}$$

If we are willing to accept the assumption that individual tree volume can be estimated by the linear regression estimate $V_i = bD_i^2H_i$, where b is the regression coefficient properly derived for the local timber, then

$$F_{v_i} = bD_i^2H_i \left(\frac{F}{0.005454D_i^2} \right) \tag{14-29}$$

$$= \left(\frac{bF}{0.005454} \right) H_i = bEH_i$$

Thus, if we use a basal area factor F of 10 and a tree volume equation such as $V = 0.01914D^2H$, the volume factor becomes

$$F_{v_i} = \frac{0.01914(10)}{0.005454}H_i \tag{14-30}$$

$$= 35.09 \ H_i$$

This means that tree diameter can be ignored in the field tally. Furthermore, every foot of height, in this case merchantable height, in a tree which qualifies represents 35.09 board feet of volume per acre. And so every 16-foot log represents 16 times 35.09 or 561 board feet per acre.

Example C is the theoretical foundation of the commonly used practice of assuming that "a one-log (16-foot logs) qualifying tree represents 600 board feet per acre, a two-log qualifying tree represents 1200 board feet per acre, and so on," as discussed by Miller (1963). This type of inventory has been termed "point sampling with diameter obviation" (Beers, 1964).

Careful study of Table 14-3a or Table 14-3b will show that there is a degree of circularity involved and that to derive certain of the formulas it is helpful to make use of relations derived separately from the table. For example, in developing the horizontal point sampling tree factor (see Example under step 2, page 232), in order to obtain the relation between the tree factor F_t and the basal area factor F, it is necessary to make use of the relation $F = 10,890k^2$, which was derived earlier. Of course, F is the "characteristic factor" for horizontal point sampling. Similarly, each other type of polyareal plot sampling has a "characteristic factor," and this factor has a simple relation to the appropriate gauge constant. These relationships, shown in Table 14-4, should be memorized or kept close at hand for ready reference.

Let us point out that, if the inventory is such that individual tree measurements of diameter and height must be recorded, and if the data are to be processed by computer, there is little advantage in poring over Table 14-3 in search of short-cut solutions. The general summarization formulas shown in Table 14-5 are all that one needs.

14-3.4 Angle Gauges for Horizontal Sampling

For horizontal point or horizontal line sampling one needs an angle gauge that will accurately "project" a small horizontal angle, generally under 5 degrees. A tree that appears larger than the projected angle is considered "in," and a tree that appears smaller than the projected angle is considered "out."

Basically, there are two different ways of "projecting" the angle.

1. By prolonging two lines of sight from the eye through two points whose lateral separation w is fixed, both of which are in the same horizontal plane and both of which are at the same fixed distance L from the eye—Fig. 14-8a.

2. By deviating the light rays from the tree through a fixed angle—Fig. 14-8b.

Types of Instruments. Instruments based on the first principle include the stick-type angle gauge, the Panama angle gauge, and the Spiegel relaskop. To construct a stick-type angle gauge, a device which clearly illustrates the principle behind all of these instruments, one simply provides a "stick" with a peep sight to position the eye and a crossarm of predetermined width, which is placed on the stick at a predetermined

Table 14-5
Summarization Equations for Use in Polyareal Plot Sampling

A. An estimate of a per-acre (hectare) characteristic, X_i, assuming m trees at *one point* (point sampling) or along a line segment *one chain* (20 meters in metric system) in length (line sampling) can be obtained from the following summarization equations, where X_1, X_2, ... X_m represent the individual tree values for the characteristic being estimated:

  1. Horizontal point sampling:

$$X_i = X_1\left(\frac{F}{BA_1}\right) + X_2\left(\frac{F}{BA_2}\right) + \cdots + X_m\left(\frac{F}{BA_m}\right)$$

where:
  F = basal area factor
  $BA_1, BA_2, \ldots, BA_m$ = basal area observed on individual trees

  2. Horizontal line sampling:

$$X_i = X_1\left(\frac{f}{D_1}\right) + X_2\left(\frac{f}{D_2}\right) + \cdots + X_m\left(\frac{f}{D_m}\right)$$

where
  f = diameter factor
  $D_1, D_2, \ldots, D_m$ = diameter observed on individual trees

  3. Vertical point sampling:

$$X_i = X_1\left(\frac{Z}{H_1^2}\right) + X_2\left(\frac{Z}{H_2^2}\right) + \cdots + X_m\left(\frac{Z}{H_m^2}\right)$$

where
  Z = quadratic height factor
  $H_1^2, H_2^2, \ldots, H_m^2$ = squared height observed on individual trees

  4. Vertical line sampling:

$$X_i = X_1\left(\frac{z}{H_1}\right) + X_2\left(\frac{z}{H_2}\right) + \cdots + X_m\left(\frac{z}{H_m}\right)$$

where:
  z = height factor
  $H_1, H_2, \ldots, H_m$ = height observed on individual trees

B. Therefore, an *average* per-acre (hectare) estimate obtained from n points visited in a sample can be calculated by dividing the sum of the n sample estimates, ΣX_i by n. For line sampling the total length of line (in chains or in number of 20 meter units) is used instead of n.

distance from the peep sight. To determine the width of the crossarm, one must know the gauge constant k for the basal area factor F to be used, and make a decision on stick length L. Then, crossarm width w will be: $w = kL$.

The Spiegel relaskop uses the same basic principle. Instead of using a stick on which to position the crossarm, the Spiegel relaskop utilizes the principle of the reflector

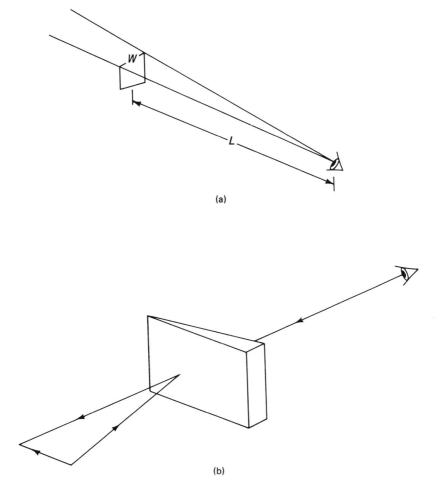

(a)

(b)

Fig. 14-8 *Projecting a horizontal angle (a) by prolonging two lines of sight, and (b) by deviating the light rays through a fixed angle.*

sight to image the scale (crossarm), which is on a wheel only a few inches from the eye, at a suitable viewing distance.

The thin prism or optical wedge is the only instrument of the second type that has been used by foresters. Briefly, a *prism* is a device made of optical glass or plastic in which the two surfaces are inclined at some angle *A* (the refraction angle), so that the deviation produced by the first surface is further increased by the second. The chromatic dispersion is also increased. However, chromatic dispersion is not a cause of appreciable error unless a telescopic device is used in conjunction with the prism.

The cruising prisms that foresters have used in horizontal point sampling generally have a refracting angle of less than 6 degrees. Such prisms are made in square, rectangular, and round shapes.

The effect of "projecting" a fixed angle with a thin prism is shown in Fig. 14-9. In Figure 14-9 the ray that is tangent to points a, b, and c on the sides of the tree to the observer's right is refracted to E. Thus, when observers at E sight through the prism to points a, b, and c, they will see these points as if they were at a', b', and c'. Of course, all visible points on each tree cross section will be displaced so that each cross section will appear to be displaced as shown in the drawing. To use the prism as a gauge, the observer looks *through* the prism at the right side of the trees—that is, at points a, b, and c (Fig. 14-9) and, at the same time, *over* the prism at the left side of the trees on the line of sight to I. We can note if the right side is, or is not, refracted past the left side. The actual "picture" that will be obtained when a round prism is used is shown in Fig. 14-10. Similar "pictures" are obtained with rectangular prisms.

Although the thin prism is the only commonly used horizontal angle gauge based on the second principle, it is quite possible to deviate the light rays from the tree through a fixed angle by using either mirrors or right-angle prisms. Compared to the thin prism, such a device would be expensive, cumbersome, and probably difficult to use. On the other hand, this instrument would be free of chromatic dispersion, a condition that appears to be difficult, or expensive, to eliminate with the thin prism. Until magnification is required for horizontal angle gauging, the elimination of chromatic dispersion is unimportant.

Prism Calibration.   Prisms for use in forest inventories are available as calibrated or noncalibrated. Although the latter are generally less expensive than calibrated prisms, they are more likely to deviate from the stated prism factor. In either case, checking or initial calibration (i.e., determination of F, the basal area factor) is imperative before field work commences.

Precise techniques for prism calibration have been described by Beers and Miller (1964) using a collimator and by Stage (1962) using a projector and screen. However, since many situations may not warrant this precision, the following procedure is commonly used.

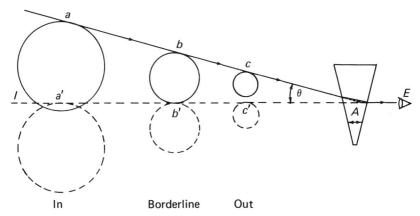

Fig. 14-9 *Representation of image deflection using a prism (cross section through prism and trees).*

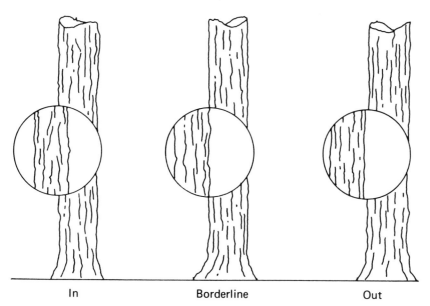

In   Borderline   Out

Fig. 14-10 *Representation of image deflection using a prism (the picture as seen by the observer when using a round prism).*

1. A flat "target" of width w is placed at some distance from the observer holding the prism.

2. The observer moves himself and/or the prism toward or away from the target until the perpendicular line of sight to the target results in the completely offset picture—obtained when one side of the target seen over the prism precisely lines up with the other side seen through the prism.

3. The perpendicular distance from the prism to the target B is carefully measured in the same units as the target width.

4. The gauge angle θ is then calculated or determined from a table of trigonometric functions since

$$\tan \theta = \frac{w}{B} \tag{14-31}$$

$$\theta = \text{arc tan} \left(\frac{w}{B} \right) \tag{14-32}$$

5. The gauge constant k is then determined from the formula

$$k = 2 \sin \frac{\theta}{2} \tag{14-33}$$

and the pertinent characteristic factor F, or f, from the formulas given in Table 14-4.

The geometry of the procedure just described is shown in Fig. 14-11, where, in addition to the generated angle, a schematic tree cross section is shown at the borderline distance R appropriate for its diameter D. Examination of Fig. 14-11 suggests several concepts worth noting.

First, the angle generated by the prism and intercepted by the flat target forms a small *right* triangle, not an isosceles triangle. The assumption of a generated isosceles triangle is appropriate for the stick-type and relascope gauges, but not for the wedge prism. Thus, by substituting w and B in the formula used for stick-type gauges to obtain the basal area factor

$$F = \frac{43,560}{1 + 4\left(\dfrac{B}{w}\right)^2} \qquad (14\text{-}34)$$

we will arrive at some error that, although small (for prisms having a small refracting angle), can be avoided by following the calibration procedure described above.

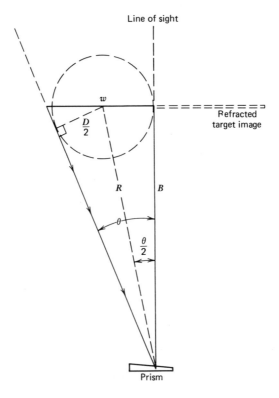

Fig. 14-11 *The geometry of prism calibration. Target of width w located at distance B, which will provide the "completely offset" picture. Dotted lines pertain to a hypothetical borderline tree located properly in the generated gauge angle θ.*

Second, from Fig. 14-11 it can be shown that the gauge constant k calculated from w and B, as explained in steps 4 and 5 above, is the same as the ratio of tree diameter D to distance R, discussed in the development of the theory of horizontal point sampling. That is, using the constant 12 to change the denominator to inches,

$$\frac{D/2}{12R} = \sin\frac{\theta}{2}$$

Therefore,

$$\frac{D}{12R} = 2\sin\frac{\theta}{2} = k$$

Regardless of whether a run-of-the-mill wedge prism or a carefully manufactured gauge is chosen, personal bias will result if the appropriate "borderline picture" is not consistently obtained. To fix this picture in mind, that is, to recognize the borderline case, the beginner should make a field check such as described by Beers and Miller (1964) or more recently by Zeide, Troxell, and Haag (1979). Briefly, this involves repeated attempts by the observer to position himself at what he interprets (from the gauge picture) the borderline distance to be. Subsequent careful measurement of tree diameter (caliper measurement) and knowledge of the gauge constant will indicate the true borderline distance. Comparison of the observer's obtained distances with the true distance will indicate the precision and if bias is present. Bias will be due either to the individual's inability to recognize the borderline picture under the conditions of the test, or to an incorrectly calibrated, poorly adjusted, or poorly manufactured gauge. It is a sound practice, even for experienced workers, to occasionally check their interpretation of the borderline picture, especially when prevailing stand conditions change radically.

Additional details regarding prisms and their calibration are given by Miller and Beers (1975).

Modification of the Gauge Angle. Since the determination of forest stand parameters is on a horizontal land area basis, the projected angle used must be corrected if the terrain is not level. Specifically, the gauge constant k must be reduced to a value k_r, where $k_r = k\,(\cos S)$, and where S is the slope angle in degrees.

For angle gauges based on the principle of projecting a horizontal angle by prolonging two lines of sight, as in Fig. 14-8a, slope correction can be achieved by reducing the crossarm width w to the value w_r, where $w_r = w(\cos S)$, or by increasing stick length L to the value L_r, where $L_r = L/\cos S$. Employing both of these concepts, Robinson (1969) describes a very useful stick-type gauge that achieves a slope-correcting variable basal area factor. The Spiegel relaskop uses an ingenious method of varying "crossarm" width. It has a "strip" scale that varies in width directly as the cosine of the slope angle. This scale, which is mounted on a weighted wheel, is seen "projected" on the target (the tree) and rotates by gravity to the appropriate position as the line of sight is raised or lowered.

Slope correction can be achieved when the thin prism is used by rotating the prism in a plane perpendicular to the line of sight through an angle equal to the slope angle.

Such rotation properly reduces the gauge angle. The practical utility of this fortuitous relationship will be discussed in Section 14-5.4. It is sufficient here to point out that by prism rotation one can change the basal area factor very conveniently. For example, a prism having a basal area factor of 10.2 can be changed to one with a factor of 10.0 by rotating it perpendicular to the line of sight through an angle of 14 percent.

$$\text{Since } k_r = k(\cos S) \tag{14-35}$$

$$\text{Leading to } F_{res} \cong F(\cos^2 C) \tag{14-36}$$

$$\cos C \cong \sqrt{\frac{F_{res}}{F}} = \sqrt{\frac{10}{10.2}} \tag{14-37}$$

$$\cong \sqrt{0.9803922} = 0.99015$$

Therefore,

$$C \cong 8°3' \text{ or } 14 \text{ percent}$$

where

k_r = resultant gauge constant after rotation
k = gauge constant of unrotated prism
C = angle of rotation
F = basal area factor (unrotated prism)
F_{res} = resultant basal area factor after rotation

Table G-9 in Beers and Miller (1966) can be used to avoid the calculations.

An additional rotation is required for slope correction, which is not directly additive to the basic rotation. Although more detail is beyond the scope of this discussion, it is worth noting that the following relationship has been found (see Beers and Miller, 1964).

$$\cos T = \cos S (\cos C) \tag{14-38}$$

where T is the total rotation in degrees required to correct for a slope of S degrees as well as to modify the gauge angle to a desired F_{res} by rotating the prism through the angle C in degrees. Angle C is found as described in the previous paragraph.

14-3.5 Angle Gauges for Vertical Sampling

For vertical point or vertical line sampling, one needs an angle gauge that will accurately "project" a large vertical angle. A tree that appears larger than the projected angle is considered *in*, and a tree that appears smaller than the projected angle is considered *out*.

If ϕ is the vertical gauge angle in degrees, then, as Fig. 14-12 shows, ϕ must be corrected for slope. But even though the vertical angle changes on slope sights, the vertical gauge constant $q = H/R$ does not change. Thus, one will project the correct vertical angle under all conditions if a device is used that maintains a constant ratio between a fixed vertical distance H and a fixed horizontal distance R.

Although several instruments have been built to achieve this directly—for example, the conometer by Hirata (1955) and Jukohscope by Kaibara (1957)—there appears to be little advantage in using such special devices. Any clinometer that has a percent scale can be used efficiently for vertical gauging as explained below.

Since the correct vertical gauge angle is projected so long as the gauge constant q is held constant, it follows that slope is automatically corrected for if q is appropriately defined. The three field situations are depicted in Fig. 14-12, which shows that:

1. On the level, $q = \tan \phi$.
2. Downhill, $q = \tan A + \tan B$.
3. Uphill, $q = \tan A - \tan B$.

Since a percent slope is the natural tangent of the vertical angle multiplied by 100, it follows that if we multiply the gauge constant q by 100, it will be expressed in percent. Consequently, such instruments (clinometers) as the Abney level and the Haga altimeter, if they have a percent scale, can be used to check trees to see if they are larger than, or smaller than, the projected vertical angle. Figure 14-13 reveals that one need remember only two simple rules to identify qualifying trees.

1. A tree will be "in" if the percent readings, summed or subtracted as appropriate, to the tip and base of a tree (or to the merchantable limit and stump height) is *greater than* 100 times the gauge constant q.
2. A tree will be "out" if the percent readings, summed or subtracted as appropriate, to the tip and base of a tree (or to the merchantable limit and stump height) is *less than* 100 times the gauge constant q.

To take a specific example, assume we have decided to use a vertical angle for gauging total height for which q is 1. Thus, $100q$ will be 100. Now, if we obtain a reading (with a percent Abney level) of $+67$ to the tip of a tree and a reading of -40 to the base of the tree, the tree will be "in" because 107 (i.e., $67 + 40$) is greater than 100. Then, on another tree, if the readings to the tree base and tip are $+15$ and $+63$, the tree will be "out" because 48 (i.e., $63 - 15$) is less than 100.

There is little question that vertical sampling, in particular, vertical line sampling, is a potentially powerful forest sampling method. Initial trials have shown that it can be efficiently applied. However, additional experience will be required to fully demonstrate its usefulness as a practical inventory technique.

14-3.6 Number of Points or Lines Required

In all types of point and line sampling, theory of the method dictates that sampling is performed *with replacement* (contrary to fixed-size plot sampling which is usually perceived as sampling *without replacement*). Thus, it can be assumed that sampling is being performed in a population that is infinite. Furthermore, the sample units

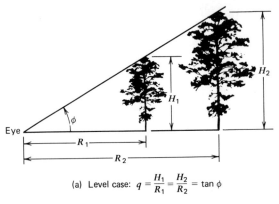

(a) Level case: $q = \dfrac{H_1}{R_1} = \dfrac{H_2}{R_2} = \tan \phi$

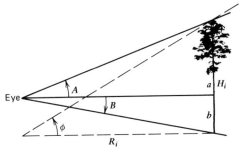

(b) Downhill case (angles A and B opposite in sign):

$$q = \frac{H_i}{R_i} = \tan \phi = \tan A + \tan B \quad (\text{Note: } \phi \neq A + B)$$

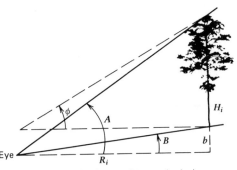

(c) Uphill case (angles A and B same in sign):

$$q = \frac{H_i}{R_i} = \tan \phi = \tan A - \tan B \quad (\text{Note: } \phi \neq A - B)$$

Fig. 14-12 *Vertical gauging of borderline trees in three field situations.*

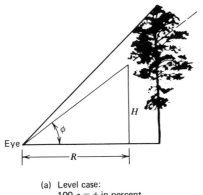

(a) Level case:
100 q = ϕ in percent

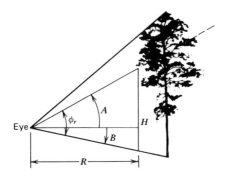

(b) Downhill case:
100 q = A in percent + B in percent

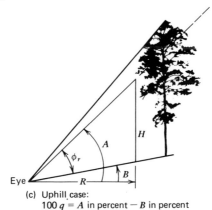

(c) Uphill case:
100 q = A in percent − B in percent

Fig. 14-13 *Selection of qualifying trees in three field situations. (When the ratio H/R is held constant, the correct vertical angles, ϕ or ϕ_r are obtained. The lines of sight of the clinometer—that is, the heavy lines—indicate that all three trees are "in." See text for elaboration.)*

(points or lines) are in a sense without area. This implies that the area of our sample has little meaning, that correction for sampling a finite population is not appropriate, and that the total area of the population has no direct effect on the calculation of sample size. Thus, for most situations the following formula can be used to determine the number of points to visit.

$$n = \frac{t^2 C^2}{a^2}$$

(14-39)

where

> n = number of points required for accuracy a, with the probability level implied by the value of t
> t = Student's t
> C = coefficient of variation (in percent) for the forest to be sampled
> a = desired precision (in percent) for the average volume (or basal area, etc.)

A crucial substitution into the formula is that for the coefficient of variation. If previous data are not available to provide good estimates, preliminary inventories might be necessary to establish reasonable approximations.

Unfortunately, when small forests are sampled (several hundred acres or less), impractical results may be obtained if the formula is used. For example, for a 40-acre tract where the coefficient of variation is expected to be approximately 70 percent and a desired precision of 10 percent is required at the .95 probability level ($t \cong 2$), then substitution into the formula leads to

$$n = \frac{(2)^2 \, (70)^2}{10^2} = 196 \text{ points}$$

Locating 196 sample points in a 40-acre woods is unnecessarily intensive. The reader should note again that forest area does not appear in the formula used and that such a problem does not exist when dealing with fixed-area plot sampling, since sampling is performed from a *finite* population and a correction factor involving forest area is part of the appropriate formula to determine number of plots.

To circumvent these impractical results for small areas, various approaches have been proposed. Often the procedure is to determine how many fixed-area plots of a certain size might be required and then to locate somewhat more (or less) sample points depending upon local experience regarding comparative accuracy of plots versus points.

Another approach involves the development of a "rule of thumb" to facilitate the decision. For example, the following "rule" has provided reasonable results in the mixed hardwood forests of Central United States.

If area in acres is	Number of points is
Less than 10	10
11–40	1 per acre
41–80	20 + 0.5 (area in acres)
81–200	40 + 0.25 (area in acres)
Over 200	Found by equation 14-39

When using *line sampling* the question to be resolved is not how many points to take but, instead, how many line segments to traverse. Very little research is available to specifically answer the question. However, it is reasonable to assume: (1) that the same techniques (equation 14-39 and a general guide for small areas) can be used for line segments as for points, and (2) that sample size for lines will be similar to that for points if approximately equal numbers of sample trees are tallied at the average point and sample line segment. Also, the length of each sample line segment must be fixed before intelligent sample size calculations can be made, since the between-line variation when using long sample lines should be less than when using short sample lines. Thus, the estimate of coefficient of variation to be substituted into equation 14-39 is influenced not only by gauge angle but also by sample line length.

Undoubtedly, considerations such as whether line segments are located randomly about the area or systematically (whether continuously or intermittently) along cruise lines will also have an effect on the statistical efficiency of line sampling. This effect, analogous to the clustering of point samples, requires further investigation. Similar unresolved questions exist for both vertical point and vertical line sampling.

14-3.7 Choosing a Suitable Gauge Constant

The choice of an appropriate gauge constant, which of course determines the gauge angle and "characteristic" factor (Table 14-4), is influenced primarily by the nature of the stands to be sampled, and secondarily by the objectives and conduct of the inventory. The choice is frequently based on local experience and other general guidelines such as: for small trees and open stands, use a small gauge constant; for large trees and dense stands, use a large gauge constant. Since small trees do not always occur in open stands and large trees do not always occur in dense stands, a compromise frequently must be made after assessing both average tree size and stand density. A more definitive approach is to first determine the average number of trees m we wish to tally per location, then after estimating, for example, the average stand basal area (in the case of horizontal point sampling), find the "suitable" characteristic factor (basal area factor in the case of horizontal point sampling) by division. Thus, for horizontal point sampling,

$$\text{Suitable } F = \frac{\text{Estimated average basal area per acre}}{m} \tag{14-40}$$

Similarly for horizontal line sampling,

$$\text{Suitable } f = \frac{\text{Average sum of diameters per acre}}{m} \qquad (14\text{-}41)$$

For vertical point sampling,

$$\text{Suitable } Z = \frac{\text{Average sum of squared heights per acre}}{m} \qquad (14\text{-}42)$$

And for vertical line sampling,

$$\text{Suitable } z = \frac{\text{Average sum of heights per acre}}{m} \qquad (14\text{-}43)$$

where

m = desired average number of trees to be tallied at a point or along a line segment

After the suitable characteristic factor is determined by the appropriate formula above, the corresponding gauge constant and gauge angle can be calculated from equations shown in Table 14-4.

The major question in this procedure obviously is—In a specific situation, what is the desired number of trees to tally at each location? Recommendations vary, but the results of a study by Sayn-Wittgenstein (1963), one of the more complete studies dealing with horizontal point sampling, indicate that the desired average tree count per sample location is somewhere between 6 and 16. Experience in the Central Hardwood and Lake States regions of the United States indicate that an average count of 10 trees gives satisfactory results. For horizontal point sampling, therefore, if average stand basal area is between 75 and 125 square feet per acre, the basal area factor to use, rounded to the nearest "convenient" number, would be 10 since substitution into equation 14-40 leads to 7.5 and 12.5, respectively.

It should be apparent that the choice of the desired average number of sample trees is somewhat arbitrary. However, several general concepts must be considered before making a decision about a suitable gauge constant.

1. For small *average* tree counts there exists a great likelihood of getting many small individual counts (say less than 5), and conversely for large *average* tree counts the likelihood is great for getting large individual counts (say more than 20).

2. Small or large tree counts can lead to inefficient samples. For very small counts the forest is "undersampled" and the inventory is economically inefficient since the at-location costs are frequently much less than the between-location costs. Therefore, measuring a few more trees at a location might materially increase the overall statistical precision but not materially affect cost. For very large tree counts the forest is "oversampled" since once a certain level of statistical precision is reached, the precision will be increased negligibly by measurement of additional trees. Then, the measurement of additional trees is largely a waste of time.

3. Estimates of forest parameters may be biased if tree counts are very high because of the increased possibility of overlooking a tree that qualifies as a sample tree, because of the increased likelihood of encountering questionable (i.e., "borderline") trees, because of the increased difficulty in viewing the tree stems because of high stand density, and because of the inclination to be less careful when more trees are to be scrutinized.

4. Regarding low tree counts, a prime consideration to note is that the effect of a mistaken action will lead to a much larger bias (in percentage) than when the tree count is high, because the base is smaller.

5. In the deliberation over what gauge constant to use, it is easily shown, but often forgotten, that the gauge constant, gauge angle, and characteristic factor all vary *inversely* as the expected tree count. Therefore, for horizontal point sampling a large basal area factor implies small tree counts, and a small basal area factor implies large tree counts.

An example of how inventory objectives can influence the size of the angle gauge can be found in the horizontal point sampling procedure commonly employed by the U.S. Forest Service in their forest survey inventories. A ten-point cluster of sample points is taken within an area approximating 1 acre for the purpose of obtaining more objective descriptions of stand conditions and treatment needs. To minimize the selection and measurement of trees from more than one point (i.e., oversampling), the gauge angle is increased over that normally applicable. In eastern United States angles of 285.37 minutes ($F = 75.0$ square feet) or 201.76 minutes ($F = 37.5$ square feet) have traditionally been used. If stands average 75 to 150 square feet of basal area, then 1 to 2 or 2 to 4 trees per point on the average will be selected, depending on the value of F (75.0 or 37.5).

A practice sometimes employed in very steep terrain in western United States exemplifies how the conduct of the inventory will, in part, influence the gauge angle chosen. To minimize the need for slope correction, to increase visibility, and to improve the volume estimate it is common to sample using half-points. Only trees on the downhill side (180 degrees) of the level line through sample point are assessed for qualification, and the diameter point scrutinized with the angle gauge is the diameter at the top of the first log; thus, if form class volume tables are being used, there is no need to determine the form class directly (for details see Bruce, 1961, and Dilworth, 1979). In this situation, then, one must reduce the gauge angle so that the basal area factor F, appropriate for full points in similar stands, is halved. A similar situation exists if line sampling is used and trees are considered for qualification on just one side of the line.

For vertical sampling the pertinent stand characteristics might be so "strange" that even crude approximations cannot be made. In such cases one can locate several fixed-size plots or horizontal point or line samples so that logical values for the sum of squared heights per acre or the sum of heights per acre (appropriate for planning a vertical point or vertical line sample) can be established.

14-3.8 Choosing the Appropriate p.p.s. System to Use

If the inventory is a large-scale operation where computer processing of the field measurements is anticipated, the p.p.s. system chosen might be selected to achieve the highest relative efficiency in estimating the stand parameter of primary interest. Several sources (Beers and Myers, 1965; Grosenbaugh, 1967) have pointed out, for example, that purely on statistical grounds, to estimate basal area and volume, horizontal point sampling should be the most efficient. Similarly, fixed-size plots should be the most efficient for estimating number of trees per acre. On the other hand, if a lesser used stand characteristic such as average tree circumference per acre is the parameter of primary interest, then horizontal line sampling should be the most efficient.

Once one of the systems is chosen, estimates of the stand parameters can be made using the appropriate formula in Table 14-5. In such large-scale inventories, then, the only "factor" that need be calculated prior to the field work is the distance multiplier (required for the checking of questionable trees). This distance multiplier is the number by which the pertinent tree dimension is multiplied to determine the tree's associated plot radius (sometimes called the *limiting distance*) or plot width. For horizontal sampling it is

$$HDM = \frac{1}{12k} \tag{14-44}$$

whereas for vertical sampling it is

$$VDM = \frac{1}{q} \tag{14-45}$$

where

HDM = horizontal distance multiplier
k = horizontal gauge constant
VDM = vertical distance multiplier
q = vertical gauge constant

Although *any* of the four types of p.p.s. sampling can be used to estimate the normal quantitative stand characteristics (volume, basal area, number of trees, etc.), in certain inventories, particularly small-scale inventories where summaries will be made directly on the field sheets, it is advantageous to ponder the type of information needed from the inventory and, with the help of Table 14-3, to select the most suitable type of p.p.s. sampling to use, and then to calculate the appropriate factors to enter on the field tally sheet. Thus, using concepts presented earlier, Table 14-6 illustrates some of the more obvious solutions to certain sampling situations. Both the opportunities open to the inventory designer and the limitations are indicated in the table. Several of these will now be discussed; repeated reference should be made to Table 14-6. Specific factor formulas can be obtained from Table 14-3.

If primary interest is in basal area per acre, horizontal point sampling can be used to estimate basal area from tree counts alone; the only factor required is the basal

Table 14-6
Type of Point or Line Sampling
and the Associated Tree Dimension Affecting the Calculation
of the Necessary Factors Suggested for Certain Inventory Situations

Per-Acre Characteristic of Primary Interest[2]	Recommended Type of Sampling	Tree Dimension Influencing Needed Factors		
		For Primary Characteristic	For Secondary Characteristic	
			Number of Trees	Basal Area
Basal area	Horizontal point	( )[1]	(D)	( )
Volume or weight	Horizontal point	(D, H)	(D)	( )
Using $V = bD^2H$	Horizontal point	(H)	(D)	( )
	Vertical line	(D)	(H)	(H)
Using local volume table	Horizontal point	(D)	(D)	( )
Diameter or circumference	Horizontal line	( )	(D)	(D)
Height	Vertical line	( )	(H)	(D, H)
Bole surface area using $S = bCH$	Horizontal line	(H)	(D)	(D)
	Vertical line	(D)	(H)	(H)
Quadratic height	Vertical point	( )	(H)	(D, H)

[1] The parentheses should be interpreted as follows: to estimate the stand characteristic indicated at the head of the column or row:

( )   Implies no tree measurements are needed, and the appropriate factor is a constant.

(D)   Implies that tree diameter must be measured, and the appropriate factor varies with diameter.

(H)   Implies that tree height must be measured, and the appropriate factor varies with height.

(D, H) Implies that both tree diameter and height must be measured, and the appropriate factor varies with both of these dimensions.

[2] V, D, H, S and C represent tree volume, diameter, height, surface area, and circumference, respectively, and b represents a properly derived regression coefficient.

area factor F. If, however, estimates of number of trees per acre are also needed, tree diameter must be measured (or estimated) and tree factors F_t, one for each diameter class, must be calculated.

If volume or weight per acre is of primary interest, it is common to use horizontal point sampling and to measure both tree diameter and height. In this case volume factors F_v are needed for each height-within-diameter class, since tree volume varies with both of these dimensions. And of course, trees per acre and basal area per acre may be obtained when the appropriate factors F_t and F are computed.

If a regression relation of the form $V = bD^2H$ relating tree volume, diameter, and height proves acceptable, then volume per acre can be estimated more efficiently by using horizontal point sampling and obviating diameter measurements (volume factors F_v are needed only for each height class), or by using vertical line sampling and obviating height measurements (volume factors z_v are needed only for each diameter class). Note, however, if number of trees per acre is also to be estimated, in both cases the obviated tree measurement must be measured and the tree factors F_t or z_t associated with those measurements calculated.

When a reliable local volume table (i.e., volume by diameter class only) is available, if horizontal point sampling is used, height measurements may be obviated in making both volume and trees per acre estimates. In this case volume factors F_v and tree factors F_t will be needed for each diameter class.

Further study of Table 14-6 will lead to other procedures for estimating the lesser used stand characteristics.

14-4 ESTIMATES ON A PER-TREE BASIS

The primary purpose of an inventory is usually to obtain estimates of forest parameters on an areal basis, that is, per acre, per hectare, or per stand (total). However, the need sometimes arises for estimates such as mean tree basal area, mean tree diameter, or mean tree height, which are on a *per-tree basis*.

With the development of point and line sampling, because of the variable probability of selection characteristics, the efficient estimation of various per-tree characteristics became quite simple, although the underlying mathematical concepts can appear devious if not carefully documented.

The following sections are presented to demonstrate several efficient estimation procedures and to attempt classification of the types of estimates that can be made.

14-4.1 Direct Estimates by Combination Sampling

Combination sampling, as used here, can be described as a technique for obtaining per-tree estimates *from tree counts only,* by combining the application of monareal (fixed-size plot) and polyareal (variable-sized plots) plot sampling on a given forest population. Evidently the first published use of the technique was by Hirata (1955) who combined vertical point sampling and circular plot sampling to estimate quadratic mean tree height. Subsequent treatment by Grosenbaugh (1958) briefly noted how other mean per-tree estimates could be made, and based on Grosenbaugh's work, Beers (1967) showed the development of the formulas to estimate quadratic and arithmetic mean tree diameter.

Combination sampling can be readily understood if one first notes that for any given forest tract the ratio

$$\frac{\text{Basal area per acre}}{\text{Number of trees per acre}}$$

represents the arithmetic mean basal area *per tree* on that tract. If horizontal point sampling is used to estimate both the numerator and denominator, tree diameters must be measured (or estimated) since the denominator requires the use of a tree factor (trees per acre represented by each tree tallied) which varies with tree diameter. On the other hand, if one uses combination sampling and estimates the numerator by horizontal point sampling and the denominator by fixed-size plot sampling, no tree measurements need be taken. To show this, assume that m_v trees have been tallied on the point sample inventory at n locations, and m_f trees have been tallied on the fixed-size circular plots at the same n locations; then an estimate of the arithmetic mean basal area per tree $\bar{b}$ is obtained by the formula developed as follows.

$$\bar{b} = \frac{F m_v / n}{\dfrac{43560}{\pi R^2} m_f / n}$$

$$= \frac{\pi R^2 k^2}{4} \frac{m_v}{m_f} \tag{14-46}$$

where

$$F = \text{basal area factor} = 10{,}890 k^2$$
$$k = \text{horizontal point sampling gauge constant}$$
$$= 2 \sin \left(\frac{\text{Gauge angle}}{2} \right) = 0.0095827 \sqrt{F}$$
$$R = \text{circular plot radius in feet}$$

It is commonly recognized and easily shown that the diameter corresponding to the arithmetic mean tree basal area is actually the quadratic mean tree diameter $\bar{d}_q$. Thus, by converting $\bar{b}$ as calculated in equation 14-46 to a diameter, we have

$$\bar{d}_q = \sqrt{\frac{\pi R^2 k^2}{4} \frac{m_v}{m_f} \frac{576}{\pi}}$$

$$= 12 R k \sqrt{\frac{m_v}{m_f}} \tag{14-47}$$

Equation 14-47 can be simplified by judicious choice of plot radius and basal area factor to achieve what has been called the "neutral" situation. For example, if one uses a basal area factor F of 10.0, the gauge constant k is $0.030303 = 1/33$. By choosing a fixed-size plot radius of 33 feet, equation 14-47 reduces to

$$\bar{d}_q = 12 \sqrt{\frac{m_v}{m_f}} \tag{14-48}$$

and one two-way table can be prepared that is applicable to other "neutral combinations" of basal area factor and plot size found to be useful. An example of such a table can be found in Beers (1967) or Beers and Miller (1973).

In similar fashion, other forms of point and line sampling can be combined with circular or rectangular fixed-size plots to estimate arithmetic mean tree diameter and arithmetic and quadratic mean tree height. A summary of selected logical combinations and corresponding formulas are shown in Table 14-7.

Table 14-7
Appropriate Equations for Estimating Certain Per-Tree Characteristics Using Combination Sampling

Characteristic and Symbol	Units	Logical Sampling Combination	Estimating Equation for		
			U.S. Units	Metric Units	
Arithmetic mean basal area ($\bar{b}$)	Square feet or square meters	Horizontal point and circular plot	$\bar{b} = \dfrac{\pi R^2 k^2}{4}\rho$	$\bar{b}$ = same as U.S.	(14-49)
Quadratic mean diameter ($\bar{d}_q$)	Inches or centimeters	Horizontal point and circular plot	$\bar{d}_q = 12Rk\sqrt{\rho}$	$\bar{d}_q = 100Rk\sqrt{\rho}$	(14-50)
Arithmetic mean diameter ($\bar{d}$)	Inches or centimeters	Horizontal line and rectangular plot	$\bar{d} = 12wk\rho$	$\bar{d} = 100wk\rho$	(14-51)
Arithmetic mean diameter ($\bar{d}$)	Inches or centimeters	Horizontal line between 2 circular plots	$\bar{d} = \dfrac{\pi R^2 k}{5.5L}\rho$	$\bar{d} = \dfrac{5\pi R^2 k}{L}\rho$	(14-52)
Arithmetic mean height ($\bar{h}$)	Feet or meters	Vertical line and rectangular plot	$\bar{h} = wq\rho$	$\bar{h}$ = same as U.S.	(14-53)
Quadratic mean height ($\bar{h}_q$)	Feet or meters	Vertical point and circular plot	$\bar{h} = Rq\sqrt{\rho}$	$\bar{h}_q$ = same as U.S.	(14-54)

R = circular plot radius in feet or meters.

w = rectangular plot half-width in feet or meters.

L = line length in chains or number of 20-meter segments.

k = horizontal sampling gauge constant = $2 \sin\left(\dfrac{\text{horizontal angle}}{2}\right)$.

q = vertical sampling gauge constant = tan (vertical angle).

ρ = the ratio: $\dfrac{\text{tree count from point or line sample}}{\text{tree count from fixed-size plot sample}}$

14-4.2 Indirect Estimates

As considered in this section, indirect per-tree estimates are those that arise as secondary to the primary purpose of the inventory. They can be categorized as *per-acre ratios* and *sample-tree averaging* types.

Per-tree Estimates as Ratios of Per-Acre Estimates. At the completion of an inventory numerous per-acre estimates have been made, and it is often tempting and reasonable to make per-tree estimates from these. For example, consider the ratio

$$\frac{\text{Mean volume per acre}}{\text{Mean number of trees per acre}}$$

It is reasonable that this ratio represents the arithmetic mean volume per tree and as such provides an unbiased estimate of the corresponding parameter. However, with regard to the *trees actually in the sample,* what type of mean is it? The following will show this ratio to be, assuming *horizontal point sampling* was used in the inventory, the weighted arithmetic mean volume of the sample trees with weights being the reciprocal of tree basal area.

$$\frac{\text{Volume per acre estimate}}{\text{Number of trees per acre estimate}} = \frac{\sum\limits_{i=1}^{m} F_{v_i}/n}{\sum\limits_{i=1}^{m} F_{t_i}/n} \tag{14-55}$$

$$= \frac{V_1\left(\dfrac{F}{B_1}\right) + V_2\left(\dfrac{F}{B_2}\right) + \cdots + V_m\left(\dfrac{F}{B_m}\right)}{\dfrac{F}{B_1} + \dfrac{F}{B_2} + \cdots + \dfrac{F}{B_m}}$$

$$= \frac{V_1\left(\dfrac{1}{B_1}\right) + V_2\left(\dfrac{1}{B_2}\right) + \cdots + V_m\left(\dfrac{1}{B_m}\right)}{\dfrac{1}{B_1} + \dfrac{1}{B_2} + \cdots + \dfrac{1}{B_m}}$$

$$= \frac{\sum\limits_{i=1}^{m} w_i V_i}{\sum\limits_{i=1}^{m} w_i}$$

$$= \text{Weighted arithmetic mean tree volume with } w_i = \frac{1}{B_i}$$

where

$$F_{v_i} = \text{volume factor} = V_i \left(\frac{F}{B_i} \right)$$

$$F_{t_i} = \text{tree factor} = \frac{F}{B_i}$$

$F = $ basal area factor
$V_i = $ individual tree volume
$B_i = $ individual tree basal area
$m = $ number of trees in the sample
$n = $ number of points visited

That equation 14-55 is the appropriate one to estimate the arithmetic mean volume per tree was pointed out by Barrett (1969). From this result one can generalize with regard to horizontal point sampling that for any tree characteristic X whether it be volume, height, or whatever,

$$\frac{\text{Estimate of arithmetic mean}}{X \text{ per tree}} = \frac{X \text{ per acre}}{\text{Number of trees per acre}} \tag{14-56}$$

$$= \frac{\sum\limits_{i=1}^{m} w_i X_i}{\sum\limits_{i=1}^{m} w_i} \tag{14-57}$$

where

$$w_i = \frac{1}{B_i} \text{ for the } m \text{ sample trees}$$

Similar statements can be made about the other types of point and line sampling. Thus, equations 14-56 and 14-57 still hold, but the weights are different.

For horizontal line sampling,

$$w_i = \frac{1}{D_i}$$

For vertical line sampling,

$$w_i = \frac{1}{H_i}$$

And for vertical point sampling,

$$w_i = \frac{1}{H_i^2}$$

where D_i and H_i represent individual tree dbh and height of the sample trees.

Interesting special cases of these results have been noted; for example, assuming horizontal point sampling, the arithmetic mean basal area per tree is estimated by forming the ratio,

$$\frac{\text{Basal area per acre}}{\text{Number of trees per acre}}$$

which is shown below to be equivalent to the harmonic mean basal area of the m sample trees. Recalling that $w_i = \dfrac{1}{B_i}$, and applying equation 14-57,

$$\text{Estimate of arithmetic} \atop \text{mean tree basal area} = \frac{\sum\limits_{i=1}^{m} w_i B_i}{\sum w_i} \tag{14-58}$$

$$= \frac{\sum\limits_{i=1}^{m} \dfrac{1}{B_i} B_i}{\sum\limits_{i=1}^{m} \dfrac{1}{B_i}}$$

$$= \frac{m}{\sum\limits_{i=1}^{m} \dfrac{1}{B_i}} \tag{14-59}$$

$$= \text{Harmonic mean basal area}$$
$$\text{of the } m \text{ sample trees}$$

The presence of the harmonic mean in this regard has been noted several times in forestry literature; according to Buckingham (1969) initial credit is given to Hirata (1956) and Bitterlich (1957). In a more general sense, however, it has long been recognized (Davies and Crowder, 1933) that the harmonic mean for a characteristic X is equivalent to the weighted arithmetic mean X if weights are specified to be the reciprocal of X, which verifies (albeit in reverse fashion) the special case just described.

For the other types of point and line sampling, similar special cases of the harmonic mean exist. That is, for horizontal line sampling,

$$\frac{\text{Sum of diameters per acre}}{\text{Number of trees per acre}} = \frac{m}{\sum\limits^{m} \dfrac{1}{D_i}} \tag{14-60}$$

$$= \text{Harmonic mean diameter}$$
$$\text{of the sample trees}$$

For vertical line sampling,

$$\frac{\text{Sum of heights per acre}}{\text{Number of trees per acre}} = \frac{m}{\sum\limits_{}^{m} \dfrac{1}{H_i}} \tag{14-61}$$

$$= \text{Harmonic mean height of the sample trees}$$

For vertical point sampling,

$$\frac{\text{Sum of (heights)}^2 \text{ per acre}}{\text{Number of trees per acre}} = \frac{m}{\sum\limits_{}^{m} \dfrac{1}{H_i^2}} \tag{14-62}$$

$$= \text{Harmonic mean squared height of the sample trees}$$

Per-Tree Estimates by Averaging the Sample Trees. Eventually the practicing forester, on completion of a horizontal point sample inventory, will wonder, "What if I take the arithmetic mean volume, height, or whatever, of the trees selected for the sample?" Since horizontal point sampling selects trees with probability proportional to basal area, intuition might lead to the conclusion that one would obtain an estimate of the arithmetic mean volume, height, and so on, weighted by tree basal area. This is in fact the case. By definition a population composed of M trees has an arithmetic mean X per tree μ, weighted by tree basal area B_i, calculated by

$$\mu_{wtd} = \frac{\sum\limits_{i=1}^{M} X_i B_i}{\sum\limits_{i=1}^{M} B_i} \tag{14-63}$$

For simplicity both numerator and denominator can be assumed on an acre basis and, therefore, can be estimated using a horizontal point sample which obtains m trees at n points by

$$\overline{X}_{wtd} = \frac{\sum\limits_{i=1}^{m} (X_i B_i)\left(\dfrac{F}{B_i}\right) \Big/ n}{mF/n}$$

$$= \frac{\sum\limits_{i=1}^{m} X_i}{m} \tag{14-64}$$

$$= \text{Unweighted arithmetic mean } X \text{ of the } m \text{ sample trees}$$

where

$$\frac{F}{B_i} = \text{Tree factor}$$

$X_iB_i = $ tree characteristic (i.e., the product of X and B) estimated in the numerator

Thus, for horizontal point sampling, calculating the unweighted arithmetic mean X of the sample trees provides an estimate of the weighted arithmetic mean X in the population, with the weight equal to the individual tree basal area. As before, similar statements can be made for the other sampling systems with only the weights changing.

For horizontal line sampling,

$$w_i = D_i$$

For vertical line sampling,

$$w_i = H_i$$

And for vertical point sampling,

$$w_i = H_i^2$$

A useful special case of the above procedure occurs in horizontal point sampling if the characteristic X_i is defined to be tree height. Taking the arithmetic mean height of the sample trees has been shown by Kendall and Sayn-Wittgenstein (1959) and later by Beers and Miller (1973), to estimate what is called *Lorey's height,* the arithmetic average tree height weighted by tree basal area. Lorey's height is frequently used, in conjunction with a measure of some mean diameter, in the estimation of total volume on a stand basis.

14-5 HORIZONTAL SAMPLING FIELD TECHNIQUES

A well-planned inventory must include implicit or explicit guides for the procedures and contingencies encountered in the field. Otherwise the inventory will not lead to consistent and reliable estimates. Field practices for fixed-size plots or strips are generally simple and well understood by the practicing forester. But, for various reasons, the field application of point and line sampling is thought by some to be devious (or at least mysterious) and fraught with rules and stipulations. In any event, careful and consistent procedures are required. Ignorance of these procedures can lead to considerable bias. The following sections provide sufficient background to avoid such bias.

14-5.1 Proper Use of Gauges

Basic to the field application of horizontal point or line sampling is the concept that once the necessary factors or constants are developed there exists a precise method for determining whether a given tree qualifies as a sample tree. Therefore, the angle gauge should be thought of as a tool to speed up the decision on the majority of trees—those which are definitely "in" or "out." In fixed-size plot sampling an occasional tree must be checked for inclusion in the plot by laying off the plot radius or plot width. Similarly, in horizontal point or line sampling an occasional tree will be questionable and must be checked; the plot dimension (plot radius or strip width) is not fixed for all trees, as it is for fixed-size plot sampling. It depends on tree diameter. The skilled worker, therefore, must be thoroughly familiar with the angle gauge and the method of checking questionable trees.

The most common gauges used for horizontal sampling are the "stick-type," the Spiegel relaskop, and the wedge prism. The picture obtained by each is discussed in Section 14-3.4.

For the first two instruments, the angle generated has its vertex at the eye of the user; therefore, *the user* must stand over the sample point so that the angle vertex is positioned vertically above the point. Directions for the use of the Spiegel relaskop are provided by the manufacturer. Application of stick-type gauges requires little more than common sense.

The use of the optical wedge, or wedge prism, will be described in greater detail since this widely used angle gauge is often misused. Wedge prisms are manufactured in square, rectangular, or round shapes and have been mounted in various ways to facilitate field use. One of the most convenient mountings is described by Beers and Miller (1964) and is known as the *Purdue point sampling block*. Other mountings are described by Bruce (1955) and Gould (1957).

Once the sampling point or location is occupied by the observer, the prism must be properly positioned before each tree is sighted critically; otherwise, bias is likely to occur. Here are rules for this positioning.

1. Since the gauge angle originates within the prism, *the prism* must be held vertically above the sample point or line when sighting; in point sampling this implies that the observer (not the prism) moves in a small circle about the point, while in line sampling the observer stands on the side of the sample line away from the tree being sighted.

2. Each tree is sighted at breast height through and over the prism.

3. The line of sight should be perpendicular to the prism bisector (i.e., the plane bisecting the refracting angle of the prism).

4. The distance from the eye to the prism is immaterial, provided a clear picture is obtained. Ten inches is the normal viewing distance.

With the prism vertically above the designated sampling point, the prism bisector can be oriented perpendicular to the line of sight by a technique analogous to "Waving the rod" in surveying, to get the proper reading on a hand-held level rod. Such a procedure is possible since, if the prism is properly oriented at right angles to the line of sight,

    a. a rotation of the prism in the vertical plane perpendicular to the line of sight will *reduce* the amount of horizontal deflection of the tree (i.e., make the images overlap more)

b. a rotation of the prism by (1) "tipping" or (2) "swinging" will *increase* the amount of horizontal deflection of the tree (i.e., make the images overlap less)

The four positions of the prism (unrotated, rotated, tipped, and swung) are shown in Fig. 14-14.

Although devices have been suggested to keep the prism over the sample point and perpendicular to the line of sight (Hutchison, 1958), it is, in most cases, sufficiently accurate to hand-hold the prism over the sample point or line as described above.

A final note with regard to gauging in *horizontal line sampling*. Although the critical viewing of trees should be made along a line of sight perpendicular to the sample line segment, many trees can be tallied as "in" when viewed over the longer oblique line of sight, for if the tree gives an overlapped picture for the oblique distance it will also give an overlapped picture for the shorter, perpendicular distance. Thus, the observer in walking the sample line should be constantly viewing trees obliquely ahead, occasionally pausing longer to scrutinize (or possibly tape-check) a near borderline tree along the perpendicular line of sight.

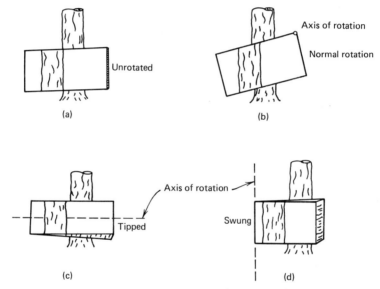

Fig. 14-14 *The effect of prism rotation on the amount of horizontal deflection. (a) Prism in unrotated position on a borderline tree. (b) Prism rotated in the vertical plane perpendicular to the line of sight—reduced horizontal deflection. (c) Prism tipped in the vertical plane parallel to the line of sight—increased horizontal deflection. (d) Prism swung in the horizontal plane—increased horizontal deflection.*

14-5.2 Trees Tallied from Two Points or Lines

Since point and line sampling are based on the concept of sampling with replacement, any tree in the forest population may qualify and be tallied from two or more sample points or lines. Provided the points or lines have been chosen in an unbiased manner, tallying a qualifying tree from two such locations will cause no bias in the subsequent estimates of mean stand parameters. In fact, *failure to do so* will result in biased estimates. If the occurrence of such double-tally trees is high, however, excessive oversampling is probably present and the inventory should be modified by either reducing the number of points (increase spacing if systematic sampling is used), increasing the gauge angle, or both.

14-5.3 Checking Questionable Trees

Any tree that is questionably "in" or "out" as determined by the angle gauge should be measured for dbh and the plot radius (or plot half-width) associated with the tree calculated from the relationship:

$$\text{Plot radius} = HDM \text{ (dbh)} \tag{14-65}$$

where *HDM* is the horizontal distance multiplier defined by the equations shown in Table 14-3. For example, if the basal area factor F is 10.0 (assuming U.S. units), then

$$HDM = \frac{33\sqrt{10}}{12\sqrt{F}} = \frac{33\sqrt{10}}{12\sqrt{10}} = 2.75$$

If the measured horizontal distance to the center of the questionable tree is greater than the calculated plot radius, the tree is "out"; otherwise it is "in." A table showing plot radii by dbh classes is frequently prepared, but is not necessary if a hand calculator is available.

A more efficient method for checking questionable trees makes use of a specially calibrated tape as described by Beers and Miller (1964). Diameters are marked on the tape at the appropriate plot radius for the basal area factor being used, thus avoiding table reference or calculations. However, with the availability of programmable hand calculators, this refinement may be unnecessary. Indeed, as a general observation, other corrections, such as for slope and boundary overlap, can be streamlined by the application of programmable calculators.

14-5.4 Slope Correction

When the inventory is conducted on sloping terrain some adjustments in procedure must be made to avoid bias, since *horizontal* land area is implicit in most stand characteristics. Although it can be shown that the errors are negligible for slopes up to 10 percent, it is recommended that one correct for all slopes, regardless of magnitude. Three unbiased methods have been proposed which result in equivalent corrections to the pertinent average stand estimates (Barrett and Nevers, 1967). The methods differ considerably, however, in their application and, under certain circumstances, one method might be preferred over the others. The decision of which to use must be made by the inventory planner. Important aspects that affect this decision are presented in the following paragraphs. The three methods have been referred to as (1) the constant gauge angle technique, (2) the adjusted constant gauge angle technique, and (3) the variable gauge angle technique (Beers, 1969).

The *constant gauge angle technique* was first proposed by Grosenbaugh (1955) and appears particularly well suited to horizontal point sampling using fixed-length stick-type angle gauges. The gauge is used on the slope to determine the number of qualifying trees in the same manner as on horizontal ground. After the tally is completed, one slope reading is made perpendicular to the contour through the sample point with an attempt made to "average out" minor depressions and hummocks. A correction is made for this "prevailing slope" by adjusting the pertinent tree characteristics (or factors) at the point by a multiplier (the secant of the prevailing slope angle in degrees). For example, if $F = 10$ square feet and 8 trees are tallied at a point where the prevailing slope is 40 percent ($21°48'$), the slope multiplier would be 1.08 (the secant of $21°48'$), leading to basal area per acre $= 8(10)(1.08) = 86.4$ square feet. Similarly, if volume per acre were being estimated, each volume factor, or the sum of the volume factors, would be multiplied by 1.08, depending on whether or not volume was desired by diameter classes.

It is important to understand that if this correction technique is used, questionable trees should be tape-checked by establishing uncorrected plot radii *on the slope*, using the horizontal distance multiplier (equation 14-65) or by employing the calibrated tape on the slope, with no correction made for slope. Thus, the constant gauge angle technique when used with point sampling establishes concentric circles on the slope having the same slope area as the appropriate circles established on level terrain.

The *adjusted constant gauge angle technique*, which appears suitable for point sampling using a prism or an adjustable stick-type gauge, was proposed by Del Hodge (1965) and independently by Bouchon (1965). As in the previous correction technique, the observer first measures the "prevailing" slope angle. Then, if a prism is used, the prism is rotated to achieve the required correction. It has been shown that the proper amount of prism rotation is found by the equation

$$\beta(\text{in degrees}) = \text{arc cos } \sqrt{\cos S} \tag{14-66}$$

or

$$\beta(\text{in percent}) = 100 \tan\left(\text{arc cos } \sqrt{\cos\left(\text{arc tan } \frac{\alpha}{100}\right)}\right) \tag{14-67}$$

where

β = angle through which the prism is rotated
S = slope angle in degrees
α = slope angle in percent

In order to resolve the status of trees judged "questionable" after the appropriate gauge angle adjustment, one must obtain an adjusted plot radius (limiting distance). The appropriate formula is

$$R_s = \frac{(HDM)\ dbh}{\sqrt{\cos S}} \tag{14-68}$$

$$= \frac{(HDM)\ dbh}{\sqrt{\cos\left(\arctan\dfrac{\alpha}{100}\right)}} \tag{14-69}$$

where

R_s = slope-measured plot radius
HDM = horizontal distance multiplier
dbh = diameter breast high of questionable tree
S = prevailing slope angle in degrees
α = prevailing slope angle in percent

The adjusted constant gauge angle technique does not employ a multiplicative correction as required in the previously described technique but, instead, adjusts for slope by reducing the gauge angle so that occasionally a tree is tallied that would not have qualified had the angle not been reduced. The technique, in theory, establishes concentric circles on the slope whose horizontal projections (ellipses) have an area appropriate for the tree dbh and gauge angle being used in the inventory.

The *variable gauge angle technique* differs from the previous two methods in that the slope to each tree, rather than one prevailing slope, influences the adjustment. Consequently, certain trees (on the same contour as the sample point) will require no adjustment in gauge angle, while others might require considerable adjustment. This technique is quite efficient for either point or line sampling and, in fact, is automatically achieved in such internal adjusting gauges as the Spiegel relaskop. It is very conveniently applied using a prism, but rather awkward when used with a stick-type gauge.

In the case of the prism, in the article which introduced the wedge prism to U.S. foresters, Bruce (1955) noted that by rotating the prism in the plane perpendicular to the line of sight through an angle equal to the slope of the line of sight the proper slope adjustment will be made.[2] Thus, although sighting at a tree over a slope distance, a simulated horizontal-distance picture will be achieved, and the tree can be evaluated as to "in" or "out." Subsequent proof that such rotation does, in fact, very closely achieve the required correction has been given by Grosenbaugh (1963b) and by Beers and Miller (1965).

The method by which the observer measures the angle and effects the prism rotation

---

[2] According to Grosenbaugh (1955), "Müller (1953) first visualized the advantage of using a wedge-prism as an angle-gauge [but] Bruce independently conceived the same idea later, and improved on it by a novel prism-rotation which automatically compensated for slope."

is a matter of individual preference. Although automatic rotation devices have been described by Bruce (1955) and Gould (1957), and undoubtedly by others, a commonly advocated procedure makes use of a flat base prism and an instrument such as the Abney level of the Suunto clinometer. By this method the slope is first determined, then the prism is rested on the clinometer and the combined unit is rotated properly through the angle equivalent to that of the slope. A similar procedure using a circular wedge prism, mounting block, and rotation scale has been described by Beers and Miller (1964). But regardless of the procedure used, the prism rotation practice must be considered rough, for in addition to the more obvious errors, if a single prism is used, a vertical displacement of the tree image results from prism rotation.[3] Therefore, the "over the prism" dbh is being compared with a "through the prism" image of a tree diameter higher than or lower than breast height, depending on the direction of prism rotation. For further detail on this phenomenon and its elimination, the reader is referred to Bruce (1955) and Beers and Miller (1964, 1966).

The point to be stressed is that by prism rotation the observer can roughly correct for the sloping line of sight, but one should tape-check trees that are close to the "questionable" status in the after-rotation picture. An efficient method of doing this is described in the following paragraphs.

Assuming horizontal point sampling is being used on sloping ground, the application of the variable gauge angle slope correction technique can be summarized as follows.

1. If a tree appears "in" or "borderline" with the unrotated prism, it can be tallied as "in."
2. If the sighted tree appears slightly "out," then
   a. Measure the angle of slope from eye level to tree breast height.
   b. Rotate the prism in the vertical plane perpendicular to the line of sight through an equivalent angle.
   c. Make a decision to tally or not to tally the tree on the basis of the corrected picture.
3. To resolve a questionable picture obtained as a result of 2c, the cruiser should
   a. Measure the tree dbh.
   b. If an ordinary distance tape is used, calculate (or find in a table) the slope-measured radius for that tree from the equation

$$R_s = (dbh)(HDM)(\sec S) \qquad (14\text{-}70)$$

$$= (dbh)(HDM)\sqrt{1 + \left(\frac{\alpha}{100}\right)^2} \qquad (14\text{-}71)$$

---

[3] The reader can verify this by sighting on a tree that has the breast high point clearly marked with a horizontal line. Prism rotation, then, will demonstrate the vertical separation of the actual and refracted image of the horizontal line.

where

R_s = slope-measured plot radius
HDM = horizontal distance multiplier
S = slope angle in degrees
α = slope angle in percent

    c. Stretch the tape along the slope from the sample point to the tree in question and determine its status with regard to the plot radius R_s.

The establishment of the proper plot radius described in 3b and 3c can be considerably streamlined if a calibrated point sampling distance tape (referred to in Section 14-5.3) has been prepared. That is, instead of step 3b, the "tape mark to hold" to establish the slope-measured plot radius can be found from the relationship

$$\text{Tape mark to hold} = (\text{tree dbh})(\sec S) \tag{14-72}$$

$$= (\text{tree dbh}) \sqrt{1 + \left(\frac{\alpha}{100}\right)^2} \tag{14-73}$$

where S and α are as previously defined. These formulas are appropriate for any gauge angle and, therefore, for all properly calibrated tapes.

The variable gauge angle technique theoretically establishes, on the slope, concentric ellipses whose individual horizontal projection is a circle having the appropriate area for each tree dbh. The slope adjustment is achieved by tallying an occasional tree that would not have qualified had the gauge angle not been reduced. The reduction of the gauge angle is accomplished either by wedge prism rotation or by the use of an internal slope-correction instrument such as the Spiegel relaskop.

If horizontal line sampling is being used, the variable gauge angle correction method as described is sound and applicable. However, to practically eliminate individual tree slope correction, the sampling line segment can be arbitrarily located perpendicular to the contours. In this way, a slope reading is necessary to determine the horizontal line length (or to "slip" the chain to achieve a fixed line length), but the lines of sight to all trees (when critically examined) will be along the contour and therefore essentially horizontal.

14-5.5 Gauging Leaning and Hidden Trees

Moderate tree lean has little effect on the field procedure. If, however, a tree leans severely to the right or left of the observer and appears to be "just in," a corrected gauge picture should be obtained. This is accomplished by rotating the angle gauge so that the vertical axis of the gauge parallels the axis of the leaning tree. Thus, if a prism is used, the sides of the prism (or prism block) should parallel the sides of the tree; if a stick-type gauge is used, the sides of the blade should parallel the tree sides; if a Spiegel relaskop is used, the sides of the instrument should parallel the tree sides.

A leaning tree that appears "out" with the unrotated gauge need not be checked, since the adjustment described above will lead to a further separation of the actual and refracted image. When a leaning tree is viewed with an unrotated prism, the effective gauge angle is reduced, and rotation of the prism is necessary to obtain the full angle.

On questionable trees and those that lean severely toward or away from the observer, a tape-check as described earlier should be made. When this is done the tree center is commonly assumed to be a point vertically above the center of the tree cross section at the ground line.

When tree lean is encountered in combination with sloping terrain, the following points should be considered.

1. If a prism is used in conjunction with the constant gauge angle slope correction technique, it should be rotated as described for leaning trees and the slope multiplier correction applied.

2. If a prism is used with the variable gauge angle slope correction technique, it should first be rotated to correct for the sloping line of sight, then rotated as described for leaning trees. Having the prism mounted on a block with a slope scale greatly simplifies this combined correction.

3. If a Spiegel relaskop is used, the slope is first adjusted for, the scales are kept locked in that position, and the entire instrument is rotated as described for leaning trees.

In dense or brush-choked stands, precautions are necessary to avoid missing or double-counting trees. Trees hidden behind other trees can be detected by swaying from side to side after each obvious tree is examined. No error is made by viewing these trees from some other point as long as the distance from the tree to the observer remains the same as the original distance to the sampling point.

Moving from the sample point is also warranted if trees are close enough together so two or more stems are superimposed while being viewed with the angle gauge.

When the breast-high point is masked by limbs or underbrush, one can check a visible point higher on the stem for qualification. If a tree qualifies for inclusion at some point above breast height it will qualify at breast height, unless the tree leans toward the observer and the distance is critical.

In any of the above situations a tape-check should be made if a tree is questionable.

14-5.6 Correcting for Boundary Overlap

In fixed-size plot sampling, if any part of a sample plot overlaps a boundary, biased estimates will result from that plot unless the area of the overlap is taken into account, or other suitable alteration in procedure is made. An analogous situation exists in point or line sampling where the plot associated with certain trees overlaps the boundary. This boundary overlap is a form of bias that can occur when any part of a tree's associated plot extends beyond the boundary of the forest tract (or forest type) being sampled. In such a situation the probability of the sample point falling within this associated plot is reduced, which, if not corrected for, leads to a certain amount of bias. By referring to the fundamental discussions on boundary overlap (Grosenbaugh,

1958, Haga and Maezawa, 1959, and Barrett, 1964), we can make the following observations: (1) bias created by overlap is negligible on large forest areas having a small proportion of edge; (2) bias is negligible if the forest outside the boundary is identical to that within the boundary; and (3) bias can be considerable on small areas, especially on long, narrow tracts, where the proportion of edge is great. Barrett and Allen (1966) in sampling a 4.84-acre woodlot for basal area observed a negative 5 percent bias using horizontal point sampling and a negative 3 percent bias using horizontal line sampling. Ashley and Beers (1970), using horizontal point sampling, found underestimates averaging 7 percent for the 20-acre woodland sampled.

Only the correction for overlap occurring in horizontal point sampling has received significant attention in forestry literature; therefore, only this case will be discussed here. Similar approaches could be used, however, for vertical point sampling, and analogous principles could be applied to horizontal and vertical line sampling.

Probably the most common way of avoiding overlap is to restrain the sample point from falling in a zone along the periphery of the tract; the width of the zone is slightly greater than the radius of the plot associated with the largest tree expected. However, by avoiding overlap bias in this way an edge-effect bias is probably being introduced, since the peripheral zone is not being sampled at all. In situations where the forest tract is large and sampling intensity is low, or where forest conditions are similar within and without the boundary, the edge-effect bias thus induced is usually of no practical significance.

Where the tract area is small, or where very precise results are needed, or where edge-effect bias is thought to be considerable, boundary bias should be corrected for. Numerous methods have been proposed to correct for boundary overlap, but the most elegant is the *mirage method* developed by Schmid (1969) and later described by Beers (1977).

The mirage method can be used for fixed-size plots as well as for point sampling and is recommended in all cases where the region beyond the boundary is accessible by the inventory personnel. To apply the mirage method in horizontal point sampling, one should follow these guidelines.

1. Determine if overlap exists. This is the case when the plot radius R_i associated with any qualifying tree exceeds the shortest (i.e., perpendicular) distance B *from the sample point to the boundary*. This can readily be checked for the qualifying tree of largest diameter D_i. Thus, if $R_i > B$, overlap exists, where $R_i = (HDM)D_i$, and

$$HDM = \text{horizontal distance multiplier}$$

$$= \frac{33\sqrt{10}}{12\sqrt{F}} \text{ (U.S. units)}$$

or

$$HDM = \frac{1}{2\sqrt{F}} \text{ (metric units)}$$

2. If overlap exists, the gauge user moves *B* units across the boundary and establishes a temporary sighting point.
3. Trees qualifying from this auxiliary point are again tallied. Thus, some trees may be tallied twice, which represents the adjustment for the original plots being undersized.

In cases where the area beyond the boundary is inaccessible another method must be used. Here the *direct weighting procedure* (Beers, 1966) appears to be the most practical.

14-6 VERTICAL SAMPLING FIELD TECHNIQUES

Both vertical point and vertical line sampling are relatively undeveloped. Although described by Hirata (1955), Strand (1957), and Grosenbaugh (1958) at a time when horizontal point sampling was still in its infancy, the application of vertical sampling has been rapidly surpassed by its horizontal companion. It is logical to assume that the introduction of the simple wedge prism for horizontal point sampling provided the impetus for its wholesale application. But a more valid reason is that tree height is much more difficult to assess with an angle gauge than is tree dbh. Thus, it is likely that vertical point and line sampling will remain little more than techniques for specialized inventories.

The most likely application of vertical sampling is in the assessment of regeneration, since the tree height being assessed can, in the near borderline situation, be measured directly with a calibrated pole. Vertical line sampling in this application has been described by Beers and Miller (1976), and a specific case using metric units is presented by Eichenberger, Parker, and Beers (1982). More detailed aspects of vertical sampling field procedure can also be found elsewhere (Beers and Miller, 1973).

14-7 PREPARATION OF SPECIAL TALLY FORMS

The efficiency of an inventory can be improved considerably by the thoughtful preparation of specific tally forms. This is particularly the case for small inventories when the data must be summarized using nonprogrammable calculators or done strictly by hand. A significant amount of time is saved by most tally forms because the data are sorted into diameter, height, and species classes as they are recorded in the field. By combining this advantage with the factor concept described in Section 14-3.3 and by giving some concentrated thought to the appropriate p.p.s. system to use (Section 14-3.8), very efficient tally forms can be prepared. Unfortunately, further treatment is not within the scope of this book. The reader is referred to Beers and Miller (1964 and 1973).

15
Growth of the Tree

Tree growth consists of elongation and thickening of roots, stems, and branches. Growth causes trees to change in weight and volume (size) and in form (shape). Linear growth of all parts of a tree results from the activities of the primary meristem; diameter growth from the activities of the secondary meristem, or cambium, which produces new wood and bark between the old wood and bark. Total and merchantable height growth, diameter growth at breast height, and diameter growth at points up the stem are elements of tree growth most commonly measured by mensurationists; from these elements, volume or weight growth of sections of the stem, or of the entire stem, may be determined. Root and branch growth is also measured in certain cases.

Tree growth is influenced by the genetic capabilities of a species interacting with the environment. Environmental influences include *climatic factors* (air temperature, precipitation, wind, and insolation); *soil factors* (physical and chemical characteristics, moisture, and microorganisms); *topographic characteristics* (slope, elevation, and aspect); and competition (influence of other trees, lesser vegetation, and animals). The sum of all these environmental factors is expressed as site quality, although competition is of less importance than the other factors since it is transient and can be changed by silvicultural treatments. When a site has favorable growing conditions, it is considered "good." When a site has unfavorable growing conditions, it is considered "poor." Because the environmental conditions favorable for one tree species may be unfavorable for another tree species, site quality must be considered by individual species. Extremes exist that provide absolute limits for all species, as exemplified by timber lines on mountains and in polar regions. But since site quality is an effect best expressed by the average reaction of the trees on an area of land, it is more readily measured for stands than for individual trees. Measures of site quality are therefore discussed in Chapter 17.

The increase in a tree (or stand) dimension should be qualified by the period of time during which the increment occurred. The period may be a day, a month, a year, a decade, and so on. When the period is a year, the increase, termed *current annual increment* (c.a.i.), is the difference between the dimensions measured at the beginning and at the end of the year's growth. Since it is difficult to measure some characteristics, such as volume, for a single year, the average annual growth for a period of years, termed *periodic annual increment* (p.a.i.), is often used in place of c.a.i. This is found by obtaining the difference between the dimensions measured at the beginning and at the end of the period, say 5 or 10 years, and dividing by the number of years in the period. If the difference is not divided by the number of years, it is termed *periodic increment*. The average annual increase to any age, termed *mean annual increment* (m.a.i.), is found by dividing the cumulative size by the age.

These increment measures are applicable to individual trees (or stands) for any measurable growth characteristic. However, they have been most commonly applied to the volume growth of stands.

15-1 GROWTH CURVES

When the size of an organism (e.g., volume, weight, diameter, or height for a tree) is plotted over its age, the curve so defined is commonly called the *growth curve* (Fig. 15-1a). Such curves, characteristically S- or sigmoid-shaped, show the cumulative size at any age. Thus, they are more descriptively termed *cumulative* growth curves. A true growth curve, which shows increment at any age, results from plotting increment over age (Fig. 15-1b).

The S-shaped form of the cumulative growth curve is evident for individual cells, tissues, and organs, and for individual plants and animals for the full life span. Also, the pattern of growth for short growing periods, such as a growing season, tends to follow the S-shaped curve.

Although the exact form of the cumulative growth curve will change when the tree dimension (height, diameter, basal area, volume, or weight) plotted over age is changed, the cumulative growth curve has characteristics that hold for all dimensions of a tree. With this in mind, an insight into tree growth can be obtained by studying Fig. 15-1a and the derived curves in Fig. 15-1b and Fig. 15-1c. During youth the growth rate increases rapidly to a maximum at the point of inflection in the cumulative growth curve, and the acceleration first increases and then drops to zero at the point of inflection in the cumulative growth curve. During maturity and senescence, the growth rate decreases with related changes in acceleration.

Curves of current annual increment, periodic annual increment, and mean annual increment may also be derived from a cumulative growth curve by computing increments from sizes read from the cumulative growth curve at chosen ages and by plotting the increments over age. Figure 15-2 shows curves of p.a.i. and m.a.i. that were derived from a cumulative height-growth curve. From Figure 15-2 we can see that m.a.i. culminates when it equals p.a.i. (this is also true when m.a.i. equals c.a.i.). A formal proof of this could be given, but the reason is obvious; m.a.i. will lag behind p.a.i. if it is smaller than p.a.i. When p.a.i. drops below m.a.i., m.a.i. must decrease; it, therefore, reaches its maximum when equal to p.a.i.

It should be noted that c.a.i. could be found analytically as the first derivative if cumulative growth is expressible as an equation.

Finally, in working with growth curves, one should realize that each species, perhaps each tree, dispenses a time of its own making. This physiological time varies from one tree species to another, and from one stage of development to another. In a sense, then, it is true that a 20-year-old aspen is older than a 20-year-old Douglas fir; the aspen develops more rapidly than the Douglas fir.

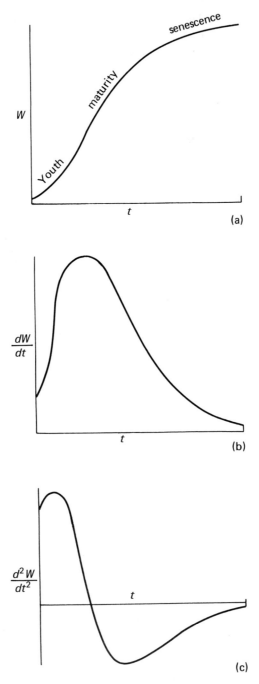

Fig. 15-1 *The curve of cumulative growth is shown in (a); growth rate in (b); and acceleration in (c). (W = size; t = age.)*

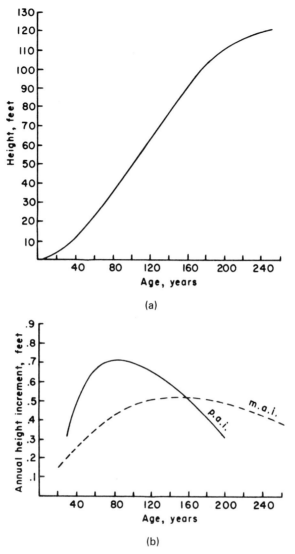

Fig. 15-2 *The cumulative height growth curve is shown in (a); the curves of periodic annual height growth (p.a.i.) and mean annual height growth (m.a.i.) are shown in (b); p.a.i. and m.a.i. curves derived from cumulative height growth curve.*

15-2 HEIGHT AND DIAMETER GROWTH

Characteristically, a cumulative growth curve of height over age for trees shows a juvenile period of less than a decade, a long maturing period when the trend is nearly linear, and a leveling off in old age. A cumulative growth curve of diameter over age shows much the same trend; there is, however, more of a tendency toward curvilinearity during the period of maturity. Since diameter is usually measured at breast height, dbh cannot be measured until a tree is over 4.5 feet tall. Consequently, since some of the early growth is before measurement begins, curves of dbh over age may not reflect some of the early growth.

Past height and diameter growth of individual trees may be determined (1) from repeated measurements of total size at the beginning and at the end of specified growing periods, and (2) from increment measurements of past growth.

15-2.1 Repeated Measurements

Height growth of an individual tree can be obtained by measuring the total (or merchantable) height of a standing tree at the initiation and at the cessation of a specified growing period and by taking the difference. Tree heights up to 75 feet can be accurately and precisely determined with telescopic measuring poles that can be purchased from forestry and engineering supply companies. However, for measuring heights over 30 feet, poles are slow and cumbersome; then it is more convenient to use a transit or some other precise tripod-mounted instrument that gives height indirectly. Repeated measurements of trees with hand-held hypsometers generally do not give sufficiently precise increment measurements.

Diameter growth of an individual tree can be obtained by measuring the diameter at the beginning and at the end of a specified period and by taking the difference. Since annual diameter increment is small, when instruments such as calipers and diameter tapes are used, measurements are frequently taken at intervals of several years. However, for short periods, even for a day, diameter growth can be obtained with more precise instruments (Chapter 2).

15-2.2 Increment Measurements

Increment measurements of past height growth can be made quickly if a reference point is marked on a tree, or on a pole by the tree. Past increment for intervals during a growing season, for a growing season, or for a specified period may be measured from this reference point. Also, past height increment may be determined by stem analysis (Section 15-4). And for species for which the internodal lengths on the stem indicate a year's growth, past height growth may be determined by measuring internodal lengths (Section 4-2).

In regions where tree growth has a seasonal or annual growth pattern, past diameter increment can be obtained from increment borings or cross-sectional cuts. Borings or cross sections can be secured at any point along the stem. However, diameter increment is most often determined at breast height from increment borings. But, when dealing with high-quality trees, it may be advisable to bore at stump height to eliminate damage in the butt log (Section 15-8). Section 4-2 discusses the taking of increment borings.

15-3   HEIGHT-DIAMETER CURVES

If two variables are each correlated with a common variable, they will appear to be correlated with each other. Since both tree height and tree diameter are correlated with age, height appears to be correlated with diameter. Such height curves, which are often plotted as free-hand curves, particularly for preparing local volume tables, appear to have a relation that can be expressed by a mathematical function. Since the curve form may vary from one forest stand to another, several functions have been developed.

1. $H = 4.5 + bD + cD^2$ (Trorey, 1932)
2. $H = 4.5 + h(1 - e^{-aD})$ (Meyer, 1940)
3. $\log H = a + b \log D$ (Stoffels and Van Soest, 1953)
4. $H = a + b \log D$ (Hendricksen, 1950)

where

H = total height in feet
D = dbh in inches
e = base of natural logarithms
a, b, c = constants

The parabolic equation $H = 4.5 + bD + cD^2$ can be used to describe the height-diameter relationships of many forest stands. However, if one desires to use a mathematical function to describe the height-diameter relationship for a particular stand, one should test to see which function is most applicable.

15-4   STEM ANALYSIS

A record of the past growth of a tree may be obtained by a stem analysis. Such a study shows how a tree grew in height and diameter and how it changed in form as it increased in size. In making a stem analysis, one counts and measures the growth rings on stem cross sections at different heights above the ground. Measurements may be taken on a standing tree by using an increment borer, if the tree is not too big or the wood too hard. It is more convenient and more accurate, however, to obtain the measurements from cut cross sections.

The procedure for making a stem analysis on cut cross sections is simple.

1. Fell tree and cut stem into sections of desired lengths.
2. Determine and record species, dbh, total height, years to attain stump height, and total age.
3. Measure and record the height of the stump, length of each section, and length of tip.
4. Measure and record average diameter at top of each section.

Table 15-1
Measurements for Stem Analysis of a 34-Year Old Red Oak

Species __Red Oak__   dbh __10.2 in.__   Total Height __54 ft__

Years to attain stump height __2__   Total Age __34__

Date __March 2, 1968__   Measured by __CIM__

Section No.	Length (feet)	Top dib (inches)	No. of Rings at Top	Distance Along Average Radius from Heart to Each 10th Ring (inches)[1]									
				1	2	3	4	5	6	7	8	9	10
1 stump	1	10.3	32	0.65	2.50	3.80	5.15						
2	16	8.7	24	1.30	3.00	4.35							
3	16	6.0	17	1.65	3.00								
4	14	2.7	9	1.35									
5	7	0	0										

[1] Double these values to give average diameters when plotting taper curves.

5. Find an average radius on each cross section and draw a line along it with a soft pencil.

6. Along each average radius count the annual rings from the cambium inward, marking the beginning of each tenth ring, or other desired period. Record the total number of rings at each cross section.

7. From the center of each cross section, measure outward toward the cambium along the average radius, recording the distance from the center to each tenth ring, or to other desired ring count. The fractional part of a decade, or other desired period, will be measured and recorded first.

Table 15-1 shows how the measurements should be recorded, and Table 15-2 how the height measurements should be summarized. Note that lengths are normally

Table 15-2
Height Summary for Stem Analysis of a 34-Year Old Red Oak

	Section No.				
	1	2	3	4	5
Length (feet)	1	16	16	14	7
Height above ground of top of section (feet)	1	17	33	47	54
Ring count, top of section	32	24	17	9	0
Years to grow section	2	8	7	8	9
Years to attain height at top of section	2	10	17	25	34

recorded to the nearest foot, diameters are recorded to the nearest $\frac{1}{10}$ inch, and radii are recorded to the nearest $\frac{1}{20}$ inch.

In making the stem analysis (Fig. 15-3), the first step is to draw a curve of *height* above ground of section tops over years to attain height at section tops (i.e., *age*) from data in Table 15-2. (This curve appears on the left side of the graph.) Next, diameters for each section (i.e., double the radial measurements) are plotted for the appropriate height from data in Table 15-1 (e.g., the three radial measurements for section 2 are doubled and plotted at 17 feet). Finally, diameters representing the same ages are connected to form the taper curves for specific ages; the terminal position of each taper curve is estimated from the curve of height over age.

15-5   AREAL AND VOLUME GROWTH

Basal area and bole surface area growth (areal growth) and volume (or weight) growth may also be of interest. Although the cumulative growth curves for both areal and volume growth are typically S-shaped, the exact form of the curves is variable.

Basal area growth may be estimated from periodic measurements of dbh. Bole surface area growth may be estimated by calculating surface area and surface area growth from periodic measurements of stem diameters at predetermined intervals along the stem. And volume growth, the most important growth determination, may

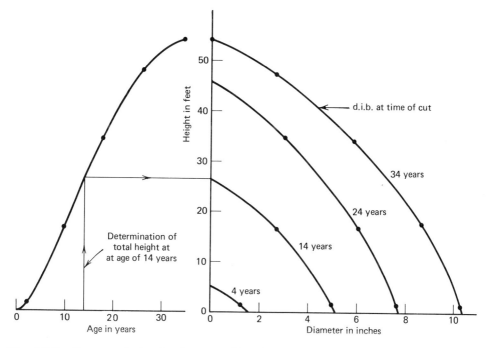

Fig. 15-3   *Stem analysis of a 34-year-old red oak; taper curves by 10-year intervals.*

be estimated by taking periodic measurements of dbh, dbh and height, or dbh, height, and form, and determining volumes at the beginning and at the end of a period from a local, standard, or form-class volume table, as appropriate, and then taking the difference. Of course, stem analysis may be used to obtain the required measurements and graphical methods to determine the volumes.

Baker (1960) has called attention to the following simple formula for rapid field computation of current annual volume increment of individual trees, which states

$$I = \frac{D \times H \times W}{100} = \frac{D \times H}{100 \times RI} \qquad (15\text{-}1)$$

where

I = current annual volume increment in cubic feet
D = diameter at breast height inside bark in inches
H = total height in feet
W = width of last ring at breast height in inches
RI = rings per inch (based on latest period of growth)

This formula is useful to determine the trees that are growing well and those that should be removed or freed from excessive competition.

Board feet or cords are not good measures of volume growth because of fundamental limitations (Chapter 7). Thus, growth estimates in board feet or cords are best made by converting cubic volume growth estimates into these units. And although weight growth has never received much attention, procedures for estimation of weight growth are not difficult. Weight and volume increments of individual trees are almost parallel in their patterns. Therefore, procedures for weight growth estimation are analogous to procedures for volume growth estimation.

15-6 EFFECTS OF ENVIRONMENTAL FACTORS ON GROWTH

The environmental factors that affect the growth of trees may be stable or transient. Stable factors—soil texture, slope, aspect, and soil nutrient level—do not change appreciably during the life of a tree. Transient factors—fluctuations in climate and competition among organisms—change cyclically or erratically during the life of a tree.

If past growth is used as a basis for growth predictions, it is important to recognize the magnitude of the effect of the transient factors. When stand competition is altered by cutting, growth will be affected; the changes for a given time are readily related to the cutting. When there are climatic variations, they also affect growth. These fluctuations, however, are not so recognizable and are not as easy to segregate from the total growth response. Yet, it may be necessary to estimate and evaluate the growth variation due to climatic changes. For example, if the effect on a tree or stand of a release cutting were under investigation, the periodic volume growth before and after the cutting could be used as the measure of the growth response, except that part of the response may be the result of variations in climate as well as a decrease in competition.

An adjustment can be made when it is necessary to eliminate the effect of climatic variations on growth. H. A. Meyer (1942) devised a procedure for adjusting diameter growth data to eliminate these effects. This procedure consists of plotting annual increments for individual trees, or averages for several trees, over time and fitting a trend line to the points. The deviations from the trend line for individual years can then be expressed as percentages of the trend line values and can be assumed to be related to changes in weather conditions. Thus, growth estimates for a period of years may be increased or decreased by the magnitude of the average percent deviation for the period. Since the variation in growth due to climate decreases as the length of the growth period increases, most growth measurements are made for periods of 5 years or more.

15-7 GROWTH PERCENTAGE

Growth percentage is a means of expressing the increment of any tree parameter in relation to the total size of the parameter at the initiation of growth. Although growth percentage is most frequently used for volume and basal area growth, it is applicable to any parameter.

In terms of simple interest, growth percent p_i is

$$p = \frac{s_n - s_0}{n s_0} (100) \tag{15-2}$$

where

s_0 = size of parameter at beginning of growth period
s_n = size of parameter at end of growth period
n = number of units of time in growth period

In this equation, average growth per unit of time is expressed as a percentage of the initial size s_0. To illustrate, if the present volume of a tree is 400 board feet and the volume 10 years ago was 300 board feet,

$$p = \frac{400 - 300}{10(300)} (100) = 3.3\%$$

In terms of compound interest, growth percent p is

$$p = \left(\sqrt[n]{\frac{s_n}{s_0}} - 1 \right) 100 \tag{15-3}$$

In this form, p may be computed by logarithms. But when compound interest tables are available, a more convenient form of the equation is

$$(1 + p)^n = \frac{s_n}{s_0} \tag{15-4}$$

The compound interest rate for the previously mentioned tree is then

$$(1 + p)^n = \frac{400}{300} = 1.333$$

$$p = 2.9\%$$

The compound interest rate is based on the premise that the increment for each unit of time is accumulated, resulting in an increasing value of s_0. Thus, as the period increases the simple and compound interest rates will diverge more and more. For short periods, however, they will be almost the same.

To avoid the use of compound interest tables, Pressler based a simple rate of interest on the average value for the period, $(s_n + s_0)/2$, which has the effect of reducing the rate to near the compound interest rate. Pressler's growth percent p_p is

$$p_p = \left(\frac{s_n - s_0}{s_n + s_0}\right)\frac{200}{n} \tag{15-5}$$

For the previous example

$$p_p = \left(\frac{400 - 300}{400 + 300}\right)\frac{200}{10} = 2.86\%$$

It is essential to remember that growth percentages are ratios between increment and initial size. Thus, percentages change as the amount of increment, and the base on which it accrued, changes. As trees grow the base of the percentage constantly increases, and the growth percentage declines even though the absolute increment may be constant or even increasing slightly. In early life the growth percentage for a tree is at its highest because the base of the ratio is small; the percentage falls as the size of the tree increases. Although young trees may grow at compound rates for limited periods, growth percent is generally an unsafe tool for predicting tree or stand growth, because of the uncertainty in extrapolating growth percent curves.

15-8 DETERMINATION OF GROWTH FROM INCREMENT CORES

Diameter growth is normally desired at breast height. And although most increment borings are made at breast height, diameter increment determined at stump height may be converted to diameter increment at breast height because the relationship between diameter at breast height d and diameter at stump height d_s for most species or species groups may be expressed by the simple linear regression equation

$$d = a + bd_s$$

Thus, diameter increment at breast height i may be obtained from diameter increment at stump height i_s as follows.

$$d + i = a + b(d_s + i_s)$$

and

$$i = a + b(d_s + i_s) - d$$

By substituting $a + bd_s$ for d,

$$i = a + b(d_s + i_s) - (a + bd_s)$$

And by simplifying,

$$i = bi_s$$

Therefore, if the diameter growth at stump height for an individual tree is multiplied by the slope coefficient b of d over d_s for the appropriate species or species group, one obtains diameter increment at breast height.

Increment borings should be taken from trees on sample plots or sample points. For example, if we used $\frac{1}{5}$-acre plots on a cruise, we might lay out $\frac{1}{20}$-acre plots within selected $\frac{1}{5}$-acre plots (say one in four) and bore trees on the $\frac{1}{20}$-acre plots. If we used horizontal point sampling ($F = 10$), we might use a 40-factor gauge at selected points to choose trees to bore. For each species or species group studied, a reliable estimate of average diameter increment by diameter classes can be obtained from a representative sample of about 100 increment measurements.

A word of warning: Boring one, two, or any predetermined number of trees per plot, rather than sampling, as explained in the previous paragraph, will result in an overestimation of growth because in open stands the sample will represent a larger proportion of the trees than in dense stands—trees in open stands grow faster than trees in dense stands.

A form for recording the field data is shown in Table 15-3. This table indicates the recommended degree of accuracy for measurements and calculations. Generally, only columns 1, 2, 3, 4, and 6 are completed in the field.

The constant K, the average ratio of diameter outside bark to diameter inside bark, is the inverse of the ratio used to calculate bark volume (Section 8-3). Thus, the equation of the straight line expressing diameter outside bark d as a function of diameter inside bark d_u is

$$d = Kd_u \tag{15-6}$$

K varies by species and, to some extent, by locality.

Average values of K are calculated as follows.

$$K = \frac{\Sigma d}{\Sigma d_u}$$

Thus, for the trees listed in Table 15-3, $K = 1.10$.

Table 15-3
Determination of Diameter Growth from Increment Cores

(1)	(2)	(3)	(4)	(5)	(6)	(7)	(8)	(9)	(10)
					Past 10 Years' Radial			Periodic	Periodic Annual dbh
Tree No.	Species	Present dbh o.b. (d)	Double Bark ($2b$)	Present dbh i.b. (d_u)	Wood Growth (L)	Past dbh i.b. (d_{up})	Past dbh o.b. (d_p)	dbh o.b. Increment (i)	o.b. Increment ($i/10$)
					(in inches)				
1	Hard maple	16.2	1.4	14.8	1.32	12.2	13.4	2.90	0.290
2	Hard maple	12.6	1.1	11.5	0.75	10.0	11.0	1.65	0.165
3	Hard maple	10.4	1.0	9.4	0.80	7.8	8.6	1.76	0.176
4	Hard maple	12.2	1.1	11.1	1.08	8.9	9.8	2.38	0.238
.		.	.	.	.	.	.	.	.
.		.	.	.	.	.	.	.	.
.		.	.	.	.	.	.	.	.
100	Hard maple	17.8	1.6	16.2	1.04	14.1	15.5	2.29	0.229
Totals:		1285.0	117.0	1168.0	98.70	970.6	1068.0	217.14	21.714
Averages:		12.85	1.17	11.68	0.987	9.71	10.68	2.171	0.2171

where:

$$d_u = d - 2b$$
$$d_{up} = d_u - 2L$$
$$d_p = K(d_{up})$$
$$i = K(2L)$$

Data are normally averaged by diameter classes. Averages for columns 5, 7, 8, 9, and 10 can be computed from the averages for columns 3, 4, and 6, and the constant K, without computing the individual values in the columns.

$$K = 1285/1168 = 1.10$$

The calculations for columns 5, 7, and 10 in Table 15-3 are self-explanatory. To calculate column 8—that is, past diameter at breast height outside bark d_p—past diameter at breast height inside bark d_{u_p} is multiplied by K (equation 15-6). To calculate column 9—that is, periodic diameter increment outside bark i—the past 10 years' radial wood growth L is doubled and multiplied by K. This is true because, if in a given period $2L$ inches of wood are laid on a given past diameter inside bark d_{u_p}, in terms of present diameter inside bark d_u, past diameter inside bark will be

$$d_{u_p} = d_u - 2L$$

and

$$2L = d_u - d_{u_p}$$

And, if in a given period i inches of wood and bark are laid on a given past diameter outside bark d_p, in terms of present diameter outside bark d, past diameter outside bark will be

$$d_p = d - i \tag{15-7}$$

and

$$
\begin{aligned}
i &= d - d_p \\
&= K(d_u) - K(d_{u_p}) \\
&= K(d_u - d_{u_p}) \\
&= K(2L)
\end{aligned}
\tag{15-8}
$$

When the data are presented as shown in Table 15-3, it is convenient to plot periodic growth or periodic annual diameter growth over present diameter outside bark, or over past diameter outside bark, and thus obtain average diameter growth by diameter classes. When the trend of the curve can be represented by a straight line, the curve can easily be fitted by the method of least squares. However, if the trend of the curve is not linear, Mawson (1982) demonstrated that the following non-linear model can be used to predict diameter growth G from diameter at breast height.

$$G = ae^{b/D} \tag{15-9}$$

where

a and b = constants
e = base of natural logarithms
D = dbh

The natural log transformation to linearize equation 15-9 is a useful alternative.

$$\ln G = \ln a + b\frac{1}{D} \tag{15-10}$$

When a mathematical function cannot be found, a free-hand curve can be fitted to the data.

Whether one plots diameter growth over present diameter outside bark or over past diameter outside bark, the final growth determinations will generally be about the same. However, the second alternative, which assumes that trees of a given diameter will have the same average diameter growth that trees of that diameter had in the past, is preferred to the first alternative, which assumes that future average diameter growth will equal past average diameter growth.

If data are available to predict the trend of diameter growth, average diameter growth may be raised or lowered to fit the trend. This, however, is usually an unnecessary refinement.

16
Stand Growth

The structure of a stand—that is, the distribution of trees by diameter classes—changes from year to year because of the growth, death, and cutting of trees. Thus, many problems of stand growth are best understood by considering a stand a population of trees, and by studying the changes in the structure of this population. For example, consider an even-aged stand for which two successive 100 percent inventories have been made (Fig. 16-1). If the periodic diameter growth of all trees was 2 inches, the periodic growth of this stand would be characterized by a displacement of the diameter distribution 2 inches to the right. The difference between the two inventory volumes represents the gross growth of the volume present at the first inventory. This is depicted in Fig. 16-1, if one omits ingrowth and disregards mortality and cut.

The importance of *ingrowth*, *mortality*, and *cut* in any expression of stand growth is illustrated in Fig. 16-1. Thus, before stand growth is considered, these important terms must be clearly defined. Although volume is the forest parameter stressed in the following definitions, the terms are equally appropriate if another characteristic such as basal area is considered.

Ingrowth is the number or volume of trees periodically growing into measurable size. (There will normally be ingrowth between any two successive inventories, particularly when measurements are made above a minimum diameter, such as 6 or 8 inches. The volume of the ingrowth may be 50 percent or more of total cubic volume growth and will be variable from one period to another.)

Mortality is the number or volume of trees periodically dying from natural causes such as old age, competition, insects, diseases, wind, and ice. (Mortality may be insignificant to catastrophic and may occur at any time during a growth period.)

Cut is the number or volume of trees periodically felled or salvaged, whether removed from the forest or not. (A cut may be light, medium, or heavy and may occur at any time during the period.)

16-1 TYPES OF STAND GROWTH

With the above definitions of ingrowth, mortality, and cut in mind, the generally accepted stand growth terms (Beers, 1962) can be defined by the following equations.

$$G_g = V_2 + M + C - I - V_1 \tag{16-1}$$

$$G_{g+i} = V_2 + M + C - V_1 \tag{16-2}$$

$$G_n = V_2 + C - I - V_1 \qquad\qquad (16\text{-}3)$$

$$G_{n+i} = V_2 + C - V_1 \qquad\qquad (16\text{-}4)$$

$$G_d = V_2 - V_1 \qquad\qquad (16\text{-}5)$$

where

G_g = gross growth of initial volume
G_{g+i} = gross growth including ingrowth
G_n = net growth of initial volume
G_{n+i} = net growth including ingrowth
G_d = net increase
V_1 = stand volume at beginning of growth period
V_2 = stand volume at end of growth period
M = mortality volume
C = cut volume
I = ingrowth volume

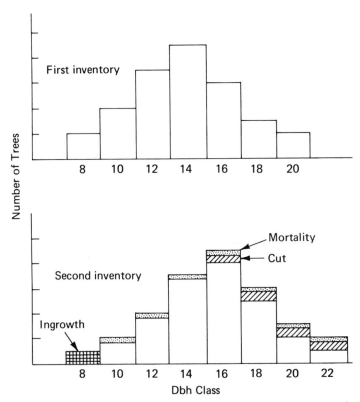

Fig. 16-1 *Schematic representation of the changes in stand structure of an even-aged stand due to growth over a 10-year period (Beers, 1962).*

It should be noted that in the above equations *mortality* and *cut* may be defined in two different ways.

1. *M* and *C* represent the volume of *M* and *C* trees at the time of their death or cutting.
2. *M* and *C* represent the volume of *M* and *C* trees at the time of the first inventory—that is, the initial volume of *M* and *C* trees.

The above growth terms are best considered in the context of repeated measurements of permanent sample plots or entire woodlands. Then it will be clear that the method of inventory generally dictates the most applicable definition of mortality and cut. For example, in inventory systems where the trees are not numbered, it is necessary to measure mortality trees at the time of the second inventory, which amounts to measuring them at the time of death, and to measure cut trees at the time of cutting. Under these conditions definition 1 would apply, and computations of growth that included *M* and *C* would then include growth put on by trees that died or were cut during the period between inventories.

If the inventory system utilizes numbered trees, as most continuous forest inventory (CFI) procedures do (Section 16-4), one can use the initial volume of cut and mortality trees and avoid measuring cut trees at the time of cutting. Under these conditions, definition 2 would apply, and computations of growth that included *M* and *C* would not include growth put on by trees that died or were cut during the period between inventories. Of course, if numbered trees are used, cut trees may be measured at the time of cutting and mortality trees at the second inventory, but this practice, which is seldom followed, requires extra care and "bookkeeping."

When gross growth of initial volume (equation 16-1) is computed using definition 2 for *M* and *C*, it includes only the growth on the trees that survived the period. Then it is often called *survivor growth*. When gross growth of initial volume is computed using definition 1 for *M* and *C*, it includes growth on trees that later died or were cut. Then it is often called *accretion*. (Note that growth on trees that later died or were cut may be an important component if cutting or mortality has been heavy, or if the interval between the inventories is long.) Marquis and Beers (1969) recommended that "the terms *survivor growth* and *accretion* be used where appropriate, and that *gross growth of initial volume* be considered a general term applicable only when either of these precise terms is not appropriate."

There is little advantage to using additional terms to qualify the growth terms expressed by equations 16-2 to 16-5. When these formulas are used, however, one must understand which definition is used for *M* and *C*.

16-1.1 Applications to Modern Continuous Forest Inventory

Equations 16-1 to 16-5 apply when tree volumes are first totaled and the resulting sums manipulated. But if we use the modern CFI approach (Section 16-4) and begin at the tree level, it is desirable to apply the following equations to calculate the various types of growth.

$$G_g = V_{s_2} - V_{s_1} \tag{16-6}$$

$$G_{g+i} = G_g + I \tag{16-7}$$

$$G_n = G_g - M \tag{16-8}$$

$$G_{n+i} = G_g + I - M \tag{16-9}$$

$$G_d = G_g + I - M - C \tag{16-10}$$

where

G_g, G_{g+i}, G_n, G_{n+i}, G_d = same variables as defined in Section 16-1
V_{s_1} = initial volume of survivor trees—that is, live trees measured at both inventories
V_{s_2} = final volume of survivor trees—that is, live trees measured at both inventories
M = initial volume of mortality trees
C = initial volume of cut trees
I = final volume of ingrowth trees

Since M and C represent the initial volume of mortality and cut trees, gross growth of initial volume G_g is correctly termed survivor growth. With this in mind, let us study Table 16-1 to find the differences between equations 16-1 to 16-5 and equations 16-6 to 16-10.

If the volume totals from Table 16-1 are used, the net growth including ingrowth is obtained from equation 16-4.

$$G_{n+i} = V_2 + C - V_1 = 749.3 + 241.4 - 744.4 = 246.3$$

Neither mortality volume nor ingrowth volume enters into the calculation. But if the growth per tree is first calculated, then the net growth including ingrowth is obtained from equation 16-9.

$$G_{n+i} = G_g + I - M = 273.8 + 34.6 - 62.1 = 246.3$$

Or in an approach that typifies the CFI procedure, G_{n+i} may be obtained by totaling the last column in Table 16-1. (Also note that G_g may be obtained by totaling the "survivor growth" column in Table 16-1.)

It should be clear that the other growth terms may be computed by the alternate equations, and that the same results will be obtained by consistent use of either equations 16-1 to 16-5 or equations 16-6 to 16-10.

Note that Table 16-1 uses the term *sound volume*. This is done to avoid the use of the terms *net* and *gross* when referring to tree or stand soundness—that is, amount of defect. This follows the recommendation of Meyer (1953) who suggested that tree or stand volume before defect deduction be termed *total* tree volume (rather than

Table 16-1
Table 16-1
Growth Data from a $\frac{1}{5}$-Acre
Permanent Sample Plot—Growth Period: 10 Years

Tree Number	Tree Status[1]	Sound Volume of						
		First Inventory	Second Inventory	Survivor Growth	Mortality	Cut	Ingrowth	Net Growth
		(board feet)						
1	20	62.1	—	—	62.1	—	—	−62.1
2	24	81.3	—	—	—	81.3	—	—
3	24	66.8	—	—	—	66.8	—	—
4	22	42.4	62.3	19.9	—	—	—	19.9
5	22	63.3	122.5	59.2	—	—	—	59.2
6	22	106.0	163.8	57.8	—	—	—	57.8
7	12	—	34.6	—	—	—	34.6	34.6
8	24	93.3	—	—	—	93.3	—	—
9	22	82.0	119.8	37.8	—	—	—	37.8
10	22	147.2	246.3	99.1	—	—	—	99.1
Plot totals		744.4	749.3	273.8	62.1	241.4	34.6	246.3
Symbol		V_1	V_2	G_g	M	C	I	G_{n+i}

[1] Tree status as used here defines the class of tree from a growth-contribution standpoint. Status at each inventory is coded as follows: 0 = not present, 1 = pulpwood size, 2 = sawlog size, 3 = cull, 4 = cut. By combining the tree classes at successive inventories, then 20 = sawlog mortality, 24 = sawlog cut, 22 = sawlog survivor tree, 12 = sawlog ingrowth from pulpwood size, and so on.

SOURCE: Beers, 1962.

gross), and that tree or stand volume after defect deduction be termed *sound* tree volume (rather than net). By employing this terminology, we can have gross and net stand growth in terms of total or sound stand volume.

16-2 ESTIMATION OF STAND GROWTH

Methods of estimating stand growth may be based on an analysis of a given stand from measured variables, or on the use of yield growth information that may be presented in tabular or equation form.

The important methods of estimating growth fall under four headings: *stand-table projection, total stand projection, yield tables,* and *derived growth and yield functions.*

16-2.1 Stand-Table Projection

To apply any stand-table projection method, the following data are needed.

1. Diameter growth information.
2. Present stand table.

3. Local volume table.

4. Information to calculate ingrowth.

5. Estimates of mortality.

Diameter growth information is most commonly obtained from increment borings (Section 15-8). However, excellent diameter growth information may be obtained from repeated measurements of permanent plots. In any case, there are three basic ways that diameter growth information may be applied to the present stand table, in conjunction with a local volume table, to obtain a growth estimate.

1. *Assume that all trees in each diameter class are located at the class midpoint, and that all trees will grow at the average rate.* Table 16-2 illustrates this approach.

- Column 2 is obtained as explained in Section 15-8.
- Column 3 = Column 1 + Column 2.
- Column 4 gives the local volume table values for the diameters given in Column 3. The values may be read from a curve of volume over tree diameter (Column 6 over Column 1), or calculated by an appropriate volume equation.
- Column 5 is obtained from inventory data.
- Column 6 gives the local volume table values for the diameters given in Column 1.
- Column 7 = Column 4 × Column 5.
- Column 8 = Column 6 × Column 5.
- Column 9 = Column 7 − Column 8.

Note that the sum of Column 9 equals the periodic gross growth of the initial volume. However, if the periodic diameter growth of the 8-inch diameter class had been over 2.00 inches, all trees in this class would have grown to measurable size, that is, to the 10-inch class, and would have been *ingrowth*. This, of course, would have increased volume production. Thus, results may be inconsistent if an attempt is made to include ingrowth. But when no attempt is made to determine ingrowth, the method gives good estimates of gross growth of the initial volume.

2. *Assume trees in each diameter class are evenly distributed through the class, and each tree will grow at the average rate.* Table 16-3 illustrates this approach. In this case a future stand table is predicted by first calculating the movement ratio M for each diameter class.

$$M = \frac{I}{C}$$

where

I = periodic diameter increment
C = diameter class interval in same units as I

Table 16-2
Calculation of 10-Year Predicted Volume Growth
Per Acre, Assuming That All Trees in Each Diameter Class Are Located
at the Class Midpoint, and That All Trees Will Grow at the Average Rate

(1)	(2)	(3)	(4)	(5)	(6)	(7)	(8)	(9)
Present dbh Class (inches)	10-Year dbh Incre- ment (inches)	Future dbh (inches)	Future Volume per Tree (cubic feet)	Present Stand Table (number)	Present Volume per Tree (cubic feet)	Future Stock Table (cubic feet)	Present Stock Table (cubic feet)	Volume Produc- tion (cubic feet)
6	2.02	8.02		41.73				
8	1.88	9.88		28.73				
10	1.74	11.74	17.0	21.73	12.5	369.4	271.6	97.8
12	1.60	13.60	24.2	17.33	18.4	419.4	318.9	100.5
14	1.46	15.46	31.9	12.87	25.6	410.6	329.5	81.1
16	1.32	17.32	40.7	9.47	34.2	385.4	323.9	61.5
18	1.18	19.18	50.1	8.27	44.1	414.3	364.7	49.6
20	1.04	21.04	62.3	5.00	55.6	311.5	278.0	33.5
22	0.90	22.90	75.3	3.47	68.5	261.3	237.7	23.6
24	0.76	24.76	89.8	2.87	83.5	257.7	239.6	18.1
26					100.1			
Total				151.47		2829.6	2363.9	465.7

Thus, the movement ratio for the 12-inch diameter class (Table 16-3) is

$$M = \frac{1.60}{2} = 0.80$$

The two digits to the right of the decimal point indicate the proportion of the trees in the class that will move one class more than indicated by the digit to the left of the decimal point. Therefore, for the 12-inch class, $0.80 \times 17.33 = 13.86$ trees move one class, and $0.20 \times 17.33 = 3.47$ trees move zero classes.

In Table 16-3 the movements for all classes are shown in Columns 7, 8, and 9, and the future stand table in Column 6. The arrows show how the trees are moved into the future stand. Columns 4 and 5 are the same as Columns 5 and 6 in Table 16-2. Columns 10, 11, and 12 are determined as follows.

- Column 10 = Column 6 × Column 5.
- Column 11 = Column 4 × Column 5.
- Column 12 = Column 10 − Column 11.

Note that the sum of Column 12 in Table 16-3 equals the periodic gross growth including ingrowth, and that ingrowth is 27.43 trees $(0.42 + 27.01)$, or 342.9 cubic feet (27.43×12.5). This is a reasonable estimate of ingrowth.

Table 16-3

Calculation of 10-Year Predicted Volume Growth Per Acre, Assuming Trees in Each Diameter Class Are Evenly Distributed Through the Class, and Each Tree Will Grow at the Average Rate

(1)	(2)	(3)	(4)	(5)	(6)	(7)	(8)	(9)	(10)	(11)	(12)
Dbh Class (inches)	10-Year dbh Increment (inches)	Movement Ratio (M)	Present Stand Table (number)	Volume per Tree (cubic feet)	Future Stand Table (number)	Number of Trees Moving 0 Classes	1 Class	2 Classes	Future Stock Table (cubic feet)	Present Stock Table (cubic feet)	Volume Production (cubic feet)
6	2.02	1.01	41.73				41.31	0.42			
8	1.88	0.94	28.73		43.03	1.72	27.01				
10	1.74	0.87	21.73	12.5	30.25	2.82	18.91		378.1	271.6	106.5
12	1.60	0.80	17.33	18.4	22.38	3.47	13.86		411.8	318.9	92.9
14	1.46	0.73	12.87	25.6	17.33	3.47	9.40		443.6	329.5	114.1
16	1.32	0.66	9.47	34.2	12.62	3.22	6.25		431.6	323.9	107.7
18	1.18	0.59	8.27	44.1	9.64	3.39	4.88		425.1	364.7	60.4
20	1.04	0.52	5.00	55.6	7.28	2.40	2.60		404.8	278.0	126.8
22	0.90	0.45	3.47	68.5	4.51	1.91	1.56		308.9	237.7	71.2
24	0.76	0.38	2.87	83.5	3.34	1.78	1.09		278.9	239.6	39.3
26				100.1	1.09				109.1		109.1
Total			151.47		151.47				3191.9	2363.9	828.0

3. *Recognize the actual position of trees in each diameter class and apply the diameter growth for individual trees in the class.* In this approach movement percentages are calculated by applying actual individual increments to individual tree diameters. A graphic solution (Wahlenberg, 1941) may be used, but a simple tabular solution is equally satisfactory and lends itself to electronic data processing. For example, in Table 16-4 tree movement percentages are computed for the 8-inch diameter class from the data used to compute Column 2 in Table 16-3. These percentages are applied to the present stand table to obtain number of trees moving. Except for this calculation, the future stand table and the future stock table are predicted as in Table 16-3.

As previously indicated, if the method depicted in Table 16-3 is used to predict growth, reasonable estimates of *ingrowth* may be determined by including, in the initial stand table, several diameter classes below the merchantable limit. If the method depicted in Table 16-2 is used, inconsistent estimates of ingrowth may be obtained by this procedure. In any case, estimates of ingrowth are unreliable for long prediction periods or for rapidly growing stands.

Mortality, which was not considered in the preceding examples (Tables 16-2 and 16-3), may be accounted for in one of the following ways.

1. By deducting predicted number of trees dying from each diameter class of the present stand table prior to projecting the present stand table.
2. By deducting predicted number of trees dying from each diameter class of the future stand table after projecting the present stand table, but before computing the future stock table.

In thrifty middle-aged stands, or in stands under intensive management, mortality will not be large and can be accurately predicted. In young stands and in old stands, mortality will often be great and, because of its erratic nature, cannot be accurately predicted. In any case allowances are made only for normal mortality resulting from old age, competition, insects, diseases, wind, and so on. No allowance is made for catastrophic mortality resulting from fire, epidemics, great storms, and so forth.

Good information on mortality may be obtained from permanent sample plots. From such information for any given stand one can determine functions relating mortality to age, diameter, stand density, and species, and thus apply the mortality information to other stands. But often when we desire to make a prediction by stand-table projection, we will lack suitable permanent sample plot data. Then, mortality estimates must be obtained from a stand inspection, which is normally made during the cruise. In the inspection, which is quite subjective, one estimates on plots or points the number of trees, by species and diameter classes, that died during some past period, say 10 years, or that will die during some future period. When the mortality information is summarized, it is expressed as percentages of the trees in each diameter class of the stand table.

A final word on stand-table projection: If accurate diameter growth information is used, any stand-table projection method will give an excellent estimate of gross growth

Table 16-4
Determination of Tree Movement
Percentages from Raw Data for 8-Inch Diameter Class

		Raw Data				Summary		
dbh Class (inches)	Present dbh (inches)	10-Year dbh Increment (inches)	Future dbh (inches)	Classes Move (number)		Classes Move (number)	Trees Moving (number)	Trees Moving (percent)
	7.1	1.5	8.6	0		0	3	30
	7.3	1.6	8.9	0		1	5	50
	7.4	1.5	8.9	0		2	2	20
	7.5	1.8	9.3	1		Total	10	100
8	7.9	2.5	10.4	1				
	8.1	1.6	9.7	1				
	8.3	1.8	10.1	1				
	8.5	2.6	11.1	2				
	8.7	1.7	10.4	1				
	8.9	2.2	11.1	2				

of the initial basal area. Of course, basal area growth is an important component of volume growth. But so is height growth. Therefore, the determination of gross growth of the initial volume also depends on the stability, during the prediction period, of the height-diameter relationship for which the local volume table was constructed. It also assumes no change in form.

It has been demonstrated that the future height-diameter relationship will not necessarily be the same as the present height-diameter relationship (Chapman and Meyer, 1949). For large areas and for uneven-aged stands this change may be slight, but for small areas and for even-aged stands the change may be substantial, even for periods of 10 to 20 years. With the exception of abnormal conditions, form changes may be safely ignored for short periods.

Thus, stand-table projection will give good results for uneven-aged stands of immature timber that are understocked. Then mortality will be small and predictable, ingrowth may be accurately predicted, and the height-diameter relationships will change only slightly. In even-aged stands, young dense stands, and overmature stands, stand-table projection will often give inaccurate results because of the change in the height-diameter relationships, and because of the high and unpredictable mortality.

Although considerable emphasis is given to the prediction of diameter growth in stand-table projections, and although diameter growth predictions are usually quite accurate, height growth and mortality predictions are often crude. Therein lies the weak link of the method. Consequently, when height growth and mortality predictions are in question, it is a waste of time to give great attention to diameter growth. In this case, a simpler system (Section 16-2.2) would save time and would give just as good results.

16-2.2 Total Stand Projection

A basic total stand projection method, termed the two-way method, was proposed by Spurr (1952). This method is based on the proposition that

$$\frac{V_f}{V_p} \cong \frac{B_f \cdot H_f \cdot F_f}{B_p \cdot H_p \cdot F_p}$$

and

$$V_f \cong V_p \left(\frac{B_f \cdot H_f}{B_p \cdot H_p} \right) \tag{16-11}$$

where

V_f = future average cubic foot volume per acre
B_f = future average basal area in square feet per acre
H_f = future average stand height in feet
F_f = future average stand form factor
V_p = present average cubic foot volume per acre
B_p = present average basal area in square feet per acre
H_p = present average stand height in feet
F_p = present averge stand form factor

Since the present stand form factor F_p will remain essentially unchanged for 10 to 20 years, the usual prediction period, F_p will be approximately equal to F_f. Therefore, the stand form factors cancel, leading to equation 16-11. And so cubic foot volume growth per acre I_V, which may be gross or net growth with or without ingrowth, depending on how B_f is determined, will be

$$I_V = V_f - V_p = V_p \left(\frac{B_f \cdot H_f}{B_p \cdot H_p} - 1 \right) \tag{16-12}$$

Present volume V_p may be determined by any desired inventory procedure. Present basal area B_p may be determined from diameter measurements of a sample of trees, but it is more efficient to use horizontal point sampling (Chapter 14). Again, present average stand height H_p may be determined from height measurements of a sample of trees, but it is more efficient to use a combination of vertical line and fixed-size plot sampling (Chapter 14), or Lorey's height using horizontal point sampling (Section 14-4.2).

The prediction of future basal area and future average stand height can be seen as the key to the accuracy of the two-way method. Since basal area growth and height growth are relatively independent, and since the variables used to estimate one are not normally used to estimate the other, each is best estimated separately. Thus, the term two-way method.

A convenient method of estimating basal area growth was developed by Spurr (1952). From information that may be collected during an inventory, gross growth of the initial basal area per acre I_B is estimated from the equation

$$I_B = \frac{\Sigma X^2(\Sigma d_u^2 - \Sigma d_{u_p}^2)}{183.3456(\Sigma d_u^2)} \tag{16-13}$$

where

X = diameter breast height in inches, o.b., for all trees tallied in stand

d_u = present diameter breast height in inches, i.b., for trees sampled for increment cores

d_{u_p} = past diameter breast height in inches, i.b., for trees sampled for increment cores

Future basal area is determined by adding basal area growth to the present basal area.

Future basal area may also be predicted by a method called *point-center extension* that utilizes increment cores and horizontal point sampling (Fender and Brock, 1963). At an appropriate number of points, a count is made of qualifying trees by the usual horizontal point-sampling procedures. Then one determines by increment cores the past periodic diameter growth of each nonqualifying merchantable tree that might grow enough to qualify at the end of the growth period. Assuming that each tree will grow the same amount in the future as it did in the past, one makes an examination with the angle gauge to determine the number of trees that will qualify at the end of the period. The check to determine the future status of any tree is made by moving a calculated distance closer to the tree and regauging; if the tree appears "in" after moving closer to the tree, it is counted as a *status changer*. Specifically,

$$\text{Distance to move toward tree in feet} = HDM(2LK) \tag{16-14}$$

Then, the future average basal area per acre in square feet B_f will be

$$B_f = (C + C_s)F \tag{16-15}$$

and

$$I_B = C_sF \tag{16-16}$$

where

HDM = horizontal distance multiplier for F value used (Table 14-3)

L = past periodic radial wood growth in inches

K = average ratio of diameter outside bark to diameter inside bark (Section 15-8)

C = present average tree count per point as determined by horizontal point sampling

C_s = average number of trees per point changing status as determined by point-center extension

F = basal area factor (Table 14-3)

Future stand height for even-aged stands is best predicted from site index curves. The present stand age and height can be used to determine site index, and the expected height at the future age can be read from the site index curves. Future stand height for uneven-aged stands can be predicted from measurements of height growth on permanent sample plots.

In fully stocked, even-aged stands past the juvenile growth period, total basal area per acre tends to remain constant, and so net basal area growth per acre will be approximately zero. Therefore, in such stands volume growth is largely associated with height growth, and the accuracy of prediction depends mainly on an accurate height growth determination.

In uneven-aged stands where the large trees are periodically cut or are periodically dying, the average height of the stand remains fairly constant. In fact, present and future average stand heights H_p and H_f may be considered approximately equal. This means that volume growth in such stands is largely associated with basal area growth, and that equation 16-12 may be simplified by canceling H_p and H_f. Thus,

$$I_V = V_p\left(\frac{B_f}{B_p} - 1\right) = \frac{V_p}{B_p}(B_f - B_p) = \left(\frac{V_p}{B_p}\right)I_B$$

And since $I_B = C_s F$,

$$I_V = \frac{V_p}{B_p}(C_s)(F) \qquad (16\text{-}17)$$

where terms are as defined for equations 16-11, 16-12, and 16-16.

Equation 16-17 exhibits a convenient growth prediction method that should not be overlooked when it is applicable.

We have now discussed stands in which growth is associated mainly with height, and stands in which growth is associated mainly with basal area. However, in partially stocked even-aged stands, and in irregular or unmanaged uneven-aged stands, volume growth may be associated with both height and basal area growth. Then care must be used to evaluate both height and basal area growth if the two-way method is used.

Finally, if one desires to determine net growth by the two-way method, mortality is deducted, in terms of basal area, from future basal area. Mortality estimates are made by the methods discussed in Section 16-2.1. If ingrowth estimates are desired, ingrowth is added, in terms of basal area, to the future basal area. Ingrowth estimates may also be made by the methods described in Section 16-2.1. However, the use of horizontal point sampling is more efficient. Then, each premerchantable tree that is "in" should be tallied as an ingrowth tree if

$$(d + 2LK) > (\text{Lower limit of smallest merchantable diameter class})$$

Then,

$$\text{Average basal area ingrowth per acre} = C_I F \qquad (16\text{-}18)$$

where

d = present diameter breast height in inches, o.b.

L, F, K = variables as defined for equations 16-14 and 16-15

C_I = average number of qualifying premerchantable trees per point that will become merchantable during growth period

16-2.3 Growth Prediction from Normal Even-Aged Yield Tables

To apply a normal yield table to determine growth, the site class and the density relative to the normal stand must be determined for individual stands that fall in different age classes, such as 10, 20, 30, or 40 years.[1] Age is determined as explained in Section 4-2. It should be mentioned that it is possible to estimate the volume of even-aged stands from normal yield tables (Section 17-4.1) if age, site index, and density relative to the normal stand are known. However, such volume estimates are seldom made in American practice, because they are not as reliable as estimates obtained from a forest inventory. But the volume estimates from normal yield tables may be compared with the results of a forest inventory, and if the two values show reasonable agreement, the yield tables may be used for prediction of future growth. Very simply,

$$I_V = p(Y_f - Y_p) \tag{16-19}$$

where

I_V = volume growth per acre

p = ratio of actual stand volume to normal stand volume (Section 17-2.1)

Y_f = future volume per acre from normal yield table

Y_p = present volume per acre from normal yield table

For periods of 10 to 20 years, a volume growth estimate by this method would give reasonable results. For longer periods, changes in stocking must be considered to avoid serious errors.

As a rule, the stocking of a young understocked stand tends to increase, or to approach the normal stand condition, while the stocking of an abnormally dense stand tends to decrease. The best bases for reliable predictions of changes in stocking are repeated observations of permanent plots. Studies based on permanent sample plots have been reported by Briegleb (1942) for Douglas fir, W. H. Meyer (1942) for Douglas fir and loblolly pine, and Watt (1950) for eastern white pine.

To illustrate the application of a normal yield table to determine growth when the trend to normal stocking is considered, assume a 10-year growth estimate is desired

---

[1] A *normal stand* is one in which all growing space is effectively occupied and yet has ample room for development of the crop trees. *Fully stocked stand* is a synonym for normal stand.

for a 70-year old Douglas fir stand with a site index of 160. From the yield table for Douglas fir (McArdle, Meyer, and Bruce, 1949), the normal volumes for trees 7 inches in diameter and over are

$$70 \text{ years } = 11,900 \text{ cubic feet}$$

$$80 \text{ years } = 13,360 \text{ cubic feet}$$

Further assume that a recent inventory has shown the average volume of the stand 7 inches in diameter and over to be 8920 cubic feet. The present stocking is thus $(8920/11,900)100 = 75$ percent of normal stocking. Assuming an increase in stocking of 4 percent per decade, the predicted stocking at age 80 will be $75 + 4$ or 79 percent of normal stocking. The estimated net growth I_V for the next 10 years is therefore

$$I_V = (13,360)0.79 - (11,900)0.75 = 1629 \text{ cubic feet}$$

16-2.4 Derived Growth and Yield Functions

Growth and yield functions may be developed for even-aged, uneven-aged, or age-indeterminable stands. Moser and Hall (1969) point out that it is preferable to present growth and yield information in formula form because of the widespread use of digital computers and of modern statistical techniques. Furthermore, modern innovations in forestry have created the need for dynamic growth and yield functions that can be revised periodically to incorporate results from new practices.

The ideal data for growth and yield functions are complete chronological records of entire stands from establishment to harvest, termed a *real growth series* (Turnbull, 1963). But in traditional practice, data has often been used from temporary plots covering a wide range of sites and ages, termed an *abstract growth series*. As a compromise between these two extremes, permanent plots may be established and remeasured at fixed intervals. This method is termed an *approximated real growth series*. The approximated real growth series will approach the real growth series.

The best independent variables to predict growth, whether it be volume, basal area, or height growth, should be selected by an objective statistical analysis in which the variables are related singly, and in combination, with the dependent growth variable.

Growth is a function of many interacting stand site factors that should be integrated over time. For even-aged stands the general form of the function at any given time might be

Growth = f(species, density, age, site quality)

Density might be expressed by basal area per unit area, number of trees per unit area, or stand density index. Site quality is usually expressed by site index.

For uneven-aged stands the general form of the function at any given time might be

Growth = f(species, density, site quality)

where neither the density nor the site quality measures are related to age.

Moser and Hall (1969) have given an appropriate methodology for deriving growth and yield models for uneven-aged stands. They point out that site index and soil characteristics have been used to express site quality in growth functions for uneven-aged stands with inconclusive results. Thus, until the effect of site can be quantified, it is best to construct separate yield functions for discrete site quality classes.

16-3 PAST GROWTH FROM REPEATED MEASUREMENTS

The determination of growth by repeated inventories of permanent sample plots, or of entire woodlands, is one of the most logical methods of determining past growth. The data obtained from a series of periodic measurements give a complete historical record of stand growth. These data can be used for growth prediction or to study the effects of cultural practices, insect attacks, weather, and other factors on growth.

The successful use of repeated measurements requires a well-designed scheme of taking measurements and of making calculations. These schemes generally fall under two headings: European *method of control* and American *continuous forest inventory system*. Both systems are designed to measure the growing stock at any given time and to give detailed estimates of growth. However, *continuous forest inventory* (CFI) is a more sophisticated management tool than the *method of control*. It provides a scientific study of individual trees and of their relationship to their immediate environment. It is an excellent channel for translating research into field practice.

One should realize that a crude estimate of net growth (as discussed in Section 16-1: $G_{n+i} = V_2 + C - V_1$) can be obtained by taking the difference between the volumes obtained on two successive, temporary sample plot inventories of an area, and adding the cut volume. Since the same sampling units are not measured in both inventories, however, the precision of the growth estimate will be somewhat poorer, and the accuracy of the growth estimate generally less, than when permanent sample plots are used. But if the permanent sample plots are not representative of the forest, the growth estimate may be biased, even though its precision may be high.

The European method of control was first presented by Gurnaud in France in 1878. The American continuous forest inventory system, which evolved from the method of control, had its genesis in 1934 (Stott, 1968). In the evolutionary process a basically different method of growth analysis has evolved: field grouping of data versus maintaining individual tree records.

Growth and stocking estimates are obtained in the method of control by grouping trees in the field by diameter classes and species, and by working up the data by simple techniques, not necessarily requiring the use of computers. Since the tree measurements are grouped, individual tree identity and location are lost. Thus, detailed summaries of growth and stocking, as obtained with CFI, are not possible.

Growth and stocking estimates are obtained in CFI by maintaining individual tree records and plot descriptions, and by making precise tree measurements in conformance to a detailed plan for computer analysis.

Although the method of control has been mainly used on woodlands where complete inventories are feasible, it is also applicable on large forests where permanent sample plots are used. And in some cases it would give the desired information at less cost than CFI. If on a large property, however, plots are small and numerous, it is more difficult to keep an accurate record of felled and dead trees with the method of control than with CFI.

16-3.1 Method of Control

The method of control depends on the following factors.

1. A well-defined procedure of measuring and remeasuring the diameters of standing trees.
2. Measuring and determining the volume of felled trees and mortality trees in the same manner as standing trees.
3. A simple method of determining ingrowth.
4. Use of permanent local volume tables (or tarifs).

Each time an inventory is taken, the diameter measurements must be made in the same manner. The point of measurement is designated by a 2-inch horizontal mark made at breast height with paint or a bark scribe. On remeasurement the old mark is renewed. If calipers are used, the graduated beam is placed on the tree touching the mark; only one measurement is normally made. But no matter what instrument is used, calipers or diameter tape, measurements are made at the height of the mark.

The diameters of cut trees and mortality trees are measured and recorded by diameter classes and species when the stand is marked for cutting, or if desired, the diameters of mortality trees may be measured at the time of an inventory. Cut and mortality trees together constitute *trees removed*.

One should note that it is possible to measure the volume of felled trees accurately and to compare this volume with the volume table. Then,

$$q = \frac{\text{Actual volume of felled trees}}{\text{Volume of felled trees from volume table}}$$

represents a correction factor that can be used for a better estimation of the volume of trees marked for cutting.

16-3.2 Determination of Diameter Increment by Diameter Classes Using Method of Control

An ingenious method, developed by French foresters, makes it possible to obtain diameter increment by diameter classes by the method of control. An example will illustrate the method (Table 16-5).

Since the calculations must begin with the highest diameter class, this class is put at the top of Table 16-5. To start, let us postpone discussion of Columns 2 and 3 and

Table 16-5

Calculation of Periodic Annual Diameter Increment by Diameter Classes by the Method of Control

(1) dbh Class (inches)	(2) Inventory Spring 1949 (number)	(3) Trees Removed (number)	(4) Inventory Spring 1949, Minus Trees Removed (number)	(5) Inventory Spring 1959 (number)	(6) Trees Rising (number)	(7) Double Rising (number)	(8) Double Effective (number)	(9) DR/DE	(10) Periodic Diameter Increment (inches)	(11) Periodic Annual Diameter Increment (inches)
32	0		0	1	0	1	1	1.000	2.00	0.200
30	2		2	4	1	4	6	0.667	1.33	0.133
28	3		3	0	3	3	3	1.000	2.00	0.200
26	3	2	1	9	0	8	10	0.800	1.60	0.160
24	10	1	9	7	8	14	16	0.875	1.75	0.175
22	10	4	6	21	6	27	27	1.000	2.00	0.200
20	23	4	19	39	21	62	58	1.069	2.14	0.214
18	37	7	30	72	41	124	102	1.216	2.43	0.243
16	73	5	68	169	83	267	237	1.127	2.25	0.225
14	194	17	177	234	184	425	411	1.034	2.07	0.207
12	249	9	240	379	241	621	619	1.003	2.01	0.201
10	418	15	403	507	380	864	910	0.949	1.90	0.190
Total	1022	64	958	1442	484	2420	2400			

$$\frac{\Sigma DR}{\Sigma DE} = 1.008$$

Average = 2.02   0.202

Number of trees ingrowth = 1442 − 958 = 484 (checks with last figure in Column 6). Trees were removed (Column 3) immediately following the 1949 inventory. Data from 143 permanent sample plots of $\frac{1}{5}$ acre located on Morgan-Monroe State Forest, Indiana. Sample area: 28.6 acres. Growth period: 10 years.

turn to Columns 4 and 5, realizing that Columns 4 and 5 include the corrected figures to be used to compute diameter increment. Consider Column 4 the first inventory figures, and Column 5 the second inventory figures. Now, in the 32-inch class, 0 trees were measured at the first inventory, and 1 tree was measured at the second inventory. Therefore, 1 tree rose from the 30- to the 32-inch class during the period between inventories. Since 1 tree moved out of the 30-inch class, this left 1 tree in the 30-inch class to be joined by 3 trees rising into the class, giving 4 trees at the second inventory. For a *specific class*,

$$\begin{pmatrix} \text{No. trees rising} \\ \text{into the class} \end{pmatrix} = \begin{pmatrix} \text{No. trees 2nd} \\ \text{inventory} \end{pmatrix} - \begin{pmatrix} \text{No. trees 1st} \\ \text{inventory} \end{pmatrix}$$
$$+ \begin{pmatrix} \text{No. trees rising} \\ \text{out of the class} \end{pmatrix}$$

Note in Table 16-5 that 484 trees rose into the 10-inch class. This represents the ingrowth and equals the difference between the totals of Columns 4 and 5.

Double rising, Column 7, is the sum of trees rising out of a class and into a class. Thus, for the 32-inch class, DR = 0 + 1 = 1; for the 30-inch class, DR = 1 + 3 = 4, and so forth. Double effective, Column 8, is the sum of trees in the first and second inventories after correction for trees removed. Thus, for the 32-inch class, DE = 0 + 1 = 1; for the 30-inch class, DE = 2 + 4 = 6, and so forth.

Column 9, DR/DE, is used to calculate the periodic diameter increment I given in Column 10.

$$I = \left(\frac{\text{DR}}{\text{DE}} \right) C \tag{16-20}$$

where C = width of diameter classes. Note that the average periodic diameter increment for all trees is obtained as follows.

$$\left(\frac{\Sigma \text{DR}}{\Sigma \text{DE}} \right) C$$

Periodic diameter increment is divided by the number of years in the period to obtain periodic annual diameter increment (Column 11). H. A. Meyer (1942) gives proof of this method, and Beers (1963) empirically substantiates it.

Diameter growth figures obtained in the above manner may be used in growth predictions in the same way as increment core data are used (Section 16-2.1). They are also valuable for silvicultural purposes.

16-3.3 Calculation of Diameter Increment in Presence of Cut and Mortality

In discussing Table 16-5 in Section 16-3.2, an explanation of how to handle trees removed by cutting or mortality was omitted. In the example, cutting, and a small amount of mortality, was assumed to take place immediately after the first inventory.

Thus, the trees removed were subtracted from the first inventory to obtain Column 4. If cutting had taken place immediately before the second inventory, we would have added the trees removed to the second inventory. If cutting had taken place halfway through the period, we would have subtracted one-half of the trees removed from the first inventory and added one-half of the trees removed to the second inventory. If cutting had taken place in the third year of a 10-year period, we would have subtracted 7/10 of the trees removed from the first inventory and added 3/10 of the trees removed to the second inventory. In formula form

$$\begin{pmatrix} \text{No. trees to subtract} \\ \text{from } i\text{th diameter class} \\ \text{of 1st inventory} \end{pmatrix} = \left(\frac{\begin{pmatrix} \text{No. years from date of} \\ \text{removal to 2nd inventory} \end{pmatrix}}{(\text{No. years in period})} \right) \begin{pmatrix} \text{No. trees removed in} \\ i\text{th diameter class} \end{pmatrix}$$

$$\begin{pmatrix} \text{No. trees to add to} \\ i\text{th diameter class of} \\ \text{2nd inventory} \end{pmatrix} = \left(\frac{\begin{pmatrix} \text{No. years from 1st inventory} \\ \text{to date of removal} \end{pmatrix}}{(\text{No. years in period})} \right) \begin{pmatrix} \text{No. trees removed in} \\ i\text{th diameter class} \end{pmatrix}$$

16-3.4 Determination of Volume Increment by Method of Control

Table 16-6 shows the calculation of volume growth by the method of control for the data used in Table 16-5. Once the columns are completed and totaled as shown in Table 16-6, the calculation of growth is simply a matter of applying the growth equations from Section 16-1. For example,

$$G_{g+i} = V_2 + (M + C) - V_1$$

$$= 136{,}988 + 8968 - 88{,}286 = 57{,}670$$

Ingrowth (No. trees) $= 1442 + 64 - 1022 = 484$

And since it is assumed that all ingrowth trees move into the smallest measurable class, in this case the 10-inch class,

$$I = 484(30) = 14{,}520 \text{ board feet}$$

A different analysis will provide information on volume growth by diameter classes. The procedure is shown in Table 16-7 for the data used in Table 16-5. Columns 2, 3, and 4 in Table 16-7 are determined exactly as Columns 4, 5, and 6 in Table 16-5. Column 6, volume difference per tree, is the difference between volumes for subsequent diameter classes; Column 7, volume increment, is the product of Columns 4 and 6. The total of Column 7 is accretion, G_g.

A careful study of Tables 16-6 and 16-7 points out the value of the method of control as a management tool. The system permits continuous analysis and examination of the makeup of a stand—the results of different cutting and cultural practices may become clearer at each inventory.

Table 16-6
Calculation of Volume Growth by the Method of Control

(1)	(2)	(3)	(4)	(5)	(6)	(7)	(8)
	Volume per Tree (board feet)	Inventory— Spring 1949		Inventory— Spring 1959		Cut and Mortality	
dbh Class (inches)		Trees (number)	Volume (board feet)	Trees (number)	Volume (board feet)	Trees (number)	Volume (board feet)
32	577	0		1	577		
30	538	2	1,076	4	2,152		
28	491	3	1,473	0			
26	442	3	1,326	9	3,978	2	884
24	386	10	3,860	7	2,702	1	386
22	329	10	3,290	21	6,909	4	1,316
20	273	23	6,279	39	10,647	4	1,092
18	217	37	8,029	72	15,624	7	1,519
16	161	73	11,753	169	27,209	5	805
14	112	194	21,728	234	26,208	17	1,904
12	68	249	16,932	379	25,772	9	612
10	30	418	12,540	507	15,210	15	450
Total		1,022	88,286	1,442	136,988	64	8,968

Type of Growth	Periodic (board feet)		Periodic Annual (board feet)	
	Total	Per Acre	Total	Per Acre
G_{g+i}	57,670	2,016	5,767	202
I	14,520	508	1,452	51
G_g	43,150	1,506	4,315	151
G_d	48,702	1,703	4,870	170

Symbols are defined in Section 16-1.

Total is for sample area of 28.6 acres.

Data from 143 permanent sample plots of ¼ acre located on Morgan-Monroe State Forest, Indiana. Sample area: 28.6 acres. Growth period: 10 years.

16-4 CONTINUOUS FOREST INVENTORY

To many foresters continuous forest inventory (CFI) implies making precise tree measurements on permanent plots in conformance to a detailed plan for computer analysis that requires individual tree and plot records. This concept, however, may be modified in various ways, but the plots always must represent the entire forest; cutting, mortality, growth, and so forth, on the plots should be representative of cutting, mortality, growth, etc., on the rest of the forest.

It should be understood that CFI will yield more than growth and volume information. It was designed as a means of periodically assessing change in the forest so

Table 16-7

Table 16-7
Calculation of Periodic Volume Growth
by Diameter Classes Using the Method of Control

(1)	(2)	(3)	(4)	(5)	(6)	(7)
dbh Class (inches)	Inventory Spring 1949, Minus Trees Removed (number)	Inventory Spring 1959 (number)	Trees Rising (number)	Volume per Tree (board feet)	Volume Difference per Tree (board feet)	Periodic Volume Increment (board feet)
32	0	1		577		
			1		39	39
30	2	4		538		
			3		47	141
28	3	0		491		
			0		49	—
26	1	9		442		
			8		56	448
24	9	7		386		
			6		57	342
22	6	21		329		
			21		56	1,176
20	19	39		273		
			41		56	2,296
18	30	72		217		
			83		56	4,648
16	68	169		161		
			184		49	9,016
14	177	234		112		
			241		44	10,604
12	240	379		68		
			380		38	14,440
10	403	507		30		
			484			
Total	958	1,442				43,150

Data from 143 permanent sample plots of ¼ acre located on Morgan-Monroe State Forest, Indiana. Sample area: 28.6 acres. Growth period: 10 years.

that management would be alerted to the need for policy changes for the forest. However, the system will not necessarily give all data needed for continuing forest management. Because of the low sampling intensity of most CFI systems, temporary samples may be needed to supplement the permanent plot data.

An efficient CFI system cannot be designed without a clear understanding of the purposes of the system. Some of the purposes include the following.

1. Provide volumes by species, tree grades, and size classes for the different timber types that are significant for differentiating products to be harvested.

2. Assess relative economic availability of trees and areas of varying qualities.

3. Provide growth information for the different timber types that will give a basis for calculating allowable cut.

4. Evaluate results of silvicultural practices, including planting, in terms of survival, quality, and growth of regeneration.

5. Evaluate need for timber stand improvement.

6. Provide a basis for determining areas to be planted, area of nonproductive land, and ratio of mortality to cut.

7. Provide values, volumes, and growth rates for depletion and other purposes in accounting for timberlands.

16-4.1 Planning Considerations for Continuous Forest Inventory

The use of stratification with optimum allocation is highly efficient for a single inventory. But since fire, wind damage, insect attacks, harvesting, and natural changes in species composition continually alter the strata, unstratified sampling is commonly employed in CFI to assess change.

Plots are commonly located in a square-grid pattern. However, random location of plots is feasible and is sometimes used. But in planning a CFI system the economy-minded forester might ask: If the total number of plot locations is reduced, and two or more plots are measured at each location instead of one plot, will the accuracy of results be reduced? Should all of the "permanent" plots be remeasured or should some be replaced by new ones? There are cases where such modifications might improve the efficiency of the CFI system; however, forest managers must answer these questions in light of their needs.

Circular plots of $\frac{1}{5}$ acre or $\frac{1}{4}$ acre are commonly used, although permanent points may also be used (Section 16-4.2). A decision on the number of plots to use is difficult, because the sample must be applicable now and in the future and must be large enough to estimate several forest parameters. But assuming a particular sampling system, and a desired degree of precision, the number of plots to use depends on the variability of the parameter being estimated and the area of the forest. In practice, sampling intensities range from about 0.03 to 0.1 percent. Within a given region, variability tends to increase as the size of the forest area increases, but not proportionately. Thus, the larger the area of the forest, the lower the sampling intensity needed for any desired precision.

Once a decision has been made on the number of plots to establish, and on the sampling system to use, plots can be efficiently located on a photo mosaic (or a map) by means of a transparent templet. To prepare the templet, a square-grid or random pattern of small holes is pricked into a transparent sheet. For example, when a 0.05 percent sample is to be used on a square-grid pattern, the spacing L in chains for $\frac{1}{5}$-acre plots would be

$$L = \sqrt{\frac{10a}{p/100}} = \sqrt{\frac{10(0.2)}{0.05/100}} = 63.25 \text{ chains (4174 feet)}$$

where

a = plot size in acres

p = percent sample

If the photo mosaic on which the plots are to be located has a scale of 1:15840, the spacing of the holes on the templet will be 3.16 inches. The completed templet is randomly placed on the mosaic so as to cover the desired ownership, and plot locations are pricked through holes that fall within the ownership. Normally, the plot locations are transferred to 9 by 9 inch contact prints that may be taken into the field.

The usual procedure for each plot is to determine, from the photograph, an azimuth and a distance to the plot center from a permanent feature, such as a survey corner, road intersection, or the like, that can be used to locate the plot for establishment and for remeasurement. This azimuth and distance is used by the field crew to locate the plot without bias. When a plot is first established, it is good practice to compare its photo and ground location, and to change the photo location if it does not agree with the ground location.

When a plot is established, its center is marked on the ground by an aluminum or treated-wood stake. The location of the stake should be carefully referenced by recording the azimuth and distance to two or more witness trees on the plot. All trees that are measured are numbered with paint or with metal tags and marked at the point of diameter breast height measurement with a short paint line.

Field data may be recorded on mark-sense cards, or prescored cards, and transferred to permanent punch cards by machine methods. Or field data may be recorded on field-tally sheets, or on tape, and transferred to cards by a key-punch operator. However, electronic data recorders that can communicate with a variety of computer hardware (Beltz and Keith, 1980) provide the most efficient system. The data are normally coded, even when on tape or in recorder memory.

A crew of two workers is satisfactory for locating, measuring, and remeasuring CFI plots. And for every four or five crews there should be a checking crew to verify the quality of work and to correct deviations from standard procedures. Approximately 10 percent of the plots should be randomly selected for checking.

Undiscovered errors can disrupt procedures and increase the costs of computing and compiling the entire inventory; the data-processing program demands error-free records. Therefore, in addition to field checks, machine-error checks should be used. In fact, machine-error checks, which compare inventories tree by tree and plot by plot, have been an important factor in encouraging better measurement, the use and development of better tools for tree measurement, standardization of field instructions, and improvement of coding systems for tree and plot descriptions. It should be emphasized that machine-error checking should be done before the field crews leave an area so corrections can be made with a minimum of effort.

The usual practice is to remeasure plots at 3- to 10-year intervals and, for large inventories, it must be decided whether to remeasure the complete system in 1 year or to measure part of the system each year during the cycle of measurement.

For more information on planning considerations for CFI see Meteer (1979).

16-4.2 Horizontal Point Sampling in Continuous Forest Inventory

There are reasons why it may be undesirable to use permanent points rather than permanent plots. But *the reasons given for not using permanent points are often the wrong ones*. We will consider the five common objections to using permanent points to emphasize this statement and to give some insight into the use of permanent points.

1. *Horizontal point sampling does not provide a representative sample.* Assuming "representative" means having the same distribution by size classes as the population, if a representative sample for estimating number of trees per acre is desired, monareal plot sampling (fixed-size plots) should be used because then a given size class can expect to be sampled in proportion to the frequency of trees in the given class. But if a representative sample for estimating basal area per acre is desired, horizontal point sampling should be used because then we are sampling with probability proportional to basal area. Thus, to estimate volume or volume growth per acre, horizontal point sampling gives a more representative sample than monareal plot sampling. And for many other important parameters, horizontal point sampling, the most logical p.p.s. system (Chapter 14) for permanent points, gives a more representative sample than monareal plot sampling.

2. *Horizontal point sampling does not provide sufficient trees in the finer tree-characteristic breakdowns.* With horizontal point sampling, contrary to monareal plot sampling, in many-aged stands the sample will be distributed rather uniformly across the diameter classes—not as many trees will be sampled in small diameter classes as with monareal plot sampling. Since small trees are less variable than large trees, fewer small trees are needed to obtain sound estimates. Furthermore, if insufficient trees are obtained, the gauge angle can be reduced so more trees will be obtained.

3. *The analysis of horizontal point sampling inventory data is too complicated; field procedures are too crude.* The formulas to obtain per-acre estimates for any tree characteristic are summarized in Table 14-3. When data from a permanent-point inventory are to be analyzed, these formulas can easily be used in computer programs for processing data. Furthermore, in the field most trees should be checked with a calibrated distance tape (Chapter 14). A gauge, commonly a prism, should be used sparingly or not at all. When this is done, field procedures can no longer be considered crude.

4. *Horizontal point sampling in CFI is too costly because more points are needed for a given accuracy than when permanent plots are used, even though fewer trees are marked and measured.* It is true that in most published studies more points than plots have been required to obtain equivalent accuracy. But in most of these studies more trees have been measured per plot than have been measured per point. If the gauge angle were reduced so that the average number of trees obtained per point equaled the average number of trees obtained per plot, we would generally need fewer points than plots.

5. *Horizontal point sampling in CFI is not desirable because volume per acre estimated from the initial inventory, added to the subsequent volume growth per acre, does not equal the volume per acre estimated from the second inventory.* In considering this objection one must examine how volume per acre for the first inventory V_1, volume per acre for the second inventory V_2, and net volume growth including ingrowth per acre G_{n+i} are obtained in horizontal point sampling. Table 16-8 gives a recommended procedure for computing volume per acre and volume growth per acre on permanent points. It will be seen from this example that there is an inherent lack of additivity when permanent points are used. Specifically,

$$V_2 \neq V_1 + G_{n+i} - C$$

Volume per acre for the first inventory, volume growth per acre, and *ingrowth* are computed from tree factors for diameters at the first inventory. Volume per acre for the second inventory is computed from tree factors for tree diameters at the second inventory. *Nongrowth* trees are included in the volume estimate for the second inventory and are computed from tree factors for tree diameters at the second inventory. Thus, it should be clear that one should not expect volume per acre for the first inventory added to volume growth per acre to equal volume per acre for the second inventory, any more than one should expect identical estimates of volume per acre from two independent samples of a small number of $\frac{1}{5}$-acre plots.

In any case, confusion on how to mark, keep track of, and process ingrowth trees and nongrowth trees (sometimes called ongrowth trees) has made foresters reluctant to use permanent sample points in CFI. However, the approach proposed by Martin (1982) tends to allay the confusion. For example, in Table 16-8 the volume growth per acre of the ingrowth tree (No. 7) was calculated as follows.

$$\text{Volume growth per acre} = \frac{F}{BA_1} v_2 = \frac{10.0}{0.005454(3.92)^2}(0.80) = 95.5$$

On the other hand, Martin aggregates trees 7 through 10 ("new" trees) and estimates the ingrowth component I as follows.

$$I = \Sigma V_n + \Sigma V_{s_2} - \Sigma V'_{s_2}$$

$$= (65.8 + 105.8 + 157.3 + 60.7)$$

$$+ (210.3 + 147.9 + 224.8 + 110.3)$$

$$- \frac{10.0}{0.005454}\left(\frac{14.21}{10.68^2} + \frac{5.49}{7.64^2} + \frac{37.94}{17.36^2} + \frac{2.58}{5.65^2}\right)$$

$$= 389.6 + 693.3 - 779.9 = 303.0$$

where

V_n = volume per acre for newly qualifying trees (Nos. 7 through 10, Table 16-8), using time 2 volume and basal area

V_{s_2} = volume per acre for survivor trees (Nos. 3 through 6, Table 16-8) at time 2, using time 2 volume and basal area

V'_{s_2} = "corrected" volume per acre for survivor trees (Nos. 3 through 6, Table 16-8), using time 2 volume and time 1 basal area

Using this procedure the estimated net increase per acre, $G_d = V_2 - V_1 = 1082.9 - 1026.6 = 56.3$, can also be estimated using equation 16-10.

$$G_d = G_g + I - M - C$$

$$= 99.2 + 303.0 - 87.3 - 258.6$$

$$= 56.3$$

Similarly, net growth including ingrowth G_{n+i} can be calculated using equation 16-4.

$$G_{n+i} = V_2 + C - V_1$$

$$= 1082.9 + 258.6 - 1026.6$$

$$= 314.9$$

Or by using equation 16-9,

$$G_{n+i} = G_g + I - M$$

$$= 99.2 + 303.0 - 87.3$$

$$= 314.9$$

It is anticipated that both the method outlined in Table 16-8 and Martin's method will lead to equivalent growth estimates in the long run. Although Martin's approach simplifies field marking and record keeping, there is a disadvantage: the contribution of an individual tree to a summary growth figure cannot be calculated for all trees since ingrowth is composed of "new tree" volume per acre and an adjustment that is a function of the corrected and uncorrected volumes of survivor trees.

The above discussion should reassure readers who have entertained the first four objections to permanent points. The fifth objection, however, is not so easily laid to rest. Indeed, this objection is closely related to a valid objection made against permanent points: the theory and methods associated with point sampling are generally unfamiliar to inventory planners, and in spite of the fact that the details of point sampling are not difficult to learn, the planners are reluctant to change to points when plots are satisfactorily serving their purposes. Another well-grounded objection is that permanent points require a check of boundary trees at each inventory, because as a tree grows its plot size increases. For permanent plots, once plot boundaries are defined, they need not be rechecked. On permanent points the field crews must go through the time-consuming procedure of looking for new qualifying trees at each inventory.

Table 16-8
Example of Volume and Volume Growth Calculations for One Permanent Sample Point ($F = 10$)

Tree Number	Tree Status[1]	First Inventory dbh (inches)	Sound Volume per Tree (cubic feet)	Sound Volume per Acre	Second Inventory dbh (inches)	Sound Volume per Tree (cubic feet)	Sound Volume per Acre	Cut per Acre	Contribution to Net Growth Including Ingrowth per Acre (cubic feet)	Survivor Growth per Acre
1	10	5.46	1.42	87.3	—	—	—	—	−87.3	—
2	24	24.06	81.64	258.6	—	—	—	258.6	—	—
3	12	10.68	13.67	219.7	11.13	14.21	210.3	—	8.7	8.7
4	11	7.64	4.69	147.3	8.25	5.49	147.9	—	25.2	25.2
5	22	17.36	37.11	225.8	17.59	37.94	224.8	—	5.0	5.0
6	11	5.65	1.53	87.9	6.55	2.58	110.3	—	60.3	60.3
7	51	3.92	—	—	4.72	0.80	65.8	—	95.5	—
8	61	—	—	—	6.57	2.49	105.8	—	—	—
9	62	—	—	—	21.75	40.59	157.3	—	—	—
10	61	—	—	—	4.76	0.75	60.7	—	—	—
Totals				1026.6			1082.9	258.6	107.4	99.2
Symbol				V_1			V_2	C	G_{n+i}	G_g

Example of volume per acre and volume growth per acre calculations for Tree Number 3:

$$\text{Volume per acre (first inventory)} = \frac{F}{0.005454 D_1^2} v_1 = \frac{10}{0.005454(10.68)^2}(13.67) = 219.7$$

$$\text{Volume per acre (second inventory)} = \frac{F}{0.005454 D_2^2} v_2 = \frac{10}{0.005454(11.13)^2}(14.21) = 210.3$$

$$\text{Volume growth per acre} = \frac{F}{BA_1} v_2 - \frac{F}{BA_1} v_1 = \frac{F(v_2 - v_1)}{0.005454 D_1^2} = \frac{10(14.21 - 13.67)}{0.005454(10.68)^2} = 8.7$$

where

D_1, D_2, v_1, and v_2 are individual tree diameters and individual tree volumes for the first and second inventories, and F is the basal area factor.

[1] Tree status defines the class of tree from the standpoint of growth contribution. Coding is as follows: 0 = dead, 1 = pulpwood size, 2 = sawlog size, 3 = cull, 4 = cut, 5 = ingrowth tree, 6 = nongrowth tree. By combining the tree classes at successive inventories, then 10 = pulpwood mortality, 24 = sawlog cut, 12 = change from pulpwood to sawlog size, 11 = pulpwood survivor tree, 51 = ingrowth tree, and 62 = nongrowth tree. An *ingrowth* tree is a tree that is "in," but below minimum diameter for measurement at the first inventory, and that grows enough during the growth period to be above minimum diameter for measurement at the second inventory. A *nongrowth* tree is a tree that is "out," regardless of size, at the first inventory and that is "in" and above minimum diameter for measurement at the second inventory.

17

Stand Structure, Density, Site Quality, and Yield

As one advances in forest mensuration from basic measurements to multi-resource inventories, the techniques become more and more allied to silviculture and forest management. This chapter covers techniques that can be considered transitional between mensuration and these fields. Although fundamental procedures and concepts are described in this chapter, the reader should search the references cited for more details on the applications of stand structure, density, site quality and yield.

17-1 STAND STRUCTURE

Stand structure is the distribution of species and tree sizes on a forest area. The structure of a stand is the result of the species' growth habits and of the environmental conditions and management practices under which the stand originated and developed. There are two typical stand structures—*even-aged* and *uneven-aged*—although under natural forest conditions there are gradations between the two. A brief treatment of certain mensurational aspects of these two structural types follows. For additional silvicultural and management detail, reference should be made to Meyer, Recknagel, Stevenson, and Bartoo (1961), Smith (1962), Davis (1966), and Daniel, Helms, and Baker (1979).

17-1.1 Even-Aged Stands

An even-aged stand is a group of trees that has originated within a short period of time. The trees in an even-aged stand thus belong to a single age class. The limits of the age class may vary, depending on the length of time during which the stand formed. A natural stand may seed-in over a period of several years. Rarely will an age class be only 1 year, except in plantations. More commonly, the age class for an even-aged stand will extend to 10 or 20 years. In some cases, a stand may appear even-aged because the trees show size uniformity. For example, stands growing slowly on poor sites may consist of trees of widely diverse ages, yet have little variation in size. Even-aged stands may be composed of shade-tolerant or shade-intolerant species, although even-aged stands of intolerant species are more common. Even-aged stands arise out of, or are perpetuated by, environmental conditions that allow trees to become established within a comparatively short, definable period. An even-aged forest may consist of several even-aged stands belonging to different age classes.

The trees in an even-aged stand are fairly consistent in height, with variations depending on their crown position as dominant, codominant, intermediate, or suppressed. The diameters in an even-aged stand show wider variation, although following a typical pattern. Most trees cluster near the average diameter, with decreasing frequencies at larger and smaller diameters. As even-aged stands grow older, the diameter class distribution changes. The total number of trees in the stand decreases, with numbers of trees appearing in larger diameter classes not previously represented. Figure 17-1 shows the diameter distribution of an even-aged stand at several ages. Several investigators have found that the diameter distributions of even-aged stands follow definite laws, and that the relationship of number of trees and diameter can be described by computed values (Baker, 1923; Meyer, 1930; Schumacher, 1930; Schnur, 1934; Anderson, 1937; and Schnur, 1937). Although the general distribution of number of trees plotted over diameter classes has usually been noted as a skewed (but nearly normal) distribution, studies by Gingrich (1967) have demonstrated that for upland forests of the central hardwood region (United States): "When the average stand diameter is large enough [8 inches, in his study] to accommodate the frequency curve in [its] positive range . . . , a normal diameter distribution exists." The stands studied did not demonstrate significant kurtosis (peakedness). From an analytical standpoint, it is worth noting that numerous studies have shown that estimated parameters of the typical even-aged structure (measures of variation, skewness, and kurtosis) are more closely related to average diameter than to stand age or a measure of site quality, though undoubtedly both of these have a considerable effect on the distribution. Thus, average tree diameter is a basic and very useful means of characterizing an even-aged stand.

A detailed discussion of stand structure analysis is beyond the scope of this discussion. However, the reader is referred to the significant contributions of Clutter and Bennett (1965), Bailey and Dell (1973), Smalley and Bailey (1974a, 1974b), Hafley and Schreuder (1977), and Hyink (1979) for state-of-the-art analysis procedures. A recent summary of current research trends in even-aged management is given by Hann and Brodie (1980).

17-1.2 Uneven-Aged Stands

A stand consisting of trees of many ages and corresponding sizes is said to be uneven-aged. The trees in an uneven-aged forest originate more or less continuously, in contrast to the single reproductive period characterizing an even-aged forest. This continuing source of new trees produces, in an undisturbed stand, trees of ages varying from germinating seedlings to overmature veterans.

In an uneven-aged forest, the trees in the crown canopy are of many heights, resulting in an irregular stand profile as viewed from a vertical cross section. The more shade-tolerant species tend to form uneven-aged stands. Cutting methods that

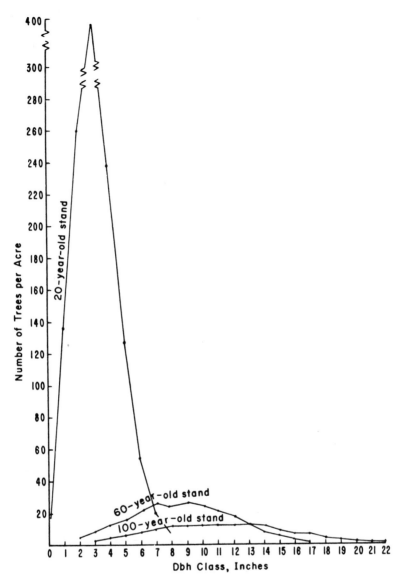

Fig. 17-1 *Examples of diameter distributions at three ages in mixed oak stands; site index 80 (Schnur, 1937).*

remove only scattered trees at short intervals maintain forest conditions favorable to shade-tolerant species and an uneven-aged stand.

The typical diameter distribution for an uneven-aged stand is a large number of small trees with decreasing frequency as the diameter increases, as shown in Fig. 17-2. The diameter distribution for small areas of uneven-aged forests may show considerably greater irregularity. As the area of the uneven-aged stand or forest increases,

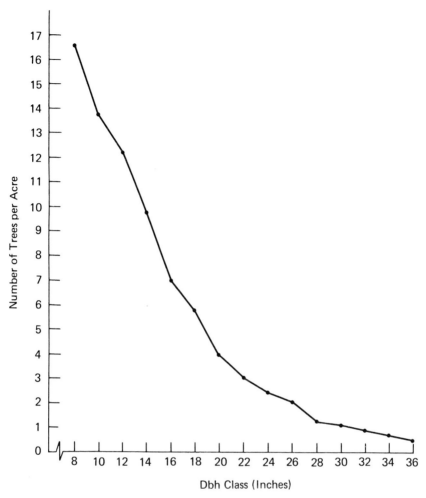

Fig. 17-2 *Diameter distribution per acre for an uneven-aged virgin stand of beech–birch–maple–hemlock (adapted from Meyer and Stevenson, 1943).*

the irregularities tend to even out and the inverse, J-shaped diameter distribution of an uneven-aged forest becomes apparent. Combining equal area of all age classes for a theoretical, fully regulated (i.e., all age classes represented by equal areas), even-aged forest will also produce the distribution typical of an uneven-aged forest. The actual forest structure on any unit area of land, however, is radically different.

Meyer (1953), basing his work on the investigations of De Liocourt (1898), studied the structure of what he termed a *balanced uneven-aged forest*. His definition of a balanced uneven-aged forest is "one in which current growth can be removed periodically while maintaining the diameter distribution and initial volume of the forest." Meyer states that a balanced uneven-aged forest will tend to have a diameter distribution whose form can be expressed by the exponential equation

$$Y = ke^{-aX} \qquad (17\text{-}1)$$

where

Y = number of trees per diameter class
X = dbh class
e = base of natural logarithms
a, k = constants for a characteristic diameter distribution

Research has shown that the typical uneven-aged distribution can be characterized by determining values for the constants a and k in equation 17-1; a determines the rate by which numbers of trees diminish in successive diameter classes, and k indicates the relative density of the stand. Thus, a can be considered the slope or rate-of-change coefficient and k the Y intercept, since it is the value of Y when $X = 0$. Meyer's work demonstrated that these constants are rather well correlated, positively. Therefore, high values of a, indicating a rapid reduction in trees for increasing diameter, are generally associated with high values of k, indicating a relatively high density of small timber, and vice versa.

A diameter distribution can be tested for conformity to the definition of a balanced structure by checking the linearity of a trend line when number of trees is plotted over diameter class on semilogarithmic paper, as shown in Fig. 17-3.

A balanced distribution implies that the number of trees in successive diameter classes follows a geometric series of the form $m, mq, mq^2, mq^3, \ldots$, where q is the ratio of the series and m is the number of trees in the largest diameter class considered. The constant q, then, is the ratio of trees in successive diameter classes.

Although the determination of a and k could proceed using the techniques of nonlinear regression made feasible by the application of modern computers, it is worthwhile to discuss the approach taken in most of the literature on the subject—that is, the application of graphical or linear least-squares techniques involving the logarithmic transformation of equation 17-1. If this is done, one obtains

$$\log Y = \log k - aX \log e \qquad (17\text{-}2)$$

or, if base e logarithms are used,

$$\ln Y = \ln k - aX \qquad (17\text{-}3)$$

Equation 17-3 is inherently simpler than 17-2 because of the exponential nature of the basic model (equation 17-1). The fitting of the least-squares line to the logarithm of Y and X is described in the following paragraphs using either base 10 logs (designated by log) or natural, base e, logs (designated by ln). Where the equations differ, the natural approach is shown in parentheses. In either case, one should be aware that a logarithmic transformation bias (Section 9-6.1) is always incurred; therefore, where feasible the nonlinear least-squares fit to the basic data, using equation 17-1, is preferable.

The transformed model (equation 17-2 or 17-3) can be handled by simple linear

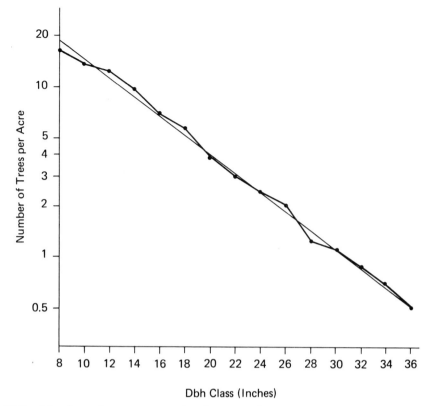

Fig. 17-3 *Number of trees per acre by diameter classes represented on semi-log paper using the data shown in Fig. 17-2. The logarithmic least-squares equation (unweighted) was log Y = 1.72242 − 0.05563X.*

regression, since by redefinition

$$Y' = b_0 + b_1 X \tag{17-4}$$

where

$$
\begin{aligned}
Y' &= \log Y \\
b_0 &= \log k \qquad \text{or} \\
b_1 &= -a \log e
\end{aligned}
\qquad
\left(
\begin{aligned}
Y' &= \ln Y \\
b_0 &= \ln k \\
b_1 &= -a
\end{aligned}
\right)
$$

The constants a and k can then be determined by

$$k = \text{antilog } (b_0) = 10^{b_0} \tag{17-5}$$

and

$$a = \frac{-b_1}{\log e} = \frac{-b_1}{0.43429} \tag{17-6}$$

Or, using natural logs,

$$(k = e^{b_0})$$ (17-7)

and

$$(a = -b_1)$$ (17-8)

It is basic to the concept of a frequency distribution that the number of trees in any diameter class of width h is given (approximately) by the equation

$$n_i = hY_i$$ (17-9)

$$= hke^{-aX_i}$$ (17-10)

where

n_i = number of trees in the ith class
X_i = midpoint of the ith class
Y_i = ordinate value on the curve corresponding to X_i
h = class width

Therefore, by definition

$$q = \frac{n_{i-1}}{n_i}$$ (17-11)

$$= \frac{hke^{-a(X_i-h)}}{hke^{-aX_i}}$$

$$q = e^{ha}$$ (17-12)

In stand structure analyses of this type, it is imperative to note that the constant k varies with both stand area and the width of the diameter class, that q is a function of class width, and that a is not dependent on either stand area or class width. The exponential equation 17-1 is usually written with the tacit assumption that the class width is 1 and that the data (number of trees in this case) are on a unit area basis. For data grouped into classes h units wide, representing totals from a stand of area A acres (or hectares), one can calculate a k' using the logarithmic least-squares approach and then find the basic k from the equation

$$k = \frac{k'}{hA}$$ (17-13)

When using the ratio q, especially in comparing stands, a common class width must be assumed. It, therefore, seems reasonable to define q as the ratio of numbers of trees in successive diameter classes of width 1, and adjust q, calculated on any other basis, back to the unity basis. For this purpose, it is easily shown that

$$q_h = (q_{h'})^{h/h'}$$ (17-14)

where q_h and $q_{h'}$, are calculated using diameter class of h and h' units, respectively.

In certain practical applications, as noted by Trimble and Smith (1976), q can be more directly obtained if one is not particularly interested in the parameters a and k. That is, from the statistical definition of the slope b_1 in equation 17-4,

$$b_1 = \frac{\log Y_i - \log Y_{i-1}}{X_i - X_{i-1}}$$

If 1-inch diameter classes are assumed ($X_i - X_{i-1} = 1$), and

$$b_1 = \log Y_i - \log Y_{i-1}$$

This leads to

$$-b_1 = \log Y_{i-1} - \log Y_i$$

and by taking antilogs,

$$10^{-b_1} = \frac{Y_{i-1}}{Y_i}$$

Thus, from equation 17-11,

$$q = 10^{-b_1} \tag{17-15}$$

or, using natural logs,

$$(q = e^{-b_1}) \tag{17-16}$$

Note, equations 17-15 and 17-16 assume 1-inch diameter classes. To determine the q assuming 2-inch classes—that is, q_2, the formulas would be

$$q_2 = 10^{-2b_1} \quad \text{and} \quad (q_2 = e^{-2b_1})$$

or, in general, for class width h,

$$q_h = 10^{-hb_1} \tag{17-17}$$

$$\text{or } (q_h = e^{-hb_1}) \tag{17-18}$$

For a numerical example we will use the data represented in Figs. 17-2 and 17-3. Using 2-inch diameter classes, the logarithmic least-squares equation was

$$\log Y = 1.72242 - 0.05563X$$

Therefore, using equations 17-5 and 17-13,

$$k' = \text{antilog } (1.72242) = 52.77$$

$$k = \frac{52.77}{(2)(1)} = 26.38$$

Using equation 17-6,

$$a = \frac{-(-0.05563)}{0.43429} = 0.1281$$

And using equation 17-12,

$$q = e^{2(0.1281)} = 1.29$$

A q of 1.29 is, therefore, appropriate for the distribution as represented (2-inch diameter classes); however, we can calculate a "standard" q using equation 17-14.

$$q_1 = (1.29)^{1/2} = 1.14$$

For details regarding the management potential of uneven-aged structure analysis, the reader is referred to the writings of H. A. Meyer and his students. Ecological applications have also been described by Leak (1964) and Schmelz and Lindsey (1965).

The application of uneven-aged structure analysis to managed forest stands has received considerable attention in recent years. In this regard, one should refer to the landmark publications of Adams and Ek (1974) and Moser (1972, 1976), and the excellent state-of-the-art publication by Hann and Bare (1979).

17-2 STAND DENSITY AND STOCKING

Measures of stand density and forest stocking are both used to depict the degree to which a given site is being utilized by the growing trees or simply to indicate the quantity of wood on an area. However, a distinction is usually made between the two terms. Gingrich (1967) describes the distinction in this way.

> *Stand density* is a quantitative measurement of a stand in terms of square feet of basal area, number of trees, or volume per acre. It reflects the degree of crowding of stems within the area. *Stocking,* on the other hand, is a relative term used to describe the *adequacy* of a given stand density in meeting the management objective. Thus, a stand with a density of 70 square feet of basal area per acre may be classified as overstocked or understocked, depending upon what density is considered desirable.

Stand density can be expressed, as above, in absolute units per unit land area according to such stand parameters as volume, basal area, crown coverage, number of trees, and so on, but often it is expressed on a relative scale as a percent of some "normal" (full or desirable) density or as a percent of the average density. When expressed in this form, it should be clear that "relative density" exists as a term transitional between "stand density" and "stocking" as defined above. Stand density is also expressed as an index that is derived by using more involved concepts (Sections 17-2.2, 17-2.3, 17-2.4, 17-2.5).

17-2.1 Relative Density

For many forestry purposes, *volume* is the ultimate expression of stand density. Ideally, then, relative density is determined by comparing the volume of an observed stand with the volume of some standard, such as the volumes of fully stocked stands for specified ages and site qualities as given in a yield table. For example, a 50-year-old stand of white pine on site index 58 land in Massachusetts was measured and its volume was found to be 4500 board feet per acre. A normal yield table showed theoretically that the full stocking for this age and site was 6290 board feet. The relative density is then $(4500/6290)(100) = 71.5$ percent. Note that relative density percentages using volume depend on the volume unit chosen. For example, board-foot density percentages will differ from cubic-foot density percentages for the same stand. Density as measured by relative volumes has several disadvantages. Volumes as expressed for the standard or fully stocked stand may be based on different merchantability limits, different log rules, or different volume units from the stand under investigation, making valid comparison difficult. In addition, stand volume estimates are too expensive if only a measure of relative density is needed.

Instead of volume, stand basal area can be used in a similar way as a density measure. The advantage of basal area is that it is easily determined and is quite consistent for fully stocked stands of specified ages and sites. Since relative density depends on the unit of volume used, even for the same stand, relative density using basal area need not be the same as that using volume.

The number of trees per unit land area can be used as another measure of stand density. At any age, there can be a wide range in the number of trees per unit land area, so that frequency by itself is of little value. For a useful descriptive measure of stand density, number of trees must be qualified by tree sizes. The stand structure as shown in a stand table describes this but is too cumbersome for practical use as a stand density description. A useful measure of density for even-aged stands based on number of trees is *Reineke's stand density index* (Reineke, 1933).

17-2.2 Stand Density Index

Stand density index (SDI) is the number of trees per acre that a stand would have at a standard average dbh. The stand density index for a stand is obtained by referring to a stand density index chart for the species. As shown in Fig. 17-4, the chart consists of a series of lines representing the relationship between number of trees per acre and average stand diameter. The chart can be constructed by plotting the logarithm of number of trees per acre over the logarithm of the average stand dbh on rectangular coordinate paper. The natural values are plotted on logarithmic paper in Fig. 17-4, yielding the same results. The number of trees and the average stand diameters are obtained from a series of sample plots. Average stand diameter is taken as the diameter of the tree of arithmetic mean basal area (i.e., quadratic mean diameter). An average relationship for all the observations will be defined by a straight line. This line represents the average stocking of all plots. The number of trees indicated by the intersection of this line and the ordinate at the standard dbh is the average stand density index (in Fig. 17-4 the 10-inch ordinate is taken as the standard). A series of parallel lines is then constructed to intersect the standard diameter ordinate at specified numbers of trees per acre. The numbers at these intersections are the stand density index values for the set of parallel lines. If the average stand density index line is taken as 100 percent, other index lines can be expressed in percentages instead of numbers of trees.

The stand density index for a stand is determined by plotting the position of the observed number of trees and the average stand dbh per acre on the stand density index chart for the species. The stand density index is indicated by the closest line to the plotted point or can be found by interpolation between the index lines. For example, the stand density index for a white pine stand of 14.6 inches average dbh, with 265 trees per acre, on Fig. 17-4 is about 480.[1] The equation form of the average stand density index line is log $N = a \log D + b$, where N = number of trees per acre and D = average stand dbh. Reineke found that the a constant was -1.605 for several species. Other investigators noted that the linear relationship expressed by the equation holds for many species and that the slope differs little, although the b constant varies considerably. The average stand density index line for Fig. 17-4 has the equation log $N = -1.598 \log D + 4.165$, based on a least-squares solution for 53 plots. Stand density index is not strongly correlated with age or site. Stands of the same age that are on the same site may have different numbers of trees and average stand diameters. This quality of independence of age or site makes stand density index an additional valuable parameter in describing a stand, especially in yield table construction (see Section 17-4).

17-2.3 Tree-Area Ratio

The tree-area ratio is a measure of density that was first proposed by Chisman and Schumacher (1940) as a measure independent of stand age and site quality and appropriate for even- or uneven-aged stands. It is based on the concept that if the space on the ground Y, occupied by a tree of diameter at breast height d, can be expressed by the relation

$$Y = b_0 + b_1 d + b_2 d^2 \tag{17-19}$$

then the total area of growing space represented on a plot or unit of ground area can be found by summing over all the trees on the plot or unit area. Thus, a measure of growing space utilization can be obtained by adding over all trees (n) on a unit area, leading to

$$\text{Tree area} = b_0 n + b_1 \Sigma d + b_2 \Sigma d^2 \tag{17-20}$$

After the constants b_0, b_1, and b_2 are obtained by least squares, the contribution to tree area of a single tree of diameter d can be found by letting $n = 1$, $\Sigma d = d$, and $\Sigma d^2 = d^2$.

Using data obtained from sample plots (preferably adjusted to the appropriate unit

---

[1] Using the technique of translation of axes (see Section 17-3.3), the stand density index can be *calculated* from the formula log SDI = log $N - a(\log D - \log 10)$. For this example, then, log SDI = log 265 + 1.598(log 14.6 $-$ 1), leading to SDI = 485.

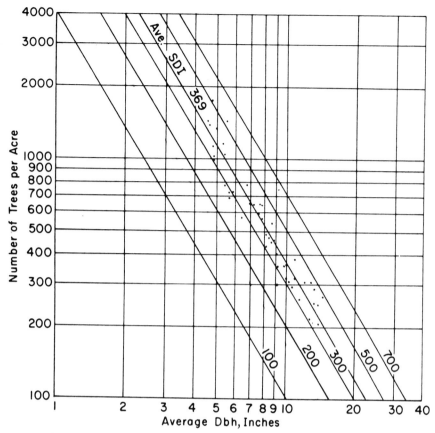

Fig. 17-4 *Stand density index (SDI) chart for white pine in southeastern New Hampshire. Equation is log N = −1.598 log D + 4.165. Based on 53 sample plots.*

area basis), the constants in equation 17-20 are obtained by the method of least squares, minimizing the sum of squared residuals.

$$\sum_{i=1}^{N}(1 - b_0X_0 - b_1X_1 - b_2X_2)^2 \tag{17-21}$$

where

X_0 = number of trees per plot (or unit area)
X_1 = sum of diameters per plot (or unit area)
X_2 = sum of squared diameters per plot (or unit area)
N = number of plots in the sample

It is important to note that "normal equations" derived from this minimization differ from the usual ones in that the column vector to the right of the equal sign becomes ΣX_0, ΣX_1, ΣX_2 rather than ΣY, $\Sigma X_1 Y$, $\Sigma X_2 Y$.

If the data used to derive the coefficients b_0, b_1, and b_2 in equation 17-20 came from stands that were deliberately chosen to be "full" or "normally" stocked, then the application of the regression equation to other stands and substitution of n, Σd, and Σd^2 will provide a tree-area figure that will reflect the proportion of *full stocking* demonstrated by that stand. On the other hand, if the data used to derive the coefficients came from stands having a range of densities, substitution into the equation for a given stand will reflect the proportion of stocking as compared to the *average stocking* of the basic data.

Although some studies (Lynch, 1958; Beers, 1960) have found that certain of the independent variables in the basic tree-area ratio model (equation 17-20) contribute nonsignificantly and might be dropped from consideration, Gingrich (1967) shows on logical grounds that all three independent variables should be retained. For further discussion, the reader should refer to the basic references cited. It should be understood that the tree-area ratio is firmly established as a useful measure of stand density, and that the final form of the regression equation is somewhat dependent on the use to which the measure is put.

17-2.4 Crown Competition Factor

A measure of density, which in final form is similar to the tree-area ratio, although considerably different in derivation, is the crown competition factor (CCF) proposed by Krajicek, Brinkman, and Gingrich (1961). The authors claim that CCF is independent of site quality and stand age and should be suitable for use in both even- and uneven-aged stands.

The development of a CCF formula for a species or species group would proceed as follows, using the example given by Krajicek et al. (1961).

1. Measurements of crown width and dbh are taken on a satisfactory number of truly open-grown trees, carefully selected to ensure that they have developed in an undisturbed, competition-free environment.

2. For this measured sample the relation of crown width (CW) to dbh (d) is found by least squares. For example, $CW = 3.12 + 1.829d$.

3. A formula for the crown area of individual trees expressed as a percent of 1 acre is derived and is called *maximum crown area* (MCA), since it indicates the maximum proportion of an acre that the crowns of trees of a given dbh can occupy. For example,

$$MCA = \left(\frac{\pi(CW)^2}{4} \right) \left(\frac{100}{43,560} \right)$$

(17-22)

$$= 0.0018 \, (CW)^2$$

By substituting the regression relating CW to dbh, we obtain

$$MCA = 0.0018 (3.12 + 1.829d)^2$$

$$= 0.0175 + 0.0205d + 0.0060d^2$$

4. By adding the MCAs for all trees on the average acre of forestland, an expression of stand density, called crown competition factor (CCF), is obtained. Thus, using the example, the CCF for a specific stand could be determined by substitution in the following equation.

$$CCF = \frac{1}{A}(0.0175\Sigma n_i + 0.0205 \ \Sigma d_i n_i + 0.0060\Sigma d_i^2 n_i) \qquad (17\text{-}23)$$

where

n_i = number of trees in the ith dbh class
d_i = midpoint of the ith dbh class
A = stand area in acres, taken to be unity if a per-acre stand table is the basis for the data to be substituted

It should be clear that any combination of MCAs that sum to 100 ($CCF = 100$) represents a closed canopy and reflects a situation where the tree crowns just touch and are sufficiently distorted to completely cover each acre of ground. However, the authors of this index (Krajicek et al., 1961) state:

> It should be emphasized that CCF is not essentially a measure of crown closure. Theoretically, complete crown closure can occur from CCF 100 to the maximum for the species (in oaks, approximately CCF 200). Instead of estimating crown closure, CCF estimates the area available to the average tree in the stand in relation to the maximum area it could use, if it were open grown.

17-2.5   Point Density

The previously described measures of density are usually employed to determine density of a stand in general or "on the average." Somewhat more specific measures of density have been developed to describe the degree of competition on a given point or tree in the stand. For an example of point density we can look at a technique developed by Spurr (1962) called the *angle-summation method*. "This method," Spurr explains, "should have its utility in such studies as the correlation between tree growth and density around that tree, the correlation between establishment of natural regeneration and point density, and the selection of plus trees in forest tree improvement research."

In the angle-summation method a point or tree is chosen on which we wish to determine the degree of competition from surrounding trees. Using the basic theory of Bitterlich's angle count method (horizontal point sampling, as described in Chapter 14), *each competing tree* is imagined to be "borderline" from the chosen point or tree and, thus, to have a specific basal area factor. Spurr's point density measure then is obtained by appropriately summing a series of basal area per acre etimates made using these trees. Other measures of point density are described and evaluated by Opie (1968), Gerrard (1969), and Johnson (1973).

17-2.6 Forest Stocking

In an excellent discussion of stocking and density, Bickford, Baker, and Wilson (1957) accept the Society of American Forester's definition of stocking as ". . . an indication of the number of trees in a stand as compared to the desirable number for best growth and management: such as well-stocked, over-stocked, partly stocked." And, although they point out that there are varying shades of meaning for the silviculturist, economist, or forest manager, the most common connotation of "forest stocking" is in the sense of "best growth." That is, the terms "overstocked" and "understocked" represent upper and lower limits of site occupancy within which there exists a degree of stocking where forest growth will be optimum. Ideally, forest managers should be able to recognize this optimum stocking point under the complete range of stand composition conditions they encounter. Many attempts have been made to quantify forest stocking so that its relationship to growth can be more readily determined. Based on the repeatedly shown concept that gross increment varies very little over a wide range of stand density, the work of Gingrich (1964, 1967) culminated in a stocking chart (Fig. 17-5) that has found considerable usage in applied forest management in the central United States.

The chart was developed using (1) a tree-area ratio equation (see Section 17-2.3) derived from data for stands that were deliberately chosen to be "fully stocked," and (2) an equation based on the crown-competition factor (see Section 17-2.4), using data from open-grown trees. Thus, in Fig. 17-5 the A line (100 percent stocking) ". . . represents a normal condition of maximum stocking for undisturbed stands of upland hardwoods of average structure," and the B line represents the lower limit of stocking for full site occupancy (Gingrich, 1967). The entire range of stocking between A and B is called "fully stocked" because the growing space can be fully utilized. It follows that stands having a combination of basal area per acre and number of trees per acre falling outside this range will demonstrate a gross increment less than the potential of the site, because of either overstocking or understocking.

Application of the stocking chart to determine a prescription of silvicultural treatment is in itself simple. For example, if stand measurements indicate a basal area per acre of 80 square feet and 175 trees per acre, an average diameter of 9.2 inches is indicated (in Fig. 17-5). Following the 9.2 line down to the point where it crosses the B level, one finds that a basal area of 65 square feet is the minimum basal area to maintain full stocking at that average diameter. Therefore, silvicultural treatment, which removes 15 square feet of basal area without materially changing the average stand diameter, leaves a stand that will still fully utilize the site and produce wood efficiently. For greater detail on application, the reader is referred to the handbooks written by Roach and Gingrich (1962, 1968) and by Roach (1977).

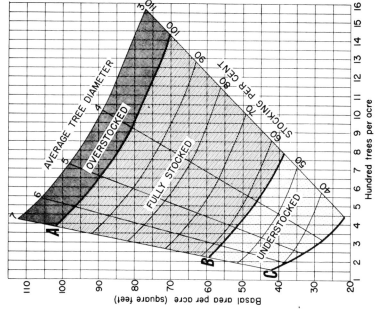

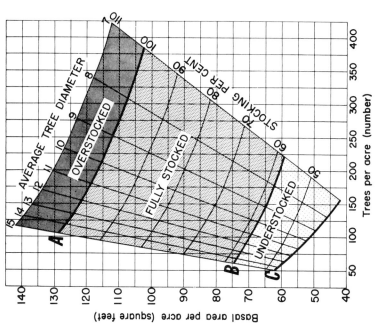

Fig. 17-5 *Relation of basal area, number of trees, and average tree diameter to stocking percent for upland hardwood stands. Tree-diameter range 7–15 (left), 3–7 (right). The area between curves A and B indicates the range of stocking where the trees can fully utilize the growing space. Curve C shows the lower limit of stocking necessary to reach the B level in 10 years on average sites. (Source: Gingrich, 1964.)*

17-3   SITE QUALITY

A knowledge of the growth responses of forest trees to the factors of the environment is important to forest management. The forester can use this knowledge to encourage the growth of desirable species, either by modifying the environment or by concentrating on sites having desirable environmental characteristics. In the past, foresters have been restricted by their inability to modify the environment on a given site in order to change stand density and structure. In the future, we may be able to extend our ability to fertilization and irrigation practices when the economic climate is propitious. At present, our application of site-growth relationships consists of encouraging the growth of a species on sites where its establishment is readily achieved or where its growth potential can be fully realized.

The relationship of the growth of forest trees to their environment, or, as it is commonly referred to, their site, is a difficult one to measure. The factors of the site and the plants themselves are interacting and interdependent, making it difficult to assign cause and effect relationships. A considerable amount of effort has been directed toward investigating the characteristics of the soil in an attempt to find some one environmental factor to serve as a reliable indicator of site quality. This approach has been found practical but frequently leaves unassigned a sizable amount of variation in the growth parameter employed. To understand fully the growth of trees in relation to environment, individual site factors cannot be studied in isolation. The interdependencies and influences of the other factors may be masked and, consequently, not recognized. Even if the primary interest is the study of one factor of the site, it must be done with recognition of the effects of the other factors.

The productivity of a site for tree growth is usually evaluated on a stand basis. Considered in this way, site quality expresses the average productivity of a designated land area for growing forest trees. A common way of expressing relative site quality is to set up from three to five classes, or ordinal ranks, such as Site I, Site II, and Site III, designating comparative productive capacities in descending rank. The characteristics of each class must be defined to enable any area to be classified. To a large extent, the definitions are in qualitative terms so that the ranking of a site is quite subjective. Wherever possible, attempts are made to introduce numerical definitions to improve the precision of site quality ranking.

Site quality can be evaluated in two general ways.

1. By the measurement of one or more of the individual site factors that are considered closely associated with tree growth. This approach evaluates site quality in terms of the environmental causal factors themselves.
2. By the measurement of some characteristics of the trees or lesser vegetation that are considered sensitive to site quality. This approach assesses site quality from the effects of the environment on the vegetation.

17-3.1   Measurement of Site Factors

Of the numerous environmental factors influencing tree growth, the important relationship between soils and tree growth has been so apparent that attention has been directed to it for a long time. An excellent summary of work in this field has been done by Coile (1952). Continuing work by other investigators has progressed along similar lines for an increasing number of species in different localities. They have shown that the soil characteristics significantly related to tree growth are not the same everywhere. Relative wetness, sandiness, depth, amount of clay in the *A* and *B* horizons, nutrient levels, soil temperature, and other characteristics have different proportionate effects, depending on the kind of soil and species involved. Evaluating site quality from soil characteristics has several advantages. The soil is comparatively stable and changes slowly. Thus, an evaluation of site quality from soil characteristics can be made regardless of the presence or absence of the forest. An area may support a dense forest or be cut over, but the site quality based on the soil will be little affected.

The usual form of a soil-site quality investigation has consisted of a multiple regression analysis with the height or site index as the dependent variable and a number of soil and other environmental characteristics as the independent variables. A commonly used approach is to use the model

$$\log H = b_0 + b_1(1/A) + b_2(B) + b_3(C) + \cdots + b_n(N)$$

where

H	= height
A	= age
$B, C, \ldots, N$	= soil or other environmental factors
$b_0, \ldots, b_n$	= constants to be determined by least squares

Environmental factors other than soil characteristics have also been studied in their relation to growth. These include aspect, slope, elevation, photoperiod, air temperature, among others. Hills (1952) has prepared a site evaluation system that considers the integrated result of all environmental factors in expressing site quality.

A major difficulty in this type of analysis pertains to the numerical coding of qualitative variables to enable the regression approach. Such a procedure can lead to difficulty because the arbitrary assignment of numbers to discrete classes and subsequent computations, by assuming a continuous scale, implies a scale of equal units that may not represent the actual relationship between the original discrete classes. The thoughtful assignment of numerical codes to qualitative variables will make the subsequent regression analysis more meaningful. For an example of this type of coding refer to Beers, Dress, and Wensel (1966) and Stage (1976).

17-3.2 Measurement of Vegetative Characteristics

The characteristics of the vegetation that can be used to express the quality of a site are the quantity of wood produced, the size characteristics of the trees, and the species of plants naturally occurring on the area.

Since the concept of site quality refers to productivity, its most direct measure is the quantity of wood grown on an area of land within a given period. It is widely assumed that the measurement of site quality in this way (using volume growth, for example) is of limited value because of the difficulty in development and application. The value of using volume growth to define site quality, however, has been reiterated by Mader (1963). The work of Stage (1966) has indicated one feasible approach based on volume growth potential estimated from relations involving stand height, age, and a relative measure of site quality. It is evident that the use of volume growth to assess site quality will develop with the increased application of data-processing equipment and analysis techniques to carefully garnered forest growth data, but apparently no specific approach has been widely adopted.

The size which trees will attain according to their age are also affected by site quality. The most useful tree-size characteristic for site evaluation is tree height. Diameter is less reliable since it is more sensitive to stand density.

The relationship of tree height to age, called *site index,* has been used for many years in evaluating site quality for even-aged stands of single species or nearly pure composition. The height of the dominant and codominant trees has usually been taken as representative of stand height for site index work. This choice is not without its limitations. It relies on subjective decisions about which are dominants and which are codominants. In addition, the crown position of a tree and its height growth may be affected by stand alterations, such as cuttings. Other stand height measures, such as the average height of a specified number of the largest trees or the height of the tallest tree, have been suggested as alternative parameters. Unless otherwise specified, site index is generally defined as the average height that the dominant and codominant trees on an area will attain at key ages such as 50 or 100 years. For example, site index 70 on a 50-year basis means that the dominants and codominants reach an average height of 70 feet in 50 years. Site index 120 on a 100-year basis means an average height of 120 feet in 100 years. Site index curves are prepared for even-aged stands, as shown in Fig. 17-6, to allow site classification for a stand at any age. To assess the site quality of an area, it is necessary only to determine the average height of the dominant and codominant trees and their average age and to locate the position of these coordinates on the site index chart for the species. The site index for the stand can be read from the closest curve. For an average height of the dominants and codominants of 65 feet and a stand age of 40 years, the sited index for white pine in the southern Appalachians from Fig. 17-6 is 75. Total age or age at dbh level can be used in the preparation of site index curves. Husch (1956) has pointed out several advantages of using age at dbh instead of total age. Measuring age at dbh eliminates the necessity of adding arbitrary corrections to convert age at increment-boring level to total age; it measures tree age after the initial period of establishment and adjustment has passed; it also utilizes a standard and conventional point for age determination.

It is important to understand that site index varies according to species. Site index charts are prepared for individual species or for typical forest types, such as the charts prepared by Schnur (1937) for mixed oak forests in the Central States. A single forest area may have different site index values, depending on the species and site index chart. To help solve this problem, regression formulas can be derived that relate the

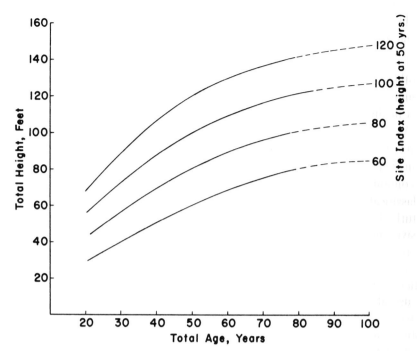

Fig. 17-6 *Site index curves for natural stands of white pine in the southern Appalachians (Doolittle and Vimmerstedt, 1960).*

site index or height of one species with that of another. Examples of this type of study are those of Foster (1959), Deitschman and Green (1965), and Norman and Curlin (1968).

The relationship of age and total height expressed by site index has enjoyed widespread popularity for several reasons. Height has been found closely correlated to the ultimate measure—volume. In addition, the two requisite measurements, height and age, are quickly and easily determined. Height growth has been considered to be only slightly affected by stand density, although some studies have shown that stand height and site index in some cases are related to stand density (Gaiser and Merz, 1951; Husch and Lyford, 1956). Finally, site index has been popular because it provides a numerical expression for site quality rather than a generalized qualitative description.

In uneven-aged stands of several species, height in relation to age cannot be used to express site quality. The height growth of a species in this type of stand is not closely related to age but more to the varying stand conditions by which it has been affected during its life. The classic concept of site index has consequently been restricted to species normally occurring in even-aged stands. McLintock and Bickford (1957) considered several alternatives for evaluating site quality in uneven-aged stands in their study of site quality for red spruce in the northeastern United States. They concluded that the relationship between height and dbh of the dominant trees in a stand was the most sensitive and reliable measure of site quality. Site index according to this concept is then defined as the height attained by dominant trees at a standard dbh. The site index for a tree of any age can be read from the height-dbh curves for the range of site indexes, as shown in Fig. 17-7. This chart utilizes a standard dbh of 14 inches.

Another approach to site quality evaluation makes use of the composition of plant communities on an area of land. This system is based on the theory that certain key species in the forest reflect the overall quality of a site for a tree species or forest type. As Westveld (1954) points out: "The concept . . . recognizes that plant communities are distinct entities developed and arranged in accordance with definite biological laws; they are not mere aggregations of plants brought together by chance. Such communities are well differentiated and are very constant for the same site." Site classification systems based on this concept have been most successful in relatively undisturbed northern forests in Finland and eastern Canada where the forests are extensive and the species few (Ilvessalo, 1937; Sisam, 1938). These systems utilize measures of presence and relative abundance of characteristic species in the minor forest vegetation, called *plant indicators,* as indicative of the quality of the site for specified tree species. An additional premise is that these shorter-lived plants are more useful site quality indicators than the trees themselves, since they are more sensitive and will, after a disturbance of the forest, return to an equilibrium with site conditions more rapidly than trees.

In summary, it has been proven that there does exist a valid correlation between plant indicators and site quality, but it must be kept on broad terms. The general applicability of plant indicator systems has several limitations. It is restricted to forests of simple composition such as occur in northern latitudes; it requires considerable ecological knowledge on the part of the forester; and it must be recognized that the lesser vegetation is affected by forest composition, stand density, and past management as well as site quality. In addition, the lesser vegetation is shallow-rooted and does not indicate conditions in deeper soil horizons, although these soil conditions affect tree growth. In spite of these limitations, indicator plants can be of general assistance in site quality evaluation.

17-3.3 Preparation of Site Index Curves

The preparation of site index curves for even-aged stands is based on average height and age measurements of the dominant and codominant trees on a series of sample plots. If temporary sample plots are used, there should be a sufficient number and distribution to cover equally the range of age and site classes found under natural conditions. For reliable relationships, a total of at least 100 plots is necessary, although more are desirable.

Using such sample data, graphical techniques have traditionally been used to construct the desired series of site index curves. In recent years regression techniques have been employed to remove the subjectivity involved in the handfitting of curves. One commonly used approach is described in the following paragraphs.

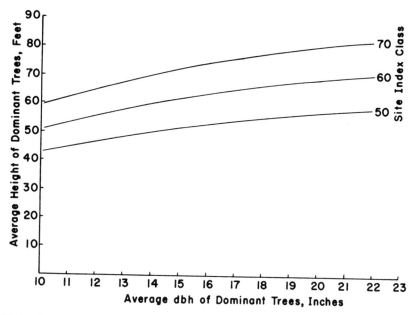

Fig. 17-7 *Site index curves for uneven-aged stands of red spruce (McLintock and Bickford, 1957).*

Using field data involving total tree height and total tree age (plot averages or individual tree data) obtained from selected dominant or codominant trees, site index curves similar to those shown in Fig. 17-6 can be derived. The procedure is as follows.

1. Transform the height and age data to logarithm of height and reciprocal of age and fit a simple linear regression, obtaining numerical values for the constants *a* and *b* in the model

$$\text{Logarithm of height} = a + b \left(\frac{1}{\text{Age}}\right) \tag{17-24}$$

The resulting equation represents the "average" site index curve for the data.

2. Locate points needed to draw a specified site index curve by moving the *Y* axis to the key age (by mathematical translation) and then, by definition, the *Y* intercept is the log of site index. For example, if the key age is 50 years, equation 17-24 is translated to become

$$\text{Log height} = \text{Log site index} + b \left(\frac{1}{\text{Age}} - \frac{1}{50}\right) \tag{17-25}$$

Thus, for a curve where the regression coefficients are

$$a = 1.950 \quad \text{and} \quad b = -4.611$$

$$\text{Log height} = \text{Log site index} - \frac{4.611}{\text{Age}} + 0.09222 \tag{17-26}$$

STAND STRUCTURE, DENSITY, SITE QUALITY, AND YIELD **341**

and the curve for site index 80, for example, can be found by substituting various ages into the following equation, solving for log of height, and then finding height.

$$\text{Log height} = \log 80 - \frac{4.611}{\text{Age}} + 0.09222$$

$$= 1.90309 - \frac{4.611}{\text{Age}} + 0.09222$$

$$= 1.99531 - \frac{4.611}{\text{Age}}$$

3. Locate other site index curves by substituting the pertinent site index number in equation 17-26 and proceeding as described in step 2.

It is often useful to calculate site index from the fitted equation rather than to read a graph. For this purpose, equation 17-25 is rearranged, by solving for the log site index, to become

$$\text{Log site index} = \text{Log height} - \frac{b}{\text{Age}} + \frac{b}{\text{Key age}} \tag{17-27}$$

For the example used in step 2, the resulting equation is

$$\text{Log site index} = \text{Log height} + \frac{4.611}{\text{Age}} - \frac{4.611}{50}$$

For a given plot one can substitute values of average tree height (as a logarithm) and average tree age and obtain logarithm of site index; then, by using antilogs, the site index can be estimated.

More recently, the fitting of more sophisticated models to the height-age relationship has become common. Brickell (1968), for example, used the model

$$H = a(1 - be^{-kA}) \frac{1}{1 - m} \tag{17-28}$$

where

H	= total tree height
A	= total tree age
e	= base of natural logarithms
a, b, k, m	= coefficients estimated by nonlinear regression procedures; a set of polymorphic site index curves were then developed from the fitted least-squares equation

Another study by Beck (1971) used a similar model. The researcher should be aware that the fitting of such nonlinear models, made feasible by the computer, is frequently justified to avoid biases commonly present in the "traditional" approach discussed in this section. Beck and Trousdell (1973) discuss the common sources of error in site index prediction—among the more serious being the assumption that the shape of the height growth curve is the same for all sites. Since this assumption is not commonly met, the use of an approach that provides flexibility in the curve shape for different sites, that is, a polymorphic approach, should be seriously considered. A linear least-squares procedure to prepare polymorphic site curves is described by Bailey and Clutter (1974).

17-4 YIELD TABLES AND FUNCTIONS

A *yield table* is a tabular presentation of volume per unit area and other stand characteristics of even-aged stands by age classes, site classes, species, and density.

Obviously, this definition is not applicable to uneven-aged stands. Volumes cannot be shown at specified ages for an uneven-aged stand, since there is no one representative average age. A type of table has been prepared for uneven-aged stands, showing the volumes produced in growth for given periods with a certain level of growing stock on land of different site qualities (Meyer, 1934). Yield records for uneven-aged stands over long periods are required for preparing this kind of table. Accumulated information of this type is limited in the United States. Consequently, little effort has been devoted to the preparation of yield tables for uneven-aged stands. The increased availability of permanent plot records portends to accelerate this area of research.

Even-aged yield tables are prepared from yield studies of the relationship between a dependent variable, such as volume, basal area, or number of trees, and independent variables describing stand conditions, such as age, site quality, and stand density. Site quality is most often measured in terms of site index, although discrete site quality classes have been used. Density has been most commonly measured in terms of basal area, although stand density index is often more convenient to use.

Yield tables are valuable in such forest management activities as regulating the cut, determining the length of rotation, and forest valuation. Growth estimates can also be derived from yield tables, as discussed in Section 16-2.3.

Although the tabular form of yield relationships has endured years of use in forest management activities, the past decade has seen a proliferation of studies that emphasize the formula form. The conclusion is inescapable—that the construction techniques employed for most of the older yield tables are made obsolete by the modern, mathematically sophisticated approaches. However, regardless of the method of analysis used, the basic nature of yield tables is worthy of discussion. Only even-aged yield tables are described here; uneven-aged yield relationships are briefly treated in Section 17-4.2.

17-4.1 Yield Tables for Even-Aged Stands

Yield tables for even-aged stands are of several types, depending on the independent variables used: *normal, empirical,* and *variable density.*

Normal yield tables show the relationship on a per unit area basis, between the two independent variables, stand age and site index, and one or more dependent variables.

An example is shown in Table 17-1. This table has been prepared for one site index and shows values for a number of dependent variables according to age. Similar tables have been prepared for other site index levels.

This type of table originated before analytical methods for handling three independent variables were available. Since normal yield tables use only two independent variables, they are conveniently constructed by graphical methods. The density variable is held constant by attempting to select sample plots of the same density. The density required has been called *full* or *normal stocking*. Full or normal stocking is supposed to describe the density of a stand that completely occupies a given site and makes full use of its growth potentialities. Since it is difficult to describe quantitatively full stocking, qualitative and somewhat subjective guides must be used. For example, the guides might be: completely closed crown canopy, no openings in the stand, and regular spacing of the trees. Such specifications leave much to the judgment of the individual in choosing so-called fully stocked stands for samples.

The stand parameter values for successive ages, shown in normal yield tables prepared from a number of sample plots, are averages derived from many stands considered fully stocked at the time they were sampled. Each stand in which these sample plots were taken may have shown varying patterns of development over its life. In past years, some stands may have been overstocked or understocked in terms of the definition of normality. When data from these samples are compiled, average relationships are developed that represent the development of a theoretically fully stocked stand over its entire life. It is quite unlikely that any existing stand will show the same pattern as is represented in a normal yield table. In reality, very few stands are encountered that can be called *fully stocked*. Both understocked and overstocked stands can be encountered, with understocking the usual case.

The volume of an existing stand may be estimated from a normal yield table, by measuring its age, site index, and relative stocking percentage or normality. Relative stocking can be measured by comparing the volume, basal area, or number of trees per acre for a stand with the yield table values shown for a stand of the same age and site index, as discussed in Section 17-2.1. There is little point in using volume to measure normality when volume is the parameter to be estimated. Basal area has been found to be the most satisfactory basis for expressing relative stocking. It is easily and quickly determined and is closely related to volume. The stocking or normality percentage times the yield table volume estimates the volume of the existing stand. This naturally assumes that relative stocking in basal area equals relative stocking in volume. This may be a tenable assumption for volume in cubic feet but can lead to serious error for board-foot estimation. Normality percentages for the same stand calculated from basal area, cubic feet, and board feet can give widely differing results.

McArdle, Meyer, and Bruce (1949) utilized the same data for a conventional yield table and prepared a form of normal yield table based on average dbh rather than age. They utilized the idea that stands of the same average dbh are similar, even though differing in age and site class. The resulting yield table then relates all of the normal yield-table variables to average stand diameter. Table 17-2 reproduces this

Table 17-1
Example of Normal Yield Table

Yield per Acre for Fully Stocked Spruce-Aspen Stands; Good Site—Index 80[5]

Spruce, Total Age (years)	Height, Average Dominant (feet)	Tolerant Softwoods, Mainly Spruce						Intolerant Hardwoods, Mainly Aspen			Entire Stand
		Number of Trees	Average dbh (inches)	Basal Area (square feet)	Volume Inside Bark			Basal Area (square feet)	Composition[3] (%)	Volume to Basal Area Factor[4] (units)	Basal Area (square feet)
					Entire Stem (cubic feet)	Merchantable Stem[1] 12 in.+ (cubic feet)	Scribner Rule[2] 12 in.+ (board feet)				
30	23	1091	1.6	16	155	0	0	88	83	16.7	104
40	34	1379	2.5	49	620	5	0	88	64	19.9	137
50	45	1344	3.3	79	1280	20	0	85	52	22.6	164
60	58	1091	4.1	100	2010	55	150	82	45	25.0	182
70	70	845	5.0	114	2740	105	600	79	41	27.1	193
80	80	643	6.0	125	3445	310	1,540	76	38	28.7	201
90	87	484	7.1	133	4060	835	3,490	71	35	30.1	204
100	93	376	8.3	140	4565	1645	7,020	66	32	31.3	206
110	98	307	9.3	145	4995	2520	11,180	60	29	32.3	205
120	102	264	10.2	150	5370	3325	15,590	53	26	33.2	203
130	106	238	10.9	153	5680	3995	18,980	45	23	34.0	198
140	109	220	11.4	156	5945	4460	21,320	36	19	34.7	192
150	112	208	11.8	158	6165	4810	23,060	26	14	35.3	184

[1] 1-foot stump, 4-inch top inside bark.
[2] 1-foot stump, 6-inch top inside bark.
[3] Hardwood composition by basal area.
[4] 4 × basal area = total cubic volume.
[5] Note: In this study, site index is defined as the height attained by the average dominant spruce at 80 years of age.

SOURCE: Macleod and Blyth, 1955.

table. To use the table, the average diameter and number of trees per acre are determined for the stand under investigation. Volume per acre is obtained by multiplying the volume per tree from the table by the number of trees per acre as found from field sampling.

Because of the several subjective decisions necessary in their preparation and use, normal yield tables have been challenged; indeed, the entire normality concept has been seriously questioned in recent years (Nelson and Bennett, 1965). To overcome the fully stocked assumption of normal yield tables, two other types of tables have been used—the empirical yield table and the variable density yield table.

An *empirical yield table* is similar to a normal yield table but is based on sample plots of average rather than full stocking. The judgment necessary for selecting fully stocked stands is eliminated, simplifying the collection of field data. Resulting yield tables show stand characteristics for the average stand density encountered in the collection of the field data.

When stand density is used as an independent variable, *variable density yield tables* result. Tables thus show the yields for various levels of stocking. This approach also has the advantage of not requiring samples to be fully stocked. Sample plots of any density can be used since the density is measured as a variable for the solution.

Mulloy (1947) has prepared variable density yield tables using stand age and stand density index as the two independent variables. Site quality has been accounted for by preparing yield relationships for discrete site classes. Table 17-3 shows the yields of even-aged stands of jack pine for a stand density index of 100. Site as a variable has been held constant for the relationship. This yield table is applicable only to the jack pine cover type in Ontario for average site conditions.

With the increased use of statistical techniques, yield studies involving more than two independent variables can be carried out. MacKinney, Schumacher, and Chaiken (1937) used age, stand density, site index, and a stand composition index as independent variables. This resulted in a more general type of variable density yield table. In a later study MacKinney and Chaiken (1939) used age, site index, and stand density as independent variables; Smith and Ker (1959) used site index, age, maximum height, average stand diameter, basal area per acre, and number of trees per acre in preparing yield equations by multiple regression.

17-4.2 Yield Functions

Most of the early normal yield tables used in America were prepared by graphical procedures described by Bruce (1926) and Reineke (1927) and were improved on by Osborne and Schumacher (1935). Departure from the graphical approach was evident in the classic study by MacKinney et al. (1937), who used the least-squares regression technique applied to a logarithmic transformation of the Pearl-Reed logistic curve. The multiple regression applications by Schumacher (1939), Smith and Ker (1959), among others, further demonstrated the superiority of sound statistical techniques over the purely graphical approach for the preparation of yield tables. The fact that regression techniques provide a "yield formula" as well as yield tables was recognized as a distinct advantage especially in early computer applications. Through all this development, however, a major problem was ignored or overlooked—because yield functions (which predict stand volume at a specified age) and growth functions (which predict volume growth over shorter periods) were often derived independently, summation of a succession of periodic *growth estimates* added to an initial volume would not necessarily lead to the final stand volume indicated by the *yield function estimate*. The application of calculus to growth and yield studies led to the resolution of this inconsistency between growth summation and terminal yield. The independent, essentially simultaneous works of Buckman (1962) and Clutter (1963) began a new era in yield studies. A brief description of their work is appropriate.

Working with even-aged red pine, Buckman (1962) emphasized that growth and yield are not independent phenomena and should not be treated as such. Furthermore, he employed methods of calculus which had been neglected in virtually all previous yield studies of this type. Beginning with the basal area *growth equation* of the form

$$Y = b_0 + b_1 X_1 + b_2 X_1^2 + b_3 X_2 + b_4 X_2^2 + b_5 X_3 \tag{17-29}$$

where

Y = periodic net annual basal area increment (i.e., dX_1/dX_2, change in basal area with respect to age)
X_1 = basal area in square feet per acre
X_2 = age in years
X_3 = site index

yield tables were prepared by iterative solution and cumulation; that is, the least-squares fit of equation 17-29 is solved for a particular site, age, and stand density. Basal area growth is then added to the stand density, 1 year is added to age, and the equation is again solved. Addition of the n successive annual growth estimates to the initial basal area provides a yield estimate n years hence.

Clutter (1963) working with loblolly pine (even-aged) clearly indicated the relationship between growth and yield models by the following definition.

> *Such models are here defined as compatible when the yield model can be obtained by summation of the predicted growth through the appropriate growth periods or, more precisely, when the algebraic form of the yield model can be derived by mathematical integration of the growth model.*

In the research reported by Clutter a yield model was first prepared of the form

$$\ln V = a + b_1 S + b_2(\ln B) + b_3 A^{-1} \tag{17-30}$$

where

$\ln V$ = logarithm to the base e of volume
A = stand age in years
S = site index in feet
B = basal area per acre in square feet

Table 17-2
Revised Douglas Fir Yield Tables (based on average diameter instead of site and age)

Average dbh of Stand[1] (inches)	Normal No. of Trees per Acre	Normal Height of Trees of Average dbh (feet)	Total Stand[2] and Entire Stand (cubic feet)	Volume Per Tree			Volume per Tree, 12 inches dbh and Over	
				5 Inches dbh and Over to a 4-Inch Top (cubic feet)	7 inches dbh and Over to a 4-inch Top (cubic feet)	12 inches dbh and Over to a 4-inch Top (cubic feet)	International ⅛-Inch Rule[3] (board feet)	Scribner Rule[4] (board feet)
2	4466	22						
3	2387	31						
4	1530	39	1.8	0.9	0.2			
5	1084	47	3.2	2.1	1.1			
6	818	55	5.1	3.8	2.6	0.3		
7	644	62	7.6	6.2	4.9	1.1	5	3
8	524	69	10.9	9.4	8.0	2.5	18	11
9	437	76	14.9	13.4	12.1	5.3	35	23
10	371	83	19.6	18.0	16.7	9.5	66	43
11	320	90	25.2	23.6	22.7	15.1	102	67
12	280	97	31.5	29.8	29.3	21.7	148	99
13	248	104	38.5	36.6	36.5	29.5	224	149

14	221	110	46.6	44.3	44.3	38.3	274	184
15	178	117	55.5	52.8	52.8	48.0	347	236
16	180	123	65	62	62	58	432	296
17	104	130	76	72	72	69	521	359
18	150	135	87	83	83	81	618	429
19	138	141	99	95	95	93	724	510
20	127	147	112	108	108	106	836	593
21	118	152	126	121	121	119	956	683
22	110	157	142	136	136	134	1075	779
23	102	162	158	152	152	150	1205	886
24	96	167	175	169	169	168	1339	999
25	91	171	193	186	186	185	1485	1125
26	85	176	213	205	205	204	1653	1262
27	80	180	234	227	227	227	1826	1405
28	76	185	256	249	249	249	2031	1562
29	72	189	279	271	271	271	2249	1730
30	68	194	302	293	293	293	2476	1905

[1] Weighted by basal area.
[2] Total stand, i.e., trees over 1.5 inches dbh.
[3] To 5-inch top.
[4] To 8-inch top.

SOURCE: McArdle, Meyer, and Bruce, 1949.

Table 17-3
Variable Density Yield Table:
Jack Pine—Medium Site—100 SDI[1]; Timigami, Ontario

Age (years)	Total Volume (cubic feet)	Current Annual Increment (cubic feet)	Mean Annual Increment (cubic feet)
10	185	23.5	18.5
20	420	24.0	21.0
30	660	15.5	22.0
40	815	11.9	20.4
50	934	11.3	18.7
60	1,047	10.3	17.4
70	1,150	10.3	16.4
80	1,253	10.3	15.7

[1] Volumes for other stand density indices can be obtained by pointing off two places in the stand density index and multiplying times the tabular values.

SOURCE: Mulloy, 1944.

and differentiated with respect to age, obtaining

$$\frac{dV}{dA} = b_2 V B^{-1}(dB/dA) - b_3 V A^{-2} \qquad (17\text{-}31)$$

where

dV/dA = rate of change of volume with respect to age or instantaneous rate of volume growth

dB/dA = rate of change of basal area with respect to age or instantaneous rate of basal area growth

Since the rate of basal area growth is not ordinarily available, regression analysis was used to obtain dB/dA as a function of age, site, and basal area. The model finally adopted was

$$dB/dA = -B(\ln B)A^{-1} + c_0 A^{-1} B + c_1 B S A^{-1} \qquad (17\text{-}32)$$

Substituting this relation for dB/dA in equation 17-31 led to the equation

$$dV/dA = -b_2 V(\ln B)A^{-1} + b_2 c_0 V A^{-1} + b_2 c_1 V S A^{-1} - b_3 V A^{-2} \quad (17\text{-}33)$$

Using the form of this equation as a model, and based on data gathered on permanent sample plots, a least-squares regression equation was obtained, thus relating volume growth with present basal area, age, site index, and volume (estimated using equation 17-30). Subsequent integration of this regression equation led to the final yield function, from which volume yield at some future time could be predicted from given initial age, basal area and volume, projected age, and site index.

Virtually all published studies of growth and yield have been undertaken for even-aged stands. The application of recent growth and yield model theory to uneven-aged stands is quite feasible, however, as shown by Moser and Hall (1969). Moser proposed a growth model (a generalization of von Bertalanffy's growth-rate function) that, when mathematically integrated, provides a yield function in which time is represented by the variable "elapsed time from an initial condition." Using annual remeasurements of permanent plot data, methods were evaluated for the development of compatible growth and yield functions for uneven-aged stands.

No serious study of growth and yield should be undertaken without prior examination of the dissertations of Turnbull (1963) and Pienaar (1965) and their published work (Pienaar and Turnbull, 1973). A major contribution of these works is that forest-growth phenomena are studied from a basic standpoint to develop the general quantitative growth theory of even-aged forest stands as well as "biomathematical growth models," as opposed to empirical and semi-empirical regression models, which abound in forestry literature.

In recent years growth and yield models have emerged as primary forest management tools. Each of the recommended approaches has an array of supporting literature, further discussion of which is inappropriate here. The interested reader is directed to summary papers such as those prepared by Evans, Burkhart, and Parker (1975), Williston (1975), and Dudek and Ek (1980).

18
Other Mensurative Considerations

This chapter will cover practical methods of estimating wildlife populations, of estimating use of outdoor recreation areas, of applying ratios of geometric magnitude, and of monitoring campsites for deterioration. It will also look realistically at the problems involved in value measurement. The chapter is not a substitute for a textbook on outdoor recreation management or on wildlife management. Instead, it is designed to describe the more important measurements with which forest resource managers are concerned.

18-1 POPULATION ESTIMATION

The most obvious way to determine the number of animals in a population, or the number of users of a recreational area, is to make a complete count. We could make a *spacial* census or a *temporal* census. A spacial census is a count of all animals or people in a specified area *at a specified point* in time. A temporal census is a count of all animals or people *passing a point* during a specified interval of time. One might also make a census of a sample of the spacial or temporal dimension over which the census is defined and obtain a statistical estimate of the complete population.

A count of bobwhite quail in a specified area on a specified date and a count of people in a specified campground on a specified date are both examples of spacial censuses. Examples of temporal censuses include counts of crows as they enter roosts at the end of the day, counts of anadromous fish as they pass through fish ladders during a given period, and counts of backpackers as they pass a specified point on a trail on a particular day.

If a complete count is taken, there generally still will be some error. Consequently, the constraints under which the data were collected should be indicated. For example, one might indicate that the entire population of woodland deer mice had been captured in a specific area. Although it is possible to capture a high proportion of the mice, the catch would not include young mice still in the nest. This should be indicated.

If a sample is taken, one should strive to obtain the best possible estimates, commensurate with funds and objectives, and to make a statement about the accuracy of the estimate.

In estimating animal populations one must have an understanding of the movement of animals (i.e., are they migratory or nonmigratory?), their normal ranges, and so on, and clearly indicate the spacial or temporal dimension of the census. In estimating recreational use there are some special problems. People use developed sites, dispersed areas, and wilderness areas. The propensity of people to visit any particular type of area is little understood. Therefore, "range" cannot be defined for recreationists even when they are put into socioeconomic groups. But, in general, recreationists tend to concentrate along and on rivers, lakes, roads, trails, and scenic and historical points of interest. And they often move in various kinds of motorized vehicles. Consequently, the problems in estimating recreational use are somewhat different than in estimating animal populations. Although there are differences, one should look for analogies. For example, the reader might consider the possibility of applying the mark-capture method of censusing animal populations (Section 18-4) to estimate recreational use.

18-2 STRIP-CENSUS METHOD OF ESTIMATING WILDLIFE POPULATIONS

Leopold (1933) described a strip-census method from the unpublished work of R. T. King for estimating ruffed grouse populations, which has been dubbed the *King method*. This method assumes that ruffed grouse will flush when a certain "critical distance" is reached between bird and observer. In practice, the observer walks a predetermined line and records the distance to each bird flushed. The average flushing distance is computed. This average distance is assumed to be one-half the effective width of a strip covered by the observer. Thus, for each grouse flushed,

$$\text{Unit area conversion factor} = \frac{H}{L(2d)}$$

And then,

$$N = \left(\frac{Hn}{L(2d)}\right)A \tag{18-1}$$

where

N = total population of study tract
H = unit area (1 hectare = 10,000 m^2; 1 acre = 43,560 ft^2)
n = total number of grouse observed
L = length of census strip in same units as H
d = average flushing distance in same units as H
A = area of tract in hectares or acres.

This method, with variations, has been used to census, in addition to ruffed grouse, white-tailed deer, snowshoe hares, quail, and woodcock in the United States (Giles, 1971), and larger mammals in the Po National Park, Upper Volta, West Africa (Heisterberg, 1977).

The most important variation of the King method was first described by Hayne

(1949). Hayne pointed out that population estimates based on average flushing distances are often significantly in error. To remedy this, although Hayne did not recognize it as such, he treated the method as a form of p.p.s. (polyareal plot sampling). Indeed, the derivation of the applicable formula is analogous to the derivation of the formulas for the application of horizontal line sampling. Here is Hayne's method.

1. Obtain plot area associated with each animal flushed in terms of flushing distance.
2. Obtain the unit area conversion factor for each animal (e.g., number of animals per hectare or per acre) by dividing the plot area associated with the animal into the unit area, and simplifying.

To obtain an estimate of the population in the study tract, one then multiplies the unit area estimate by the total area of the study tract. Thus, for the ith grouse flushed,

$$\text{Unit area conversion factor} = \frac{H}{L(2d_i)}$$

and then,

$$N = \frac{H}{2L}\left(\frac{1}{d_1} + \frac{1}{d_2} + \frac{1}{d_3} + \cdots + \frac{1}{d_n}\right)A \tag{18-2}$$

where

N = total population of study tract
H = unit area (1 hectare = 10,000 m²; 1 acre = 43,560 ft²)
L = length of census strip in same units as H
$d_1, d_2, d_3, \ldots, d_n$ = flushing distances in same units as H
A = area of study tract in hectares or acres

If a strip census is to be conducted, it is recommended that equation 18-2 be used to work up the data. In addition, one must understand that certain fundamental assumptions must be satisfied if satisfactory results are to be obtained. These are:

- The distance at which the animals will flush or be seen on the approach of an observer will vary with individual animals.
- The animals are scattered randomly over the study tract with regard to flushing or sighting distance, and the animals will not move out of the observer's path without being seen.

18-3 FECAL-COUNT METHOD OF ESTIMATING WILDLIFE POPULATIONS

Fecal counts have been used for many years to determine the presence or absence of ungulates, rabbits, small mammals, and gallinaceous birds. But in certain cases pellet group counts can be used to estimate the size of wildlife populations.

The most widespread application of pellet group counts for estimating wildlife populations is the estimation of deer numbers. If one knows the number of pellet groups per unit area, the average number of pellet groups deposited per deer per day, and number of days of deer use, one can estimate the deer population from the following formula.

$$N = \left(\frac{\Sigma y_i \left(\frac{1}{a}\right)}{nD(13)} \right) A \tag{18-3}$$

where

N = total population of deer on study tract
y_i = number of pellet groups on the ith plot
a = plot area in hectares or acres
n = total number of plots taken
D = days of deer use
A = area of tract in hectares or acres

The number 13 in the denominator of equation 18-3 is the estimated average number of pellet groups deposited per deer per day. Although the defecation rate varies with the succulence of forage, food intake, and age, 13 pellet groups per day is a practical value for field use in the late fall and winter with natural populations of white-tailed or mule deer. [McCain (1948) reworked the data on white-tailed deer from Bennett, English, and McCain (1940) to obtain this value for 1 deer-day. The value has been documented for mule deer (Smith, 1964).]

The days of deer use are best obtained in areas where preservation of pellet groups is optimal. Consequently, the method is best adapted to locations where days of use can be taken as the interval between the average date of leaf abscission and the date of the survey, or between the date of last snowfall and the date of the survey.

Circular and rectangular sample plots with areas ranging from 0.005 to 0.05 hectares (0.012 to 0.12 acres) have been used to make fecal counts, but 0.004 hectare (0.01 acre) circular plots are recommended. Systematic or random sampling may be used, but systematic sampling is generally applied, because a systematic sample is simpler to lay out (Chapter 12).

White (1968) found in a study of white-tailed deer on the Crane Ammunition Depot in Martin County, Indiana, that the mean number of pellet groups per plot varies with cover type (i.e., open, brush, and timber). Consequently, sampling is more efficient if it is proportionate to the area of the cover types (Chapter 12).

Some wildlife scientists feel that it is more important to use pellet group counts to determine the relationship of the amount of deer dung per unit area (i.e., amount of deer use) to range use and condition than to determine population. For example, yearly use trends in a wintering ground, or in a particular cover type, can be determined from pellet group counts.

18-4 RATIO METHODS OF ESTIMATING WILDLIFE POPULATIONS

It is interesting to speculate on when the concept of ratio first appeared. The idea that one herd of ungulates is twice as large as another and the idea that one lake is one-half as large as another both involve notions of ratio that probably developed early in the history of the human race. One idea has to do with the ratio of numbers, the other with the ratio of geometric magnitudes. The ratio of numbers is often used to estimate wildlife populations, while the ratio of geometric magnitudes may be used to make comparisons that normally do not deal with populations. Consequently, the ratio of numbers will be discussed in this section; the ratio of geometric magnitudes will be discussed in Section 18-5.

Two basic ratio methods of estimating wildlife populations are the *mark-recapture method* and the *change-of-composition method*. Both methods generally involve estimating the proportions of two classes of animals in a population at Time 1, then changing the proportions by adding or removing known numbers of known classes and estimating the new proportions of the classes at Time 2. (Although the proportions of more than two classes might be estimated, they seldom are.) The classes of animals, which must remain constant between Time 1 and Time 2, might be marked and unmarked, male and female, juvenile and adult, game and nongame species, and so on. Losses and additions to a population other than those carried out for the purpose of the population estimate must be nonselective with respect to the classes of animals used for the population estimation, and they must not occur while the specified additions or removals are in progress.

Of all the ratio methods of estimating populations, the *mark-recapture method* is the oldest. Petersen (1896), who was probably the first to use the method, used it to study the yearly immigration of young plaice into Limfjord from the German Sea. Lincoln (1930), who often gets credit for being the first to use the method, used it to estimate waterfowl abundance on the basis of banding returns. Consequently, this ratio estimator is called the *Lincoln index* by many people and the *Petersen estimator* by others. Since most ratio methods can be considered to be variations of the *change-of-composition method*, this method will be discussed first.

To derive the equation for the *change-of-composition method*, we first write two structural equations (Rupp, 1966) that show the relationship between the desired quantities.

$$P_2 N_2 = P_1 N_1 + C_1$$

$$N_2 = N_1 + C_1 + C_2$$

where

N_1 = total population at Time 1
N_2 = total population at Time 2
C_1 = net number of animals of Class 1 added or removed between Times 1 and 2

C_2 = net number of animals of Class 2 added or removed between Times 1 and 2

P_1 = proportion of N_1 that consists of a class of animals (e.g., Class 1)

P_2 = proportion of N_2 that consists of a class of animals (must be same class used to determine P_1, e.g., Class 1)

Now, substituting $(N_1 + C_1 + C_2)$ from the second equation for N_2 in the first equation and solving for N_1, we obtain

$$N_1 = \frac{C_1 - P_2(C_1 + C_2)}{P_2 - P_1} \tag{18-4}$$

When P_1, P_2, C_1, and C_2 are known or can be estimated, and when $P_1 \neq P_2$, this equation is suitable for estimating population. But if $P_1 = P_2$, no population estimate is possible. Also note that, if animals are added between Time 1 and Time 2, C_1 and C_2 are plus; if animals are removed, C_1 and C_2 are minus.

Equation 18-4 is most commonly used to estimate deer populations. However, it may be used with other game animals and with fish. Rupp (1966) gave a good example of the use of the formula in estimating a legal-sized trout population, which was shown by sampling to consist of 70 percent brook trout and 30 percent brown trout at Time 1. Between Time 1 and Time 2 anglers caught and removed 200 brook trout and 30 brown trout. Sampling at Time 2 showed the composition of the population to be 55 percent brook trout and 45 percent brown trout. Then, $C_1 = -200$, $C_2 = -30$, $P_1 = 0.70$, $P_2 = 0.55$, and the estimate of the total trout population at Time 1 is

$$N_1 = \frac{-200 - 0.55(-200 - 30)}{0.55 - 0.70} = 490$$

Since the *mark-recapture method* is simply a special application of the change-of-composition method, equation 18-4 may be used to estimate populations when the mark-recapture method is applied. For example, assume that 78 deer were captured, marked, and released on the Crane Naval Ammunition Depot, Indiana, in October 1975, and that during the 1975 hunting season 38 marked deer and 1405 unmarked deer were killed. In this case we put the marked deer in Class 1 and the unmarked deer in Class 2. Note that the marking of 78 animals constitutes a removal from Class 2 and an addition to Class 1. Also note that there were no Class 1 animals at Time 1. Thus,

$C_1 = 78 - 38 = 40$ = net change in Class 1 animals during period

$C_2 = -78 - 1405 = -1483$ = net change in Class 2 animals during period

$P_1 = 0/N_1 = 0$ = proportion of N_1 that consists of Class 1 animals

$P_2 = 38/(1405 + 38) = 0.02633$ = proportion of N_2 that consists of Class 1 animals

And so by equation 18-4 the estimated deer population N_1 just before the 78 deer were marked is

$$N_1 = \frac{40 - 0.02633[40 + (-1483)]}{0.02633 - 0}$$

$$= \frac{78}{0.02633} = 2962$$

Since the numerator in the above formula equals the total number of animals marked M, and since $P_1 = 0$, for the application of the mark-recapture method equation 18-4 may be rewritten

$$N_1 = \frac{M}{P_2} \tag{18-5}$$

It is illuminating to discuss the mark-recapture method in terms of the *urn model*. Think of an urn containing W white balls and M black balls. Then the expected ratio of the number of white balls drawn w to the number of black balls drawn m will be the same as W to M. That is,

$$\frac{W}{M} = \frac{w}{m} \quad \text{and} \quad \frac{W + M}{M} = \frac{w + m}{m}$$

If we denote the population of balls $W + M$ as N, and the number of balls drawn $w + m$ as n, we obtain

$$\frac{N}{M} = \frac{n}{m}$$

Think now of a wildlife population N_1 that is analogous to the population of balls in the urn. In this population M animals are marked (analogous to the black balls in the urn). When n animals are captured, observed, or killed (analogous to the drawing of balls from the urn), of which m are marked, the wildlife population N_1 may be estimated from the formula

$$N_1 = \frac{Mn}{m} \tag{18-6}$$

For the example consisered above, equation 18-6 gives the same result as equation 18-4: $N_1 = 78(1405 + 38)/38 = 2962$. Indeed, equation 18-6 is identical to equation 18-5, which was derived from equation 18-4, since $n/m = 1/P_2$.

It should be emphasized that all ratio methods of estimating wildlife populations assume that the probability of observing, capturing, or killing any group of animals is proportional to their frequency in the population. This may not be true, particularly when a population is divided into subpopulations and population estimates are desired for each subpopulation (e.g., a deer population might be divided into adult bucks, buck fawns, adult does, and doe fawns). Nevertheless, ratio methods are useful in censusing a variety of animal populations.

18-5 RATIO OF GEOMETRIC MAGNITUDES

The *coefficient of condition* K is an interesting ratio of magnitudes that has been employed by fishery investigators (Lagler, 1956).

$$K = \frac{W}{L^3} \tag{18-7}$$

where

W = weight of fish in grams
L = length of fish in centimeters

This coefficient is a good measure of the relative robustness of fish and therefore has been widely used to indicate the suitability of a specific environment for a species, by comparison with regional averages, and to measure the effects of environmental improvement on a species.

Investigators have found that K will change with age, sex, and season. Consequently, when values of K are to be compared, the values should be for fish of the same length, age, and sex, and should be collected on the same date.

The *diversity index I* is another useful ratio of geometric magnitudes (Patton, 1975). This index can be used to express and compare the irregularity of lake shorelines, of opening or clearcut boundaries, or of watershed boundaries. Specifically, the diversity index is the ratio of the perimeter of an opening (lake, clearcut, watershed) to the circumference of a circle with the same area as the opening. The formula is

$$I = \frac{P}{C}$$

Since in terms of area, $C = 2\sqrt{\pi A}$,

$$I = \frac{P}{2\sqrt{\pi A}} = \frac{0.282P}{\sqrt{A}} \tag{18-8}$$

where

P = perimeter of the opening in meters or feet
A = area of the opening in meters or feet
C = circumference of a circle of area A

The diversity index may also be expressed as a percentage.

$$I(\%) = (I - 1)100$$

For example, if a "forty" (a square 40-acre area) was clearcut, $I = 1.13$, and I(percent) $= (1.13 - 1)100 = 13$ percent. This means that the square clearcut has 13 percent more perimeter than a 40-acre circle.

Since game (particularly low-radius game) is a phenomenon of edges, it is valuable to express "edge" in terms of the diversity index. And since for low-radius game the more types the better, a statement on the number of types that are on the edge makes the index more meaningful. For example, a clearcut ($I = 1.89$) which includes 5 types might be denoted 1.89(5).

18-6 RECREATIONAL USE ESTIMATION

Many managers of outdoor recreation areas do not have reliable methods of estimating the number of visitors using their facilities. As Madden and Knudson (1978) point out, many estimates of use are reported with no statement about the accuracy of the estimates. With this point in mind, it is illuminating to consider the methods that are used to estimate recreation use.

1. Guesses
2. Observational estimates
3. Self-counting systems
4. Direct-counting systems
5. Indirect-counting systems

The first two methods are more widely used than most people realize, or will admit. Also, they almost always indicate increasing use and are often gross overestimations.

Information for the self-counting method can be obtained from user permits, registration books, guest books, organizational records, and self-administered questionnaires. The accuracy of such counts depends on cooperation and truthfulness.

Direct counting involves enumeration by an observer, or by a remote-sensing device such as a terrestrial camera or television camera, of recreationists on a given site. Its effectiveness depends on the selection of appropriate times to make the counts and on the feasibility of counting all or nearly all of the recreationists present at the times selected.

Indirect counting involves counting or measuring things that are correlated with the number of recreationists on an area. These counts must be inexpensive so that they can be taken on a regular basis. They include concession receipts, water consumption, axle counts, lift and tow ticket sales, turnstile counts, and so on. Direct counts are taken to calibrate estimating equations that use the indirect counts.

Regarding units of measurement with which to report recreation use, the Bureau of Outdoor Recreation (1973) stated the following.

> *Each federal land managing agency will report annually to the Bureau of Outdoor Recreation, in accordance with the Land and Water Conservation Fund Act of 1965, as amended, on recreation use at each management unit, using the recreation visitor-hour as the standard unit of measure. When available and appropriate, agencies also should include recreation visit and activity-hour data.*

A recreation visitor-hour is the presence for recreation purposes of one or more persons for continuous, intermittent, or simultaneous periods of time aggregating 60 minutes.

A recreation activity-hour is a recreation visitor-hour attributable to a specific recreation activity.

A recreation visit is the entry of any person into a site or area of land or water for recreation purposes.

Other units of measurement, which derive from the visitor-hour and the activity-hour, are the visitor-day and the activity-day. A *visitor-day* is 12 visitor-hours. An *activity-day* is the average number of hours of participation per day in a given activity. For example, for a hypothetical recreational area an activity day might be 1.9 hours for hiking, 2.7 hours for boating, and 4.3 hours for fishing.

Methods of estimating the desired statistic (visitor-hours, activity-hours of swimming, activity-hours of hiking, number of visits per day, etc.) will vary with the type of site: developed site, dispersed area, or wilderness area. A *developed site* is one that has facilities and attractions that concentrate use on a small area (e.g., a swimming beach, a picnic area, a campground, etc.). A *dispersed area* is an area that has less use per unit area than a developed site (e.g., areas in forests and wildlife refuges, and along and on rivers and lakes). A *wilderness area* is a large area with access by trails only. It is beyond the scope of this book to discuss the estimation of visitor use in other than developed sites. And for such sites the discussion will be limited to the use of traffic counters. For more information on attendance methodology see Madden and Knudson (1978).

In using traffic counters one determines the relationship between the desired statistic and axle counts by simultaneously measuring both. For a given site the traffic counter should be placed across a road at a point where most of the traffic will pass.

To derive a predicting equation, Madden and Knudson (1978) recommend that one take a random sample of about 45 one-hour observations for the season of interest (e.g., high-use season, off season, etc.). Then, to obtain data to prepare an equation to estimate visits per hour, one should read the traffic counter at the start and end of the sampling hour, and count the number of first-time visitors (for the day) in each vehicle (car, bus, bicycle, etc.) that pass during the hour. The count should not include employees, service drivers, or others who are not first-time visitors. When the data have been collected, simple linear regression may be used (but may not be the best method) to determine the relationship between number of axle counts per hour and number of first-time visitors per hour. Then both sides of the equation are multiplied by 12 to obtain an equation to predict number of first-time visitors per 12-hour day from number of axle counts per 12-hour day.

One might also take a random sample of 10 or more 12-hour days (James and Ripley, 1963). On the sample days the number of people visiting an area and the use levels of recreational facilities may be determined hourly during the 12-hour period. This procedure may be used to establish the relationship between axle counts and visitor-hours, axle counts and activity-hours for specific activities (swimming, picnicking, hiking, etc.), as well as between axle counts and number of first-time visits per 12-hour day. It should be understood that a linear equation will not necessarily give the best fit for all relationships. Other equation forms might be better.

If there are no major changes in facilities or use patterns in a developed site, an equation should give satisfactory results for 3 to 5 years (James, 1966).

18-7 MONITORING CAMPSITES FOR SIGNS OF DETERIORATION

The carrying capacity of a campsite may be defined as the number of visitor-hours that the site can provide each year without major biological or physical deterioration, or without appreciable impairment of the recreational experience. Consequently, in monitoring a campsite the manager should look for biological, physical, and social clues that indicate that use exceeds capacity. By measuring the changes in the vegetation at the same time each year, the manager can determine if there is a trend toward deterioration. On-site opinion polls may be used to determine how biological and physical deterioration affect the recreational experience.

It is useful to establish condition classes, or other criteria, as a reference base for monitoring a given type of campsite. The following modification of a classification system given by Frissell (1978) illustrates a standard for judging recreation impact on wilderness campsites.

Condition Class	Visible Indicators	Management
1	Ground vegetation flattened but not permanently injured.	No management action necessary.
2	Ground vegetation worn away around fireplaces or centers of activity.	Future use should be carefully monitored to detect adverse change.
3	Ground vegetation lost on most of site, but humus and litter still present in all but a few areas.	Modification of current use patterns and intensities may be needed to prevent further change.
4	Bare mineral soil widespread. Tree roots exposed on the surface.	Withdraw use from these sites and allow recovery.
5	Soil erosion obvious. Trees reduced in vigor or dead.	Sites should be closed permanently and alternate ones located.

For a developed campsite that has roads, trails, parking spurs, and facility pads, complete with tables, fireplaces, etc., at each unit, the criteria would be different.

But one must always remember: "Although physical characteristics may define a site's initial durability, the decision to limit use rather than let the site deteriorate, intensify management, or even pour more concrete is dictated by human objectives, not ecological imperatives" (Wagner, 1974).

Plots or transects may be used to monitor the ground vegetative cover. However, stereo photos taken from permanent camera stations offer advantages over plots and transects. They give permanent three-dimensional records that can be compared to "standard stereograms" to judge campsite condition. Stereoscopic pictures differ so entirely from flat pictures in the impressions they produce that they open up a wholly new range of possibilities. In addition to giving the effect of depth they have another quality not possessed by two-dimensional photographs; they reproduce surface texture with striking naturalness. They give you the feeling of being on the ground. Details on how to make stereograms are given in Chapter 13.

18-8   VALUE MEASUREMENT

Forest resource managers may be called on to measure, or determine, outdoor recreation preferences, the esthetic impact of timber harvests, the reasons people participate in outdoor recreation, and so forth. If one becomes involved in such studies, one will often be forced to go beyond what is known in order to achieve the desired objectives. In fact, considering the various methodological problems and the special difficulties associated with such measurements, progress made in this area is quite minimal.

The reason the problem is complex is because it involves the measurement of *value* (i.e., the relative worth, merit, or importance of a thing). It is essential to understand the problems posed by value measurement.

Measurement presupposes something to be measured. Unless we know what that something is, no measurement has any significance. Value measurement, therefore, requires definition. But even after a value has been defined, how can we be sure, when measurement comes along, that what is defined and what is measured are the same thing? For example, can we really expect a subject to give figures (or ranks) that accurately reflect the relative attractiveness of various sites? And given that attractiveness is an abstract notion and that people do not necessarily know their subconscious evaluation of things, can any statement from subjects about their inner feelings have a knowable level of validity?

So a major problem with value measurements is defining values and then developing quantitative expressions for the values. Indeed, the hope seems to be that some theory of measurement or some special treatment of the problem will be forthcoming. But one must face up to the difficulties involved and realize that given some uses of value, the measurement of value is either theoretically impossible or extremely difficult.

Also, accuracy is a highly relative term for value measurements. For example, statisticians often define precision in terms of a confidence interval. This interval has a specific probability of including the "true" measurement. Each set of observations is the basis for weaving a net to catch the truth. The confidence interval tells us the probability of a successful catch. It is quite possible to determine the confidence interval in the estimation of timber volumes; however, it becomes uncertain in the measurement of wildlife populations and recreation use, and almost impossible in the measurement of values.

In closing this chapter we would emphasize that techniques for measuring value are not new. They extend back to the eighteenth century. However, since the end of World War II, there has been an explosive growth of research in this area. If one desires to explore the field, a good starting point is the Canadian Outdoor Recreation Demand Study (Parks Canada, 1976).

Appendix

Table A-1
Some Converting Factors for Common Units of Measurement

A. Length

	mm	cm	m	km	in.	ft	yd	rod	chain	mile
1 millimeter	1	0.1	0.001	—	0.0394	0.0033	—	—	—	—
1 centimeter	10	1	0.01	10^{-5}	0.3937	0.032808	0.010936	—	—	—
1 meter	1,000	100	1	10^{-3}	39.3701	3.28084	1.09361	0.19884	0.04971	0.00062
1 kilometer	10^6	10^5	1,000	1	39,370.1	3,280.84	1,093.61	198.839	49.7097	0.62137
1 inch	25.4	2.54	0.0254	—	1	0.08333	0.02778	0.00505	—	—
1 foot	304.8	30.48	0.3048	$.3048 \times 10^{-3}$	12	1	0.33333	0.060606	0.01515	0.000189
1 yard	—	91.44	0.9144	$.9144 \times 10^{-3}$	36	3	1	0.18182	0.04545	0.000568
1 rod	—	502.92	5.0292	0.00503	198	16.5	5.5	1	0.25	0.003125
1 chain[a]	—	2,011.68	20.1168	0.02012	792	66	22	4	1	0.0125
1 mile[b]	—	—	1,609.344	1.6093	63,360	5,280	1,760	320	80	1

[a] There are 100 links per chain; each link = 0.66 feet or 7.92 inches; 1 furlong = 10 chains

[b] 1 international nautical mile = 1852 meters.

Table A-1 (continued)

B. Area

	cm²	m²	hectare	km²	in.²	ft²	yd²	chain²	acre	mile²
1 centimeter²	1	0.0001	10^{-8}	—	0.15500	0.001076	1.196×10^{-4}	—	—	—
1 meter²ᵃ	10,000	1	0.0001	10^{-6}	1,549.997	10.76387	1.19599	0.00247	0.000247	3.861×10^{-7}
1 hectareᵃ	10^8	10,000	1	0.01	15.50×10^6	107,638.7	11,959.9	24.7104	2.471044	0.0038610
1 kilometer²	10^{10}	10^6	100	1	—	10,763,867	1,195,985	2,471.04	247.104	0.38610
1 inch²	6.4516	6.4516×10^{-4}	—	—	1	0.006944	0.7716×10^{-5}	—	—	—
1 foot²	929.0304	0.092903	—	—	144	1	0.11111	0.000230	0.000023	—
1 yard²	8,361.27	0.836131	—	—	1,296	9	1	0.002066	0.000207	—
1 chain²ᵇ	—	404.687	0.040469	0.0000405	627,264	4,356	484	1	0.1	0.000156
1 acreᶜ	—	4,046.87	0.404687	0.0040469	6,272,640	43,560	4,840	10	1	0.001562
1 mile²	—	258,999.7	258.9997	2.590	—	27,878,400	3,097,600	6,400	640	1

ᵃ 1 hectare = 100 are.

ᵇ 1 square chain = 10,000 square links.

ᶜ 640 acres = 1 section; 36 sections = 1 township (6 × 6 miles).

Basal area per unit land area converting factors:

1 square foot per acre = 0.2296 square meters per hectare.

1 square meter per hectare = 4.3560 square feet per acre.

Table A-1 (continued)

C. Volume

	cc	liter[b]	m³	in.³	ft³	yd³	oz[a]	pint[a]	qt[a]	gal[a]	qt[c]
1 cubic centimeter	1	0.001	0.000001	0.061024	—	—	0.033814	0.002113	0.001057	—	0.00091
1 liter	1,000.027	1	0.001	61.0237	0.035315	0.0013	33.8147	2.11342	1.05671	0.264178	0.90810
1 cubic meter	10^6	999.973	1	61,023.7	35.3145	1.308	—	—	1,056.68	264.170	908.08
1 cubic inch	16.3871	0.016387	16.38×10^{-6}	1	0.000579	21.43×10^6	0.554113	0.034632	0.017316	0.004329	0.01488
1 cubic foot	28,317	28.316	0.028317	1,728	1	0.03704	957.5	59.8442	29.9221	7.48052	25.714
1 cubic yard	764,555	764.56	0.76456	46,656	27	1	—	—	—	—	—
1 ounce[a]	29.5737	0.29573	29.57×10^{-6}	1.80469	0.00104	—	1	0.0625	0.03125	0.00781	0.02686
1 pint[a]	473.179	0.473167	—	28.875	—	—	16	1	0.5	0.125	0.42968
1 quart[a]	946.359	0.94633	0.000946	57.75	0.03342	—	32	2	1	0.25	0.85937
1 gallon[a]	3,785.43	3.78533	0.003785	231	0.13368	—	128	8	4	1	3.43747
1 quart[c]	1,101.23	1.1020	0.001	67.2006	0.03889	—	—	—	1.16365	0.29091	1

[a] U.S. fluid measure. British imperial gallon = 277.24 in.³ = 1.2009 U.S. gallon.
[b] Volume of 1 kg of water at 4°C and 760 mm pressure.
[c] U.S. dry measure. 32 quarts = 1 bushel.

Volume per unit land area converting factors:
1 cubic foot per acre = 0.06997 cubic meters per hectare.
1 cubic meter per hectare = 14.29 cubic feet per acre.

Table A-1 (continued)

D. Mass

	grain[a]	ounce[a]	pound[a]	ton[a,b]	gram	kilogram	metric ton
1 grain[a]	1	0.002286	0.000143	—	0.64799	—	—
1 ounce[a]	437.5	1	0.0625	—	28.34953	0.028350	—
1 pound[a]	7,000	16	1	0.0005	453.592	0.453592	0.0004536
1 ton[a,b]	—	32,000	2,000	1	—	907.1849	0.90718
1 gram	15.43236	0.035274	0.002205	—	1	0.001	—
1 kilogram	15,432.4	35.27396	2.204622	0.001102	1,000	1	0.001
1 metric ton	—	—	2,204.62	1.102311	—	1,000	1

[a] Avoirdupois system.

[b] U.S. ton; 1 imperial ton = 2240 pounds.

Mass per unit land area converting factors:

1 U.S. ton per acre = 2.241 metric tons per hectare.
1 metric ton per hectare = 0.44609 U.S. tons per acre.

Table A-2
Areas of Some Plane Figures

Figure	Diagram	Formula
Rectangle		$A = l\,w$
Parallelogram		$A = b\,h$
Triangle		$A = \dfrac{b\,h}{2}$ or $A = \sqrt{S(S-a)(S-b)(S-c)}$ where $S = \frac{1}{2}(a+b+c)$
Trapezoid		$A = \frac{1}{2}(a+c)\,h$
Circle		$A = \pi r^2$ or $A = \dfrac{\pi D^2}{4}$ where $r = \dfrac{D}{2}$ Circumference $= 2\pi r = \pi D$
Circular sector		$A = \dfrac{\theta r^2}{2}$ where θ is in radians $A = \dfrac{\theta \pi r^2}{360}$ where θ is in degrees
Circular segment		$A = (\tfrac{1}{2})\, r^2\,(\theta - \sin\theta)$ where θ is in radians
Ellipse		$A = \pi a b$ Perimeter $= \pi(a+b)\left[1 + \frac{1}{4}\left(\dfrac{a-b}{a+b}\right)^2 + \frac{1}{64}\left(\dfrac{a-b}{a+b}\right)^2 + \dots\right]$
Parabola		$A = (\tfrac{2}{3})\, l\, d$ Length of arc $= l\left[1 + \frac{2}{3}\left(\dfrac{2d}{l}\right)^2 - \frac{2}{5}\left(\dfrac{2d}{l}\right)^4 + \dots\right]$

Table A-3
Volume and Surface Areas of Some Solids

The following symbols are used in the formulas:

V	= volume	L	= slant height
A_b	= area of the base	P_b	= perimeter of lower base
A_m	= area of a midsection parallel to A_b and A_u	P_u	= perimeter of upper section
		P_r	= perimeter of right section
A_u	= area of an upper section	S_l	= lateral surface area
A_r	= area of a right section	S_t	= total surface area
h	= altitude	d	= distance from base to intermediate section

Solid	Diagram	Formula
Prismoid		$V = \frac{h}{6}(A_b + 4\,A_m + A_u)$ $S_l = \frac{1}{2}(P_b + P_u)\,L$ $S_t = S_l + A_b + A_u$
Prism or cylinder		$V = A_b\,h$ $S_l = P_r L$ $S_t = S_l + 2\,A_b$
Pyramid or cone		$V = \frac{1}{3}\,A_b\,h$ $S_l = \frac{1}{2}\,P_b\,L$ $S_t = S_l + A_b$
Frustum of cone or pyramid		(treat as prismoid or) $V = \frac{h}{3}\,(A_b + \sqrt{(A_b)(A_u)} + A_u)$ $S_l = \frac{1}{2}(P_b + P_u)\,L$ $S_t = S_l + A_b + A_u$
Sphere		$V = \frac{4}{3}\,\pi\,r^3$ $S_t = 4\,\pi\,r^2$

Table A-3 (continued)

Solid	Diagram	Formula
Paraboloid		$V = \frac{1}{2} A_b h$ $S_l = \left(\frac{2\pi r}{12 h^2}\right)\left[(r^2 + 4h^2)^{\frac{3}{2}} - r^3\right]$ $S_t = S_l + A_b$
Frustum of paraboloid		$V = \frac{A_b + A_u}{2} d$ $S_l = $ surface $acb - gck$ $S_t = S_l + A_b + A_u$
Neiloid		$V = \frac{1}{4} A_b h$ $S_l = 2\pi \int_0^h X^{\frac{3}{2}}\sqrt{1 + \frac{9X}{4}}\ dX$ $S_t = S_l + A_b$
Frustum of neiloid		$V = \frac{d}{6}(A_b + 4 A_m + A_u)$ $S_l = $ surface $acb - gck$ $S_t = S_l + A_b + A_u$

Areas of Circles in Square Feet for Diameters in Inches

Diameter	0.0	0.1	0.2	0.3	0.4	0.5	0.6	0.7	0.8	0.9
0	0.000	0.000	0.000	0.000	0.001	0.001	0.002	0.003	0.003	0.004
1	0.005	0.007	0.008	0.009	0.011	0.012	0.014	0.016	0.018	0.020
2	0.022	0.024	0.026	0.029	0.031	0.034	0.037	0.040	0.043	0.046
3	0.049	0.052	0.056	0.059	0.063	0.067	0.071	0.075	0.079	0.083
4	0.087	0.092	0.096	0.101	0.106	0.110	0.115	0.120	0.126	0.131
5	0.136	0.142	0.147	0.153	0.159	0.165	0.171	0.177	0.183	0.190
6	0.196	0.203	0.210	0.216	0.223	0.230	0.238	0.245	0.252	0.260
7	0.267	0.275	0.283	0.291	0.299	0.307	0.315	0.323	0.332	0.340
8	0.349	0.358	0.367	0.376	0.385	0.394	0.403	0.413	0.422	0.432
9	0.442	0.452	0.462	0.472	0.482	0.492	0.503	0.513	0.524	0.535
10	0.545	0.556	0.567	0.579	0.590	0.601	0.613	0.624	0.636	0.648
11	0.660	0.672	0.684	0.696	0.709	0.721	0.734	0.747	0.759	0.772
12	0.785	0.799	0.812	0.825	0.839	0.852	0.866	0.880	0.894	0.908
13	0.922	0.936	0.950	0.965	0.979	0.994	1.009	1.024	1.039	1.054
14	1.069	1.084	1.100	1.115	1.131	1.147	1.163	1.179	1.195	1.211
15	1.227	1.244	1.260	1.277	1.294	1.310	1.327	1.344	1.362	1.379
16	1.396	1.414	1.431	1.449	1.467	1.485	1.503	1.521	1.539	1.558
17	1.576	1.595	1.614	1.632	1.651	1.670	1.689	1.709	1.728	1.748
18	1.767	1.787	1.807	1.827	1.847	1.867	1.887	1.907	1.928	1.948
19	1.969	1.990	2.011	2.032	2.053	2.074	2.095	2.117	2.138	2.160
20	2.182	2.204	2.226	2.248	2.270	2.292	2.315	2.337	2.360	2.382
21	2.405	2.428	2.451	2.474	2.498	2.521	2.545	2.568	2.592	2.616
22	2.640	2.664	2.688	2.712	2.737	2.761	2.786	2.810	2.835	2.860
23	2.885	2.910	2.936	2.961	2.986	3.012	3.038	3.064	3.089	3.115
24	3.142	3.168	3.194	3.221	3.247	3.274	3.301	3.328	3.355	3.382
25	3.409	3.436	3.464	3.491	3.519	3.547	3.574	3.602	3.631	3.659
26	3.687	3.715	3.744	3.773	3.801	3.830	3.859	3.888	3.917	3.947
27	3.976	4.006	4.035	4.065	4.095	4.125	4.155	4.185	4.215	4.246
28	4.276	4.307	4.337	4.368	4.399	4.430	4.461	4.493	4.524	4.555
29	4.587	4.619	4.650	4.682	4.714	4.746	4.779	4.811	4.844	4.876
30	4.909	4.942	4.974	5.007	5.041	5.074	5.107	5.140	5.174	5.208
31	5.241	5.275	5.309	5.343	5.378	5.412	5.446	5.481	5.515	5.550
32	5.585	5.620	5.655	5.690	5.726	5.761	5.796	5.832	5.868	5.904
33	5.940	5.976	6.012	6.048	6.084	6.121	6.158	6.194	6.231	6.268
34	6.305	6.342	6.379	6.417	6.454	6.492	6.529	6.567	6.605	6.643
35	6.681	6.720	6.758	6.796	6.835	6.874	6.912	6.951	6.990	7.029
36	7.069	7.108	7.147	7.187	7.227	7.266	7.306	7.346	7.386	7.426
37	7.467	7.507	7.548	7.588	7.629	7.670	7.711	7.752	7.793	7.834
38	7.876	7.917	7.959	8.001	8.042	8.084	8.126	8.169	8.211	8.253
39	8.296	8.338	8.381	8.424	8.467	8.510	8.553	8.596	8.640	8.683
40	8.727	8.770	8.814	8.858	8.902	8.946	8.990	9.035	9.079	9.124
41	9.168	9.213	9.258	9.303	9.348	9.393	9.439	9.484	9.530	9.575
42	9.621	9.667	9.713	9.759	9.805	9.852	9.898	9.945	9.991	10.038
43	10.085	10.132	10.179	10.226	10.273	10.321	10.368	10.416	10.463	10.511
44	10.559	10.607	10.655	10.704	10.752	10.801	10.849	10.898	10.947	10.996
45	11.045	11.094	11.143	11.192	11.242	11.291	11.341	11.391	11.441	11.491
46	11.541	11.591	11.642	11.692	11.743	11.793	11.844	11.895	11.946	11.997
47	12.048	12.100	12.151	12.203	12.254	12.306	12.358	12.410	12.462	12.514
48	12.566	12.619	12.671	12.724	12.777	12.830	12.882	12.936	12.989	13.042
49	13.095	13.149	13.203	13.256	13.310	13.364	13.418	13.472	13.527	13.581
50	13.635	13.690	13.745	13.800	13.854	13.909	13.965	14.020	14.075	14.131

Table A-4(b)
Areas of Circles: (1) in Square Feet for Circumferences and Diameters in Inches, and (2) in Square Meters for Circumferences and Diameters in Centimeters

Diameter in./cm	Circum-ference in./cm	Area ft²	Area m²	Diameter in./cm	Circum-ference in./cm	Area ft²	Area m²
1	3.14	0.005	—	51	160.22	14.186	0.240
2	6.28	0.022	—	52	163.36	14.748	0.212
3	9.42	0.049	0.001	53	166.50	15.321	0.221
4	12.57	0.087	0.001	54	169.65	15.904	0.229
5	15.71	0.136	0.002	55	172.79	16.499	0.238
6	18.85	0.196	0.003	56	175.93	17.104	0.246
7	21.99	0.267	0.004	57	179.07	17.721	0.255
8	25.13	0.349	0.005	58	182.21	18.348	0.264
9	28.27	0.442	0.006	59	185.35	18.986	0.273
10	31.42	0.545	0.008	60	188.50	19.635	0.283
11	34.56	0.660	0.010	61	191.64	20.295	0.292
12	37.70	0.785	0.011	62	194.78	20.966	0.302
13	40.84	0.922	0.013	63	197.92	21.648	0.312
14	43.98	1.069	0.015	64	201.06	22.340	0.322
15	47.12	1.227	0.018	65	204.20	23.044	0.332
16	50.26	1.396	0.020	66	207.34	23.758	0.342
17	53.41	1.576	0.023	67	210.49	24.484	0.352
18	56.55	1.767	0.025	68	213.63	25.220	0.363
19	59.69	1.969	0.028	69	216.77	25.967	0.374
20	62.83	2.182	0.031	70	219.91	26.725	0.385
21	65.97	2.405	0.035	71	223.05	27.494	0.396
22	69.12	2.640	0.038	72	226.19	28.274	0.407
23	72.26	2.885	0.042	73	229.34	29.065	0.418
24	75.40	3.142	0.045	74	232.48	29.867	0.430
25	78.54	3.409	0.049	75	235.62	30.680	0.442
26	81.68	3.687	0.053	76	238.76	31.503	0.454
27	84.82	3.976	0.057	77	241.90	32.338	0.466
28	87.96	4.276	0.062	78	245.04	33.183	0.478
29	91.11	4.587	0.066	79	248.18	34.039	0.490
30	94.25	4.909	0.071	80	251.33	34.907	0.503
31	97.39	5.241	0.075	81	254.47	35.785	0.515
32	100.53	5.585	0.080	82	257.61	36.674	0.528
33	103.67	5.940	0.086	83	260.75	37.574	0.541
34	106.81	6.305	0.091	84	263.89	38.484	0.554
35	109.96	6.681	0.096	85	267.04	39.406	0.567
36	113.10	7.069	0.102	86	270.18	40.339	0.581
37	116.24	7.467	0.108	87	273.32	41.282	0.594
38	119.38	7.876	0.113	88	276.46	42.237	0.608
39	122.52	8.296	0.119	89	279.60	43.202	0.622
40	125.66	8.727	0.126	90	282.74	44.179	0.636
41	128.81	9.168	0.132	91	285.88	45.166	0.650
42	131.95	9.621	0.138	92	289.03	46.164	0.665
43	135.09	10.085	0.145	93	292.17	47.173	0.679
44	138.23	10.559	0.152	94	295.31	48.193	0.694
45	141.37	11.045	0.159	95	298.45	49.224	0.709
46	144.51	11.541	0.166	96	301.59	50.266	0.724
47	147.65	12.048	0.173	97	304.73	51.318	0.739
48	150.80	12.566	0.181	98	307.88	52.382	0.754
49	153.94	13.095	0.189	99	311.02	53.456	0.770
50	157.08	13.635	0.196	100	314.16	54.542	0.785

Table A-5
Circular- and Square-Plot Dimensions

Area		Radius of Circular Plot		Side of Square Plot		Diagonal of Square Plot	
English System							
Acres	**Square Feet**	**Feet**	**Chains**	**Feet**	**Chains**	**Feet**	**Chains**
1.00	43,560	117.75	1.784	208.71	3.162	295.16	4.472
0.50	21,780	83.26	1.262	147.58	2.236	208.71	3.162
0.25	10,890	58.88	0.892	104.36	1.581	147.58	2.236
0.20	8,712	52.66	0.798	93.34	1.414	132.00	2.000
0.10	4,356	37.24	0.564	66.00	1.000	93.34	1.414
0.05	2,178	26.33	0.399	46.67	0.707	66.00	1.000
0.01	435.6	11.78	0.178	20.87	0.316	29.52	0.447
0.001	43.56	3.72	0.056	6.60	0.100	9.33	0.141
Metric System							
Hectares	**Square Meters**	**Meters**		**Meters**		**Meters**	
1.00	10,000	56.41		100.00		141.42	
0.50	5,000	39.89		70.71		100.00	
0.25	2,500	28.21		50.00		70.71	
0.20	2,000	25.23		44.72		63.24	
0.10	1,000	17.84		31.62		44.72	
0.05	500	12.62		22.36		31.62	
0.01	100	5.64		10.00		14.14	
0.001	10	1.78		3.16		4.47	

Table A-6
Values of *t*

D.F.	Probability of a Larger Value of *t*, Sign Ignored							
	.5	.4	.3	.2	.1	.05	.02	.01
1	1.000	1.376	1.963	3.078	6.314	12.706	31.821	63.657
2	.816	1.060	1.386	1.886	2.920	4.303	6.965	9.925
3	.765	.978	1.250	1.638	2.353	3.182	4.541	5.841
4	.741	.941	1.190	1.533	2.132	2.776	3.747	4.604
5	.727	.920	1.156	1.476	2.015	2.571	3.365	4.032
6	.718	.906	1.134	1.440	1.943	2.447	3.143	3.707
7	.711	.896	1.119	1.415	1.895	2.365	2.998	3.499
8	.706	.889	1.108	1.397	1.860	2.306	2.896	3.355
9	.703	.883	1.100	1.383	1.833	2.262	2.821	3.250
10	.700	.879	1.093	1.372	1.812	2.228	2.764	3.169
11	.697	.876	1.088	1.363	1.796	2.201	2.718	3.106
12	.695	.873	1.083	1.356	1.782	2.179	2.681	3.055
13	.694	.870	1.079	1.350	1.771	2.160	2.650	3.012
14	.692	.868	1.076	1.345	1.761	2.145	2.624	2.977
15	.691	.866	1.074	1.341	1.753	2.131	2.602	2.947
16	.690	.865	1.071	1.337	1.746	2.120	2.583	2.921
17	.689	.863	1.069	1.333	1.740	2.110	2.567	2.898
18	.688	.862	1.067	1.330	1.734	2.101	2.552	2.878
19	.688	.861	1.066	1.328	1.729	2.093	2.539	2.861
20	.687	.860	1.064	1.325	1.725	2.086	2.528	2.845
22	.686	.858	1.061	1.321	1.717	2.074	2.508	2.819
24	.685	.857	1.059	1.318	1.711	2.064	2.492	2.797
26	.684	.856	1.058	1.315	1.706	2.056	2.479	2.779
28	.683	.855	1.056	1.313	1.701	2.048	2.467	2.763
30	.683	.854	1.055	1.310	1.697	2.042	2.457	2.750
40	.681	.851	1.050	1.303	1.684	2.021	2.423	2.704
60	.679	.848	1.046	1.296	1.671	2.000	2.390	2.660
120	.677	.845	1.041	1.289	1.658	1.980	2.358	2.617
∞	.674	.842	1.036	1.282	1.645	1.960	2.326	2.576

SOURCE: Table A-6 is adapted from Table III of Fisher and Yates, 1938. *Statistical Tables for Biological, Agricultural and Medical Research*, published by Oliver & Boyd, Ltd., by permission of the authors and publishers.

References Cited

Adams, D. M., and Ek, A. R. 1974. Optimizing the management of uneven-aged forests. *Canadian Jour. For. Res.* 4:274–87.

Adams, E. L. 1971. Effect of moisture loss on red oak sawlog weight. *Northeast. For. Exp. Sta. Res. Note* NE-133.

Adams, E. L. 1976. The adjusting factor method for weight-scaling truckloads of mixed hardwood sawlogs. *U.S.F.S. Northeast. For. Exp. Sta. Res. Paper* NE-344.

Aho, P. E., and Roth, L. F. 1978. Defect estimation for white fir in the Rogue River National Forest. *USDA Forest Serv. Res. Paper* PNW-240.

Alexander, M. E. 1978. Estimating fuel weights of two common shrubs in Colorado lodgepole pine stands. *U.S.F.S. Rocky Mtn. For. and Range Exp. Sta. Res. Note* RM-354.

Anderson, R. T. 1937. The application of Fourier's series in forest mensuration. *Jour. Forestry* 35:293–99.

Ashley, M. D., and Beers, T. W. 1970. Boundary-line overlap in horizontal point sampling. *Purdue Univ. Agr. Exp. Sta. Res. Bull.* no. 865.

Ashley, M. D., and Roger, R. E. 1969. Tree heights and upper-stem diameters. *Photogramm. Eng.* 35:136–46.

Avery, T. E. 1975. *Natural resource measurements.* 2nd ed. New York: McGraw-Hill Book Co.

Avery, T. E., and Meyer, M. P. 1959. Volume tables for aerial timber estimating in northern Minnesota. *U.S.F.S. Lake States For. Exp. Sta. Paper* no. 78.

Bailey, R. L., and Clutter, J. L. 1974. Base-age invariant polymorphic site curves. *Forest Sci.* 20:155–59.

Bailey, R. L., and Dell, T. R. 1973. Quantifying diameter distributions with the Weibull function. *Forest Sci.* 19:97–104.

Baker, F. S. 1923. Notes on the composition of even-aged stands. *Jour. Forestry* 21:712–17.

Baker, F. S. 1960. A simple formula for gross current annual increment. *Jour. Forestry* 58:488–89.

Barnes, R. M. 1963. *Motion and time study.* New York: John Wiley & Sons.

Barrett, J. P. 1964. Correction for edge effect bias in point-sampling. *Forest Sci.* 10:52–55.

Barrett, J. P. 1969. Estimating averages from point-sample data. *Jour. Forestry* 67:185.

Barrett, J. P., and Allen, P. H. 1966. Angle gauge sampling a small hardwood tract. *Forest Sci.* 12:83–89.

Barrett, J. P., and Nevers, H. P. 1967. Slope correction when point sampling. *Jour. Forestry* 65:206–207.

Bartoo, R. A., and Hutnik, R. J. 1962. Board foot volume tables for timber tree species in Pennsylvania. *Penn. State Forest School Res. Paper* no. 30.

Baskerville, G. L. 1972. Use of logarithmic regression in the estimation of plant biomass. *Canadian Jour. Forestry* 2:49–53.

Beck, D. E. 1971. Polymorphic site index curves for white pine in the southern Appalachians. *U.S.F.S. Southeast For. Exp. Sta. Paper* SE-80.

Beck, D. E., and Trousdell, K. B. 1973. Site index: accuracy of prediction. *U.S.F.S. Southeast For. Exp. Sta. Res. Paper* SE-108.

Beers, T. W. 1960. An evaluation of site index for mixed oak in the Central Hardwood Region. Ph.D. thesis. Purdue Univ.

Beers, T. W. 1962. Components of forest growth. *Jour. Forestry* 60:245–48.

Beers, T. W. 1963. Empirical substantiation of the (double rising)/(double effective) method of diameter growth estimation. *Jour. Forestry* 61:278–80.

Beers, T. W. 1964. Cruising for pulpwood by the ton without concern for tree diameter: point sampling with diameter obviation. *Purdue Univ. Extension Mimeo* F-49.

Beers, T. W. 1966. The direct correction for boundary-line slopover in horizontal point sampling. *Purdue Univ. Agric. Exp. Sta. Res. Prog. Rep.* 224.

Beers, T. W. 1967. Rapid estimation of forest parameters using monareal and polyareal combination sampling. *Proc. Indiana Acad. Sci.* 76:251–57.

Beers, T. W. 1969. Slope correction in horizontal point sampling. *Jour. Forestry* 67:188–92.

Beers, T. W. 1973. Revised composite tree volume tables for Indiana hardwoods. *Purdue Univ. Agric. Exp. Sta. Res. Prog. Rep.* 417.

Beers, T. W. 1974. Optimum upper-log viewing distance. *Purdue Univ. Agric. Exp. Sta. Bull.* no. 39.

Beers, T. W. 1977. Practical correction for boundary overlap. *Southern Jour. of Appl. For.* 1:16–18.

Beers, T. W. 1979. *Random numbers, means, regression, and the programmable calculator.* W. Lafayette, Ind.: T & C Enterprises.

Beers, T. W. 1980. Log volume calculations with Hewlett-Packard programmable calculators. *Purdue Univ. Agric. Exp. Sta. Jour. Paper* no. 8259.

Beers, T. W., Dress, P. E., and Wensel, L. C. 1966. Aspect transformation in site productivity research. *Jour. Forestry* 64:691–92.

Beers, T. W., and Miller, C. I. 1964. Point sampling: research results, theory, and applications. *Purdue Univ. Agric. Exp. Sta. Res. Bull.* no. 786.

Beers, T. W., and Miller, C. I. 1965. Refined application of point sampling. *Purdue Univ. Agric. Exp. Sta. Jour. Paper* no. 2447.

Beers, T. W., and Miller, C. I. 1966. Horizontal point sampling tables. *Purdue Univ. Agric. Exp. Sta. Res. Bull.* no. 808.

Beers, T. W., and Miller, C. I. 1973. *Manual of forest mensuration.* W. Lafayette, Ind.: T & C Enterprises.

Beers, T. W., and Miller, C. I. 1976. Line sampling for forest inventory. *Purdue Univ. Agric. Exp. Sta. Res. Bull.* no. 934.

Beers, T. W., and Myers, C. C. 1965. The use of permanent points—good or bad? *Proc. Conference on C.F.I. Michigan Tech. Univ.* pp. 182–97.

Behre, C. E. 1935. Factors involved in the application of form-class volume tables. *Jour. Agric. Res.* 51:669–713.

Bell, J. F., and Gourley, R. 1980. Assessing the accuracy of a sectional pole, Haga altimeter, and alti-level for determining total height of young coniferous stands. *South. Jour. Appl. For.* 4:136–38.

Beltz, R. C., and Keith, G. C. 1980. Electronic technology speeds forest survey. *South. Jour. Appl. For.* 4:115–18.

Belyea, H. C. 1931. *Forest measurement.* New York: John Wiley & Sons.

Bennett, L. J., English, P. F., and McCain, R. 1940. A study of deer populations by use of pellet group counts. *Jour. Wildl. Mgt.* 4:398–403.

Besley, L. 1967. Importance, variation and measurement of density and moisture. *Wood Measurement Conf. Proc. Tech. Rep.* no. 7. University of Toronto: Faculty of Forestry.

Bickford, C. A., Baker, F. S., and Wilson, F. G. 1957. Stocking, normality, and measurement of stand density. *Jour. Forestry* 55:99–104.

Bitterlich, W. 1947. Die winkelzählmessung. (Measurement of basal area per hectare by means of angle measurement.) *Allg. Forst-u. Holzwirts. Zfg.* 58:94–96.

Bitterlich, W. 1948. Die Winkelzhählprobe. (The angle-count sample plot.) *Allg. Forst-u. Holzwirts. Zfg.* 59(1/2):4–5.

Bitterlich, W. 1957. Harmonische mittel in winkelzahlproben. *Allg. Forstztg.* 68(13/14) July.

Bitterlich, W. 1959. *Sektorkluppen aus Leichtmetall* (Calipering forks made of light alloys). Wien: Holz-Kurier 14:15–17.

Bitterlich, W. 1978. Single tree measurement by the tele-relaskop—a highly efficient tool for forest inventories. Jakarta, Indonesia: 8th World Forestry Congress.

Bouchon, J. 1965. Les prismes relascopiques. *Rev. For. Fran.* 17:365–73.

Bower, D. R., and Blocker, W. W. 1966. Accuracy of bands and tape for measuring diameter increments. *Jour. Forestry* 64:21–22.

Boyer, W. D. 1968. Foliage weight and stem growth of longleaf pine. *U.S.F.S. Southern For. Exp. Sta. Res. Note* 50–86.

Braathe, P., and Okstad, T. 1967. Trade of pulpwood based on weighing and dry matter samples. *XIV IUFRO Congress.* Sect. 41. Vol. IX:236–42.

Brickell, J. E. 1968. A method for constructing site index curves from measurements of tree age and height—its application to inland Douglas-fir. *U.S.F.S. Intermt. For. and Range Exp. Sta. Paper* INT-47.

Briegleb, P. A. 1942. Progress in estimating trend of normality percentage in second-growth Douglas-fir. *Jour. Forestry* 40:785–93.

Brinker, R. C., and Wolf, P. R. 1977. *Elementary surveying.* 6th ed. New York: IEP—a Dun-Donnelley Publisher.

Brown, J. K. 1976. Predicting crown weights for 11 Rocky Mountain conifers. In *Forest biomass studies*, XV IUFRO Congress. Gainesville, Florida.

Brownlee, K. A. 1967. *Statistical theory and methodology in science and engineering.* 2nd ed. New York: John Wiley & Sons.

Bruce, D. 1926. A method of preparing timber yield tables. *Jour. Agric. Res.* 32:543–57.

Bruce, D. 1955. A new way to look at trees. *Jour. Forestry* 53:163–67.

Bruce, D. 1961. *Prism cruising in the western United States and volume tables for use therewith.* Portland, Oreg.: Mason, Bruce & Girard, Consulting Foresters.

Bruce, D. 1974. Changing to the metric system. *Jour. Forestry* 72:746–48.

Bruce, D. 1975. Evaluating accuracy of tree measurements made with optical instruments. *Forest Sci.* 21:421–26.

Bruce, D. 1976. Metrication: what's next? *Jour. Forestry* 74:737–38.

Bruce, D. 1979. *Effects of metrication on analysis and summaries of forest measurements.* Fort Collins, Colo.: Proc. Forest Resource Inventories Workshop, pp. 812–29.

Bruce, D., and Schumacher, F. X. 1950. *Forest mensuration.* 3rd ed. New York: McGraw-Hill Book Co.

Buckingham, F. M. 1969. The harmonic mean in forest mensuration. *Forestry Chron.* 45:104–106.

Buckman, R. E. 1962. Growth and yield of red pine in Minnesota. *U.S. Dept. Agric. Tech. Bull.* no. 1272.

Bureau of Outdoor Recreation. 1973. *Outdoor recreation: a legacy for America.* Washington, D.C.: U.S. Dept. of the Interior.

Cable, D. R. 1958. Estimating surface area of ponderosa pine foliage in central Arizona. *Forest Sci.* 4:45–49.

Calkins, H. A., and Yule, J. B. 1935. *The Abney level handbook.* U.S.F.S.

Campbell, R. A. 1962. A guide to grading features in southern pine logs and trees. *U.S.F.S. Southeast. For. Exp. Sta. Paper* no. 156.

Campbell, R. A. 1964. Forest Service log grades for southern pine. *U.S.F.S. Southeast. For. Exp. Sta. Res. Paper* SE-11.

Carron, L. T. 1968. *An outline of forest mensuration with special reference to Australia.* Canberra: Australian National University Press.

Chamberlain, E. B., and Meyer, H. A. 1950. Bark volume in cordwood. *Tappi* 33:554–55.

Chapman, H. H., and Meyer, W. H. 1949. *Forest mensuration.* New York: McGraw-Hill Book Co.

Chatfield, C. 1975. *The analysis of time series: theory and practice.* London: Chapman and Hall.

Chehock, C. R., and Walker, R. C. 1975. *Sample weight scaling with 3-P sampling for multi-product logging.* Atlanta, Ga.: U.S.F.S. State and Private For. S.E. Area.

Chisman, H. H., and Schumacher, F. X. 1940. On the tree-area ratio and certain of its applications. *Jour. Forestry* 38:311–17.

Clark, J. F. 1906. Measurement of sawlogs. *Forestry Quart.* 4:79–93.

Clutter, J. L. 1963. Compatible growth and yield models for loblolly pine. *Forest Sci.* 9:354–71.

Clutter, J. L., and Bennett, F. A. 1965. Diameter distribution in old-field slash pine plantations. *Georgia For. Res. Council Rep.* 13.

Cochran, W. G. 1977. *Sampling techniques.* 3rd ed. New York: John Wiley & Sons.

Cody, J. B. 1976. Merchantable weight tables for New York State red pine plantations. *College of Environ. Sci. and Forestry* (Syracuse, New York) *Applied For. Res. Inst. Res. Note* 23.

Coile, T. S. 1952. Soil and growth of forests. *Advances in Agronomy* 4:330–98.

Crow, T. R. 1978. Common regressions to estimate tree biomass in tropical stands. *Forest Sci.* 24:110–14.

Cummings, W. H. A. 1941. A method of sampling the foliage of a silver maple tree. *Jour. Forestry* 39:382–84.

Cunia, T. 1965. Continuous forest inventory, partial replacement of samples and multiple regression. *Forest Sci.* 11:480–502.

Curtis, K. S. 1976. *Optical distance measurement.* Indianapolis, Ind.: Indiana Society of Professional Land Surveyors, Inc.

Curtis, R. O. 1967. A method of estimation of gross yield of Douglas-fir. *Forest Sci.* Monograph 13.

Curtis, R. O., and Bruce, D. 1968. Tree heights without a tape. *Jour. Forestry* 66:60–61.

Daniel, T. W. 1955. Bitterlich's Spiegelrelaskop—a revolutionary general use forestry instrument. *Jour. Forestry* 53:844–46.

Daniel, T. W., Helms, J. A., and Baker, F. S. 1979. *Principles of silviculture.* 2nd ed. New York: McGraw-Hill Book Co.

Daubenmire, R. F. 1945. An improved type of precision dendrometer. *Ecology* 26:97–98.

Davies, G. R., and Crowder, W. F. 1933. *Methods of statistical analysis in the social sciences.* New York: John Wiley & Sons.

Davis, K. P. 1957. *American forest management.* New York: McGraw-Hill Book Co.

David, K. P. 1966. *Forest management: regulation and valuation.* 2nd ed. New York: McGraw-Hill Book Co.

Dawkins, H. C. 1957. Some results of stratified random sampling of tropical high forest. *Seventh British Commonwealth Forestry Conf.* Item 7 (iii).

Deitschman, G. H., and Green, A. W. 1965. Relations between western white pine site index and tree height of several associated species. *U.S.F.S. Intermt. For. and Range Exp. Sta. Res. Paper* INT-22.

Del Hodge, J. 1965. Variable plot cruising. A short-cut slope correction method. *Jour. Forestry* 63:176–80.

De Liocourt, F. 1898. De l'aménagement des sapiniéres. *Bull de la Société Forestiére de Franche-Compté el Belfort.* Besancon, France.

Diller, O. D., and Kellog, L. F. 1940. Local volume table for yellow poplar. *U.S.F.S. Central States For. Exp. Sta. Tech. Note* no. 1.

Dilworth, J. R. 1979. *Log scaling and timber cruising.* Corvallis, Oreg.: O.S.U. Book Stores, Inc.

Doebelin, E. O. 1966. *Measurement systems: applications and design.* New York: McGraw-Hill Book Co.

Donnelly, D. M., and Barger, R. L. 1977. Weight scaling for southwestern ponderosa pine. *U.S.F.S. Rocky Mtn. For. and Range Exp. Sta. Res. Paper* RM-181.

Doolittle, W. T., and Vimmerstedt, J. P. 1960. Site curves for natural stands of white pine in southern Appalachians. *U.S.F.S. Southeast. For. Exp. Sta. Res. Note* no. 141.

Drew, A. P., and Running, S. W. 1975. Comparison of two techniques of measuring surface area of conifer needles. *Forest Sci.* 21:231–32.

Dudek, A., and Ek, A. R. 1980. A bibliography of worldwide literature on individual tree-based forest stand growth models. Dept. For. Resources. *Univ. of Minn. Staff Paper Series* no. 12.

Eichenberger, J. K., Parker, G. R., and Beers, T. W. 1982. A method for ecological forest sampling. *Purdue Univ. Agric. Exp. Sta. Res. Bull.* no. 969.

Eller, R. C., and Keister, T. D. 1979. The Breithaupt Todis dendrometer. *Southern Jour. App. Forestry* 3:29–32.

Ellis, B. 1966. *Basic concepts of measurement.* Cambridge: At the University Press.

Enghardt, H., and Derr, H. J. 1963. Height accumulation for rapid estimates of cubic volume. *Jour. Forestry* 61:134–37.

Estola, J. D. 1979. 3-P random numbers and a hand-held programmable calculator. *Resource Inventory Notes.* (Jan. 1979).

Evans, T. F., Burkhart, H. E., and Parker, R. C. 1975. Site and yield information applicable to Virginia's hardwoods: a review. *Virginia Poly. Inst. and State Univ.* FWS-2-75.

FAO. 1976. *Harvesting man-made forests in developing countries.* FAO: TF-INT 74 (SWE) Rome.

Fender, D. E., and Brock, G. A. 1963. Point center extension: a technique for measuring current economic growth and yield of merchantable forest stands. *Jour. Forestry* 61:109–14.

Ferree, M. J. 1946. The pole caliper. *Jour. Forestry* 44:594–95.

Fisher, R. A., and Yates, F. 1938. Statistical tables for biological, agricultural and medical research. Edinburgh: Oliver and Boyd, Ltd.

Fogelberg, S. E. 1953. *Volume charts based on absolute form class.* Ruston: Louisiana Tech Forestry Club of Louisiana Polytech. Inst.

Forbes, R. D., and Meyer, H. B. (Eds). 1955. *Forestry handbook.* (Soc. Amer. Foresters.) New York: The Ronald Press Co.

Forest Products Laboratory. 1958. Overall work plan for development of log and bolt grades for hardwoods. *U.S.F.S. For. Prod. Lab. Rep.* TGUR-16.

Forest Products Laboratory. 1966. Hardwood log grades for standard lumber. *U.S.F.S. For. Prod. Lab. Res. Paper* FPL-63.

Foster, R. W. 1959. Relation between site indexes of eastern white pine and red maple. *Forest Sci.* 5:279–91.

Frayer, W. E. 1966. Weighted regression in successive forest inventories. *Forest Sci.* 12:464–72.

Freese, F. 1962. Elementary forest sampling. *U.S. Dept. Agric. Handbook* no. 232.

Freese, F. 1973. A collection of log rules. *USDA Forest Serv. Gen. Tech. Rep.* FPL 1.

Fries, J. 1965. Eigenvector analyses show that birch and pine have similar form in Sweden and in British Columbia. *Forestry Chron.* 41:135–39.

Fries, J., and Matern, B. 1965. *On the use of multivariate methods for the construction of tree taper curves.* Stockholm: Paper no. 9, IUFRO Advisory Group of Forest Stat. Conf.

Frissell, S. S. 1978. Judging recreation impacts on wilderness campsites. *Jour. Forestry* 76:481–83.

Fritts, H. C., and Fritts, E. C. 1955. A new dendrograph for recording radial changes of a tree. *Forest Sci.* 1:271–76.

Furnival, G. M. 1961. An index for comparing equations used in constructing volume tables. *Forest Sci.* 7:337–41.

Gaiser, R. N., and Merz, R. W. 1951. Stand density as a factor in measuring white oak site index. *Jour. Forestry* 49:572–74.

Garay, L. 1961. *An introduction to tariff volume tables.* Univ. of Washington: For. Biom. Res. Group Rev. Paper no. 1.

Garland, H. 1968. Using a polaroid camera to measure trucked hardwood pulpwood. *Pulp and Paper Mag. Canada* 69(8):86–87.

Garrison, G. A. 1949. Uses and modifications of the "moosehorn" crown closure estimator. *Jour. Forestry* 47:733–35.

Gerrard, D. J. 1969. Competition quotient: a new measure of the competition affecting individual forest trees. *Mich. State Agric. Exp. Sta. Res. Bull.* no. 20.

Gevorkiantz, S. R. 1950. Converting International 1/4-inch gross sawlog scale to peeled total volume in cubic feet. *U.S.F.S. Lake States For. Exp. Sta. Tech. Note* no. 329.

Gevorkiantz, S. R., and Olsen, L. P. 1955. Composite volume tables for timber and their application in the Lake States. *U.S. Dept. Agric. Tech. Bull.* no. 1104.

Giles, R. H., Jr. (ed.). 1971. *Wildlife management techniques.* 3rd ed. (revised). Washington, D.C.: The Wildlife Society.

Gingrich, S. F. 1964. Criteria for measuring stocking in forest stands. *Proc. Soc. Amer. Foresters:* 198–201.

Gingrich, S. F. 1967. Measuring and evaluating stocking and stand density in upland hardwood forests in the Central States. *Forest Sci.* 13:38–53.

Girard, J. W. 1933. Volume tables for Mississippi bottomland hardwoods and southern pines. *Jour. Forestry* 31:34–41.

Giurgiu, V. 1968. *Dendrometrie.* Bucarest, Romania; Editura Agrosilvica.

Godman, R. M. 1949. The pole diameter tape. *Jour. Forestry* 47:585–89.

Gould, E. M., Jr. 1957. The Harvard forest prism holder for point sampling. *Jour. Forestry* 55:730–31.

Grosenbaugh, L. R. 1948. Improved cubic volume computation. *Jour. Forestry* 46:299–301.

Grosenbaugh, L. R. 1952a. Shortcuts for cruisers and scalers. *U.S.F.S. Southern For. Exp. Sta. Occ. Paper* no. 126.

Grosenbaugh, L. R. 1952b. Plotless timber estimates—new, fast, easy. *Jour. Forestry* 50:32–37.

Grosenbaugh, L. R. 1954. New tree-measurement concepts: height accumulation, giant tree, taper and shape. *U.S.F.S. Southern For. Exp. Sta. Occ. Paper* no. 134.

Grosenbaugh, L. R. 1955. Better diagnosis and prescription in southern forest management. *U.S.F.S. Southern For. Exp. Sta. Occ. Paper* no. 145.

Grosenbaugh, L. R. 1958. Point-sampling and line-sampling: probability theory, geometric implications, synthesis. *U.S.F.S. Southern For. Exp. Sta. Occ. Paper* no. 160.

Grosenbaugh, L. R. 1963a. Optical dendrometers for out-of-reach diameters: a conspectus of some new theory. *Forest Sci.* Monograph 4.

Grosenbaugh, L. R. 1963b. Some suggestions for better sample-tree measurement. *Proc. Soc. Amer. Foresters:* 36–42.

Grosenbaugh, L. R. 1964. STX-FORTRAN 4 PROGRAM—for estimates of tree populations from 3P sample-tree-measurements. *U.S.F.S. Pac. S.W. For. and Range Exp. Sta. Paper* PSW-13.

Grosenbaugh, L. R. 1965. Three-pee sampling theory and program "THRP" for computer generation of selection criteria. *U.S.F.S. Pac. S.W. For. and Range Exp. Sta. Paper* PSW-21.

Grosenbaugh, L. R. 1966. Tree form: definition, interpolation, extrapolation. *Forestry Chron.* 42:444–57.

Grosenbaugh, L. R. 1967. The gains from sample-tree selection with unequal probabilities. *Jour. Forestry* 65:203–6.

Grosenbaugh, L. R. 1973. Metrication and forest inventory. *Jour. Forestry* 71:84–85.

Grosenbaugh, L. R. 1976. Approximate sampling variance of adjusted 3P sampling estimates. *Forest Sci.* 22:173–76.

Grosenbaugh, L. R. 1979. 3P sampling theory, examples, and rationale. *U.S. Dept. Interior. Bureau of Land Mgt. Tech. Note* no. 331.

Guttenberg, S. 1973. Evolution of weight-scaling. *Southern Lumberman* 227:12.

Guttenberg, S., Fassnacht, D., and Siegel, W. C. 1960. Weight-scaling southern pine sawlogs. *U.S.F.S. Southern For. Exp. Sta. Occ. Paper* no. 177.

Guttenberg, S., and Reynolds, R. R. 1953. Cutting financially mature loblolly and shortleaf pine. *U.S.F.S. Southern For. Exp. Sta. Occ. Paper* no. 129.

Hafley, W. L., and Schreuder, H. T. 1977. Statistical distributions for fitting diameter and height data in even-aged stands. *Canadian Jour. For. Res.* 7:481–87.

Haga, T., and Maezawa, K. 1959. Bias due to edge effect in using the Bittlerlich method. *Forest Sci.* 5:370–76.

Hamilton, D. A. 1978. Specifying precision in natural resource inventories. *U.S.F.S. Rocky Mtn. For. and Range Exp. Sta. Gen. Tech. Rep.* RM-55.

Hamilton, G. J. 1974. Metrication in British forestry. *Jour. Forestry* 72:749–52.

Hanks, L. F. 1976. Hardwood tree grades for factory lumber. *U.S.F.S. Northeast For. Exp. Sta. Res. Paper* NE-333.

Hann, D. W., and Bare, B. B. 1979. Uneven-aged forest management: state of the art (or science?). *U.S.F.S. Gen. Tech. Rep.* INT-50.

Hann, D. W., and Brodie, J. D. 1980. Even-aged management: basic managerial questions and available or potential techniques for answering them. *U.S.F.S. Gen. Tech. Rep.* INT-83.

Hann, D. W., and McKinney, R. K. 1975. Stem surface area equations for four tree species of New Mexico and Arizona. *U.S.F.S. Intermt. For. and Range Exp. Sta. Res. Note* INT-190.

Hansen, M. H., and Hurwitz, W. N. 1943. On the theory of sampling from finite populations. *Annals Math. Stat.* 14:333–62.

Hansen, M. H., Hurwitz, W. N., and Madow, W. G. 1953. *Sample survey methods and theory.* 2 vols. New York: John Wiley & Sons.

Hardy, S. S., and Weiland, G. W. 1964. Weight as a basis for the purchase of pulpwood in Maine. *Univ. Maine Agric. Exp. Sta. Tech. Bull.* no. 14.

Hayne, D. W. 1949. An examination of the strip census method for estimating animal populations. *Jour. Wildl. Mgt.* 13:145–47.

Heisterberg, J. F. 1977. Flora and fauna of Po National Park, Upper Volta, West Africa. Master's thesis. W. Lafayette, Ind.: Purdue University.

Henricksen, H. A. 1950. Hojde-diameter diagram med logaritmisk diameter (Height-diameter diagram with logarithmic diameter). *Dansk Skovforen. Tidsskr.* 35:193–202.

Herrick, A. M. 1940. A defense of the Doyle rule. *Jour. Forestry* 38:563–67.

Herrick, A. M. 1955. How to grade hardwood sawlogs. *Purdue Univ. Agric. Ext. Ser. Bull.* no. 346.

Herrick, A. M. 1956. The quality index in hardwood sawtimber management. *Purdue Univ. Agric. Exp. Sta. Bull.* no. 632.

Herrick, D. E., and Jackson, W. L. 1957. The butt log tells the story. *U.S.F.S. Central States For. Exp. Sta. Note* no. 102.

Hills, G. A. 1952. The classification and evaluation of site for forestry. *Ontario Dept. Lands and Forests. Res. Rep.* no. 24.

Hirata, T. 1955. Height estimation through Bitterlich's method, vertical angle count sampling. *Jap. Jour. Forestry* 37:479–80.

Hirata, T. 1956. Harmonic means in Bitterlich's sampling. *Tokyo Univ. For., Misc. Inf.* no. 11, July.

Hool, J. N., and Beers, T. W. 1964. Time-dependent correlation coefficients from remeasured forest plots. *Purdue Univ. Agric. Exp. Sta. Res. Prog. Rep.* 156.

Hummel, F. C. 1951. Instruments for the measurement of height, diameter and taper on standing trees. *Forestry Abs.* 12:261–69.

Hummel, F. C. 1955. *The volume/basal area line; a study in forest mensuration.* London: For. Comm. Bull. no. 24.

Husch, B. 1947. A comparison between a ground and aerial photogrammetric method of timber surveying. Master's thesis. Syracuse: New York State College of Forestry.

Husch, B. 1956. Use of age at dbh as a variable in the site index concept. *Jour. Forestry* 54:340.

Husch, B. 1962. Tree weight relationships for white pine in southeastern New Hampshire. *Univ. New Hampshire Agric. Exp. Sta. Tech. Bull.* no. 106.

Husch, B. 1963. *Forest mensuration and statistics.* New York: The Ronald Press Co.

Husch, B. 1971. Planning a forest inventory. *FAO Forest Products Studies* no. 17, Rome.

Husch, B. 1980. *How to determine what you can afford to spend on inventories.* La Paz, Mexico: IUFRO Workshop on Arid Land Resource Inventories.

Husch, B., and Lyford, W. H. 1956. White pine growth and soil relationship in southeastern New Hampshire. *Univ. New Hampshire Agric. Exp. Sta. Tech. Bull.* no. 95.

Husch, B., Miller, C. I., and Beers, T. W. 1972. *Forest mensuration.* 2nd ed. New York: The Ronald Press Co.

Hutchison, O. K. 1958. Use the "wedge" with a staff compass. *U.S.F.S. Central States For. Exp. Sta. Note* no. 123.

Hyink, D. M. 1979. A generalized method for the projection of diameter distributions applied to uneven-aged forest stands. Ph.D. thesis. Purdue Univ.

Ilvessalo, Y. 1937. Perä-pohjolan luonnon normaalien metsiköiden kasvu ja kckitys (Growth of natural normal stands in central north Finland). *Comm. Inst. Forestalis Fenniae* 24:1–168.

Imprens, I. I., and Schalck, J. M. 1965. A very sensitive electric dendrograph for recording radial changes of a tree. *Ecology* 46:183–84.

IUFRO. 1959. *The standardization of symbols in forest mensuration.* International Union of Forest Research Organizations.

Jackson, W. L. 1962. Guide to grading defects in ponderosa and sugar pine logs. *U.S.F.S. Pac. S.W. For. and Range Exp. Sta.,* 34 pp.

James, G. A. 1966. *Instructions for using traffic counters to estimate recreation visits and use on developed sites.* Asheville, N.C.: U.S. Forest Service Southeast. For. Exp. Sta.

James, G. A., and Ripley, T. H. 1963. Instructions for using traffic counters to estimate recreation visits and use. *U.S.F.S. Southeast. For. Exp. Sta. Res. Paper* SE-3.

Jeffers, J. N. R. 1956. Barr and Stroud dendrometer, Tupe F. P. 7. (London) *Rep. For. Res. For. Comm.,* 1954/55:127–36.

Jensen, V. S. 1940. Cost of producing pulpwood on farm woodlands of the Upper Connecticut River Valley. *U.S.F.S. Northeast For. Exp. Sta. Occ. Paper* no. 9.

Johnson, E. W. 1972. Basic 3-P sampling. *Auburn Univ. Agric. Exp. Sta. For. Dept. Series* no. 5.

Johnson, E. W. 1973. Relationship between point density measurements and subsequent growth of southern pines. *Auburn Univ. Agric. Exp. Sta. Bull.* no. 447.

Johnson, F. A., Lowrie, J. B., and Gohlke, M. 1971. 3P sample log scaling. *U.S.F.S. Pac. N.W. For. and Range Exp. Sta. Res. Note* PNW-162.

Jonson, T. 1910. Taxatoriska undersökningar om skogsträdens form. I. Granens stamform (Forest mensurational investigations on forest tree form. Spruce stem form). *Skogs-vårdsf. Tidskr.* 8:285–328.

Jonson, T. 1911. Taxatoriska undersökningar om skogsträdens form. II. Tallens stamform (Forest mensurational investigations on forest tree form. Pine stem form). *Skogsvårdsf. Tidskr.* 9:285–329.

Jonson, T. 1912. Taxatoriska undersökningar öfver skogsträdens form. III. Form-bestämning a stående träd (Forest mensurational investigations concerning forest tree form. Form determination of standing trees). *Svenska Skogsvårdsf. Tidskr.* 10:235–75.

Kaibara, I. 1957. *Jukohscope*. Taican Corp., Tokyo.

Kallio, E., Lothner, D. C., and Marden, R. M. 1973. A test of a photographic method for determining cubic-foot volume of pulpwood. *USDA Forest Serv. Res. Note* NC-155.

Keen, E. A. 1950. The relascope. *Empire For. Rev.* 29. no. 3.

Keilson, J., and Waterhouse, C. 1978. *Proc. 1st conf. on radiocarbon dating with accelerators.* University of Rochester. p. 391.

Kendall, M. G. 1973. *Time-series.* London: C. Griffin and Co.

Kendall, R. H., and Sayn-Wittgenstein, L. 1959. An evaluation of the relascope. *Dept. North. Affairs and Nat. Res. Canada. For. Res. Div. Tech. Note* no. 77.

Kinerson, R. S. 1973. A transducer for investigation of diameter growth. *Forest Sci.* 18:230–32.

Koch, P. 1972. Utilization of southern pine. *U.S.F.S. Agric. Handbook 420.* Southern For. Exp. Sta.

Kozak, A., and Smith, J. H. G. 1966. Critical analysis of multivariate techniques for estimating tree taper suggests that simpler methods are best. *Forestry Chron.* 42:458–63.

Krajicek, J. E., Brinkman, K. A., and Gingrich, S. F. 1961. Crown competition—a measure of density. *Forest Sci.* 7:36–42.

Labau, V. J. 1967. Literature on the Bitterlich method of forest cruising. *U.S.F.S. Pac. N.W. For. and Range Exp. Sta. Res. Paper* PNW-47.

Lagler, K. F. 1956. *Freshwater fishery biology.* 2nd ed. Dubuque, Iowa: Wm. C. Brown Co.

Lahiri, D. B. 1951. A method of sample selection providing unbiased ratio estimates. *Bull. Inst. International de Statistique* 33:133–40.

Larson, P. R. 1963. Stem form development of forest trees. *Forest Sci.* Monograph 5.

Leak, W. B. 1964. An expression of diameter distribution for unbalanced, uneven-aged stands and forests. *Forest Sci.* 10:39–50.

Leary, R. A., and Beers, T. W. 1963. Measurement of upper-stem diameter with a transit. *Jour. Forestry* 61:448–50.

Lemmon, P. E. 1957. A new instrument for measuring over-story density. *Jour. Forestry* 55:667–69.

Leopold, A. 1933. *Game management.* New York: Scribner's.

Lexen, B. 1943. Bole area as an expression of growing stock. *Jour. Forestry* 39:624–31.

Liming, F. G. 1957. Homemade dendrometers. *Jour. Forestry* 55:575–77.

Lincoln, F. C. 1930. Calculating waterfowl abundance on the basis of banding returns. *U.S.D.A. Circ.* no. 118 (May, 1930).

Lockard, C. R., Putnam, J. A., and Carpenter, R. D. 1963. Grade defects in hardwood timber and logs. *U.S. Dept. Agric. Handbook* no. 244.

Loetsch, F., and Haller, K. A. 1964. *Forest inventory. Vol. I.* Munich: BLV Verlagsgesellschaft.

Loetsch, F., Zöhrer, F., and Haller, K. E. 1973. *Forest Inventory. Vol. II.* Munich: BLV Verlagsgesellschaft.

Lussier, L. F. 1961. Work sampling applied to logging. A powerful tool for performance analyses and operations control. *Pulp and Paper Mag. Canada* 62:130–40.

Lynch, D. W. 1958. Effects of stocking on site measurement and yield of second-growth ponderosa pine in the Inland Empire. *U.S.F.S. Intermt. For. and Range Exp. Sta. Res. Paper* no. 56.

Maass, A. 1939. Tallens form bedömd av diametern 2.3 meter från marken (Stem form of pine as determined by diameter 2.3 meters above ground). *Svenska skogsvfören. Tidskr.* 37:120–40.

MacKinney, A. L., and Chaiken, L. E. 1939. Volume, yield and growth of loblolly pine in the Mid-atlantic Region. *U.S.F.S. Appalachian For. Exp. Sta. Tech. Note* no. 33.

MacKinney, A. L., Schumacher, F. X., and Chaiken, L. E. 1937. Construction of yield tables for nonnormal loblolly pine stands. *Jour. Agric. Res.* 54:531–45.

MacLeod, W. K., and Blyth, A. W. 1955. Yield of even-aged, fully-stock spruce-poplar stands in northern Alberta. *Dept. North. Affairs and Nat. Res. Canada. Tech. Note* no. 18.

Madden, D. B., and Knudson, D. M. 1978. *Attendance methodology.* Indianapolis, Ind.: Indiana Department of Natural Resources.

Mader, D. L. 1963. Volume growth measurement—an analysis of function and characteristics in site evaluation. *Jour. Forestry* 61:193–98.

Madgwick, H. A. I. 1964. Estimation of surface area of pine needles with special reference to *Pinus resinosa. Jour. Forestry* 62:636.

Maeglin, R. R. 1967. Effect of tree spacing on weight yields for red and jack pine. *Jour. Forestry* 63:647–50.

Marden, R. M., Lothner, D. C., and Kallio, E. 1975. Wood and bark percentages and moisture contents of Minnesota pulpwood species. *U.S.F.S. North Central For. Exp. Sta. Res. Paper* NC-114.

Marquardt, D. W. 1963. An algorithm for least-squares estimation of nonlinear parameters. *Jour. Soc. Indust. Appl. Math.* 11:431–41.

Marquis, D. A., and Beers, T. W. 1969. A further definition of some forest growth components. *Jour. Forestry* 67:493.

Martin, G. L. 1982. A method for estimating ingrowth on permanent horizontal sample points. *Forest Sci.* 28:110–114.

Matérn, B. 1958. On the geometry of the cross-section of a stem. *Meddel. Statens Skogs-forskningsinst* 46:1–28.

Mawson, J. C. 1982. Diameter growth estimation on small forests. *Jour. Forestry* 80:217–19.

McArdle, R. E., Meyer, W. H., and Bruce, D. 1949. The yield of Douglas fir in the Pacific Northwest. *U.S. Dept. Agric. Tech. Bull.* no. 201. Rev.

McCain, R. 1948. A method for measuring deer range use. *Trans. Thirteenth N. Am. Wildlife Conf.* 431–41.

McClure, J. P., Cost, N. D., and Knight, H. A. 1979. Multiresource inventories—a new concept for forest survey. *U.S.F.S. Southeast For. Exp. Sta. Res. Paper* SE-191.

McLintock, T. F., and Bickford, C. A. 1957. A proposed site index for red spruce in the northeast. *U.S.F.S. Northeast. For. Exp. Sta. Paper* no. 93.

Meteer, J. W. 1979. Planning inventories based on permanent plot systems. *Forest Resource Inventories Workshop Proceedings. Colorado State Univ.*, pp. 354–61.

Mesavage, C. 1965. Three-P sampling and dendrometry for better timber estimating. *South. Lumberman* 211 (Dec. 15):107–109.

Mesavage, C. 1967. Random integer dispenser. *U.S.F.S. Southern For. Exp. Sta. Res. Note* no. 50-49.

Mesavage, C. 1969. Measuring bark thickness. *Jour. Forestry* 67:753–54.

Mesavage, C., and Girard, J. W. 1946. *Tables for estimating board foot volume of timber*. Washington, D.C.: U.S. Forest Service.

Mesavage, C., and Grosenbaugh, L. R. 1956. Efficiency of several cruising designs on small tracts in North Arkansas. *Jour. Forestry* 54:569–76.

Meyer, H. A. 1940. A mathematical expression for height curves. *Jour. Forestry* 38:415–20.

Meyer, H. A. 1942. Methods of forest growth determination. *Pennsylvania State College School of Agriculture. Agric. Exp. Sta. Bull.* no. 435.

Meyer, H. A. 1953. *Forest mensuration*. State College, Pa.: Renns Valley Publishers.

Meyer, H. A., Recknagel, A. B., Stevenson, D. D., and Bartoo, R. A. 1961. *Forest management*. 2nd ed. New York: The Ronald Press Co.

Meyer, H. A., and Stevenson, D. D. 1943. The structure and growth of virgin beech-birch-hemlock forests in northern Pennsylvania. *Jour. Agric. Res.* 67:465-84.

Meyer, W. H. 1930. Diameter distribution series in even-aged forest stands, Yale Forest. *Yale Univ. School of Forestry Bull.* no. 28.

Meyer, W. H. 1933. Approach of abnormally stocked stands of douglas fir to normal conditions. *Jour. Forestry* 31:400–406.

Meyer, W. H. 1934. Growth in selectively cut ponderosa pine forests of the Pacific Northwest. *U.S. Dept. Agric. Tech. Bull.* no. 407.

Meyer, W. H. 1942. Yield of even-aged stands of loblolly pine in northern Louisiana. *Yale Univ. School of Forestry Bull.* no. 51.

Miller, C. I. 1952. The determination of simple methods for obtaining form-class of individual trees. Purdue University Agricultural Experiment Project no. 644 (mimeographed).

Miller, C. I. 1959. Comparison of Newton's, Smalian's, and Huber's formulas. Dept. of Forestry and Conservation, Purdue University (mimeographed).

Miller, C. I. 1963. Faster point sampling. *Jour. Forestry* 61:299–300.

Miller, C. I. 1973. Stereo models for measuring the space scene. *Photogrammetric Engineering* 39:599–604.

Miller, C. I., and Beers, T. W. 1975. Thin prisms as angle gauges in forest inventory. *Purdue Univ. Agric. Exp. Sta. Res. Bull.* no. 929.

Miller, C. I., and Beers, T. W. 1981. Exercises in forest resource measurements. *Purdue Univ. Dept. of For. and Nat. Res.*

Moffitt, F. H. 1967. *Photogrammetry*. 2nd ed. Scranton, Pa.: International Textbook Co.

Moffitt, F. H., and Bouchard, H. 1975. *Surveying*. 6th ed. New York: Intext Educational Publishers.

Moser, J. W., Jr. 1972. Dynamics of an uneven-aged forest stand. *Forest Sci.* 18:184–91.

Moser, J. W., Jr. 1976. Specifications of density for the inverse J-shaped diameter distribution. *Forest Sci.* 22:177–80.

Moser, J. W., Jr., and Beers, T. W. 1969. Parameter estimation in nonlinear volume equations. *Jour. Forestry* 67:878–79.

Moser, J. W., Jr., and Hall, O. F. 1969. Derived growth and yield functions for uneven-aged forest stands. *Forest Sci.* 15:183–88.

Müller, G. 1953. Das baumzahbrohr. *Allg. Forstztg.* 64:249–51.

Mulloy, G. A. 1944. Empirical stand density yield tables. *Dept. Mines and Res. Ottawa, Canada, Silv. Res. Note* no. 73.

Mulloy, G. A. 1947. Empirical stand density yield. *Dept. Mines and Res. Ottawa, Canada, Silv. Res. note* no. 82.

Nash, A. J. 1948. The Nash scale for measuring tree crown widths. *Forestry Chron.* 24:117–20.

National Bureau of Standards. 1977. *The international system of units* (SI). Special Publication 330. Washington, D.C.: U.S. Dept. Commerce.

Nelson, T. C., and Bennett, F. A. 1965. A critical look at the normality concept. *Jour. Forestry* 63:107–109.

Norman, E. L., and Curlin, J. W. 1968. A linear programming model for forest production control at the AEC Oak Ridge Reservation. *Oak Ridge National Lab. Rep.* no. ORNL-4349.

Nylinder, P. 1958. *Variations in weight of barked spruce pulpwood.* Stockholm: Uppsats. Instn. Virkeslära K. Skogshögsh no. 15.

Nylinder, P. 1967. *Weight measurement of pulpwood.* Stockholm: Rapp. Instn. Virkeslära K. Skogshögsk no. R57.

Olson, D. F. 1971. Sampling leaf biomass in even-aged stands of yellow poplar (*Liriodendron tulipifera* L.). In *Forest biomass studies.* XV IUFRO Congress, Gainesville, Florida.

Opie, J. E. 1968. Predictability of individual tree growth using various definitions of competing basal area. *Forest Sci.* 14:314–23.

Ore, O. 1948. *Number theory and its history.* New York: McGraw-Hill Book Co.

O'Regan, W. G., and Arvanitis, L. G. 1966. Cost effectiveness in forest sampling. *Forest Sci.* 12:406–14.

Osborne, J. G. 1942. Sampling errors of systematic and random surveys of cover type areas. *Jour. Amer. Stat. Assn.* 37:256–64.

Osborne, J. G., and Schumacher, F. X. 1935. The construction of normal-yield and stand tables for even-aged timber stands. *Jour. Agric. Res.* 51:547–64.

Pardé, J. 1955. Un dendrometre Blume-Leiss (The Blume-Leiss hypsometer). *Rev. For. Fran.* 7:207–10.

Pardé, J. 1961. Dendrometrie. Imprimerie Louis-Jean Gap.

Parks Canada. 1976. *Canadian outdoor recreation demand study. Vol. 2: Technical notes.* Waterloo: Ontario Research Council on Leisure.

Patrone, G. 1963. *Lezioni di dendrometri*. Florence: B. Cappini and Co.

Patton, D. R. 1975. A diversity index for quantifying habitat "edge." *Wildl. Soc. Bull.* 3:171–73.

Petersen, C. G. J. 1896. The yearly immigration of young plaice into the Limfjord from the German Sea. *Rept. Danish Biol. Sta.* 6:1–77.

Phipps, R. L., and Gilbert, G. E. 1960. An electric dendrograph. *Ecology* 41:389–90.

Pienaar, L. V. 1965. Quantitative theory of forest growth. Ph.D. thesis. Univ. of Washington.

Pienaar, L. V., and Turnbull, K. J. 1973. The Chapman-Richards generality of Von Bertalanffy's growth model for basal area growth and yield in even-aged stands. *Forest Sci.* 19:2–22.

Prodan, M. 1965. *Holzmesslehre*. Frankfurt on the Main, Germany: J. D. Sauerlander's Verlag.

Raspopov, I. M. 1955. K metodike izucenija proekcii kron derevjev (A method of studying the crown projection of trees). *Bot. Z.* 40:825–27.

Reineke, L. H. 1927. A modification of Bruce's method of preparing timber yield tables. *Jour. Agric. Res.* 35:843–56.

Reineke, L. H. 1932. A precision dendrometer. *Jour. Forestry* 30:692–97.

Reineke, L. H. 1933. Perfecting a stand density index for even-aged forests. *Jour. Agric. Res.* 46:627–38.

Rinehart, R. P., Hardy, C. C., and Rosenau, H. G. 1978. Measuring trail conditions with stereo photography. *Jour. Forestry* 76:502–503.

Roach, B. A. 1977. A stocking guide for Allegheny hardwoods and its use in controlling intermediate cuttings. *U.S.F.S. Res. Paper* NE-373.

Roach, B. A., and Gingrich, S. F. 1962. Timber management guide for upland Central hardwoods. *U.S.F.S. Central States For. Exp. Sta.*, 33 pp.

Roach, B. A., and Gingrich, S. F. 1968. Even-aged silviculture for upland central hardwoods. *U.S. Dept. Agric. Handbook* no. 355.

Robinson, D. W. 1969. The Oklahoma state angle gauge. *Jour. Forestry* 67:234–36.

Rocky Mtn. Forest and Range Experiment Station. 1978. Integrated inventories of renewable resources: proceedings of the workshop. *U.S.F.S. Rocky Mtn. For. and Range Exp. Sta. Gen. Tech. Rep.* RM-55.

Row, C., and Fasick, C. 1966. Weight scaling tables by electronic computer. *For. Products Jour.* 16(8):41–45.

Row, C., and Guttenberg, S. 1966. Determining weight-volume relationships for sawlogs. *For. Products Jour.* 16(5):39–47.

Rupp, R. S. 1966. Generalized equation for the ratio method of estimating population abundance. *Jour. Wildl. Mgt.* 30:523–26.

Sayn-Wittgenstein, L. 1963. An attempt to find the best basal area factor for point sampling. *Canada Dept. Forestry, For. Res. Br. Ottawa.* Inservice report.

Schiffel, A. 1899. Form and Inhalt der Fichte (Form and volume of spruce). *Mitt. aus d. forstl. Versuchsan. Österreiche* 24.

Schmelz, D. V., and Lindsey, A. A. 1965. Size-class structure of old-growth forests in Indiana. *Forest Sci.* 11:259–64.

Schmid, P. 1969. Stichproben am Waldrand. *Mitt. Schweiz. Anst. Forstl. Versuchswes.* 45, 3:234–303.

Schnur, G. L. 1934. Diameter distribution for old-field loblolly pine stands in Maryland. *Jour. Agric. Res.* 49:731–43.

Schnur, G. L. 1937. Yield, stand, and volume tables for even-aged upland oak forests. *U.S. Dept. Agric. Tech. Bull.* no. 560.

Schroeder, J. G., Taras, M. A., and Clark, A. 1975. Stem and primary products weights for longleaf pine sawtimber trees. *U.S.F.S. Southeast. For. Exp. Sta. Res. Paper* SE-139.

Schroeder, J. G., Campbell, R. A., and Rodenbach, R. C. 1968a. Southern pine log grades for yard and structural lumber. *U.S.F.S. Southern For. Exp. Sta. Res. Paper* SE-39.

Schroeder, J. G., Campbell, R. A., and Rodenbach, R. C. 1968b. Southern pine tree grades for yard and structural lumber. *U.S.F.S. Southern For. Exp. Sta. Res. Paper* SE-40.

Schumacher, F. X. 1930. Yield, stand, and volume tables for Douglas-fir in California. *Calif. Agric. Exp. Sta. Bull.* no. 491.

Schumacher, F. X. 1939. A new growth curve and its application to timber yield studies. *Jour. Forestry* 37:819–20.

Schumacher, F. X., and Hall, F. S. 1933. Logarithmic expression of timber-tree volume. *Jour. Agric. Res.* 47:719–34.

Scott, C. T. 1979. *Midcycle updating: some practical suggestions.* Fort Collins, Colo.: Proc. Forest Resource Inventories Workshop, pp. 362–70.

Seip, H. K. 1964. Tremaling. *Skogbruk og Skogindustri, Vol. 3*, Norway.

Shepperd, W. D. 1973. An instrument for measuring tree crown width. *U.S.F.S. Rocky Mtn. For. and Range Exp. Sta. Res. Note* RM-229.

Shigo, A. I., and Larson, E. vH. 1969. A photo guide to the patterns of discoloration and decay in living northern hardwood trees. *U.S.F.S. Northeast. For. Exp. Sta. Res. Paper* NE-127.

Shiue, C. J. 1960. Systematic sampling with multiple random starts. *Forest Sci.* 6:42–50.

Sisam, J. W. B. 1938. Site as a factor in silviculture. Its determination with special reference to the use of indicator plants. *Dominion Forest Service: Canada. Silv. Res. Note* no. 54.

Smalley, G. W., and Bailey, R. L. 1974a. Yield tables and stand structure for loblolly pine plantations in Tennessee, Alabama, and Georgia highlands. *U.S.F.S. Southern For. Exp. Sta. Res. Paper* SO-96.

Smalley, G. W., and Bailey, R. L. 1974b. Yield tables and stand structure for shortleaf pine plantations in Tennessee, Alabama, and Georgia highlands. *U.S.F.S. Southern For. Exp. Sta. Paper* SO-97.

Smith, A. D. 1964. Defecation rates of mule deer. *Jour. Wildl. Mgt.* 28:435–44.

Smith, D. M. 1962. *The practice of silviculture.* 7th ed. New York: John Wiley & Sons.

Smith, J. H. G., and Ker, J. W. 1959. Empirical yield equations for young forest growth. *British Columbia Lumberman.* Sept.

Smith, J. H. G., Ker, J. W., and Csizmazia, J. 1961. Economics of reforestation of Douglas-fir, western hemlock, and western red cedar in the Vancouver Forest District. *Univ. of British Columbia Forestry Bull.* no. 3.

Smith, J. R. 1970. *Optical distance measurement.* London: Crosby Lockwood and Sons.

Society of American Foresters. 1971. *Terminology of forest science, technology practice and products.* Washington: Multilingual Forestry Terminology Series no. 1.

Spurr, S. H. 1952. *Forest inventory.* New York: The Ronald Press Co.

Spurr, S. H. 1962. A measure of point density. *Forest Sci.* 8:85–96.

Stage, A. R. 1962. A field test of point-sample cruising. *U.S.F.S. Intermountain For. and Range Exp. Sta. Res. Paper* no. 67.

Stage, A. R. 1966. Simultaneous derivation of site-curve and procedures. *Proc. Soc. Amer. Foresters*: 134–36.

Stage, A. R. 1976. An expression for the effect of aspect, slope, and habitat type on tree growth. *Forest Sci.* 22:457–60.

Stevens, S. S. 1946. On the theory of scales of measurement. *Science* 103:677–80.

Stoffels, A., and Van Soest, J. 1953. Principiele vraagstukken bije proefperken. 3, Hoogte-regressie (The main problem in sample plots. 3, Height regression). *Ned. Boschb.-Tijdschr.* 25:190–99.

Stott, C. B. 1968. A short history of continuous forest inventory east of the Mississippi. *Jour. Forestry* 66:834–37.

Strand, L. 1957. "Relaskopisk" hoyde-og kubikkmassebestemmelse (Relascopic height and cubic volume determination). *Norsk Skogbruk* 3:535–38.

Swan, D. A. 1959. Weight scaling in the Northeast. *Amer. Pulpw. Assn. Tech. Release* no. 59-R5.

Swank, W. T., and Schreuder. 1974. Comparison of three methods of estimating surfa᷎ area and biomass for a forest of young eastern white pine. *Forest Sci.* 20:91–100.

Taras, M. A. 1956. Buying pulpwood by weight. *U.S.F.S. Southeast. For. Exp. Sta. Res. Paper.* no. 74.

Taras, M. A., and Clark, A. 1977. Above ground biomass of longleaf pine in a natural sawtimber stand in southern Alabama. *U.S.F.S. Southeast. For. Exp. Sta. Res. Paper* SE-162.

Tardif, G. 1965. Some considerations concerning the establishment of optimum plot size in forest survey. *Forestry Chron.* 41:93–102.

Thomson, G. W., and Deitschman, G. H. 1959. Bibliography of world literature on the Bitterlich method of plotless cruising. *Iowa State Univ. Agric. and Home Eco. Exp. Sta.*

Timson, F. G. 1974. Weight and volume variation in truckloads of logs hauled in the Central Appalachians. *U.S.F.S. N.E. For. Exp. Sta. Res. Paper* NE-300.

Tomlinson, R. W., and Burger, T. C. 1977. *Electronic distance measuring instruments.* 3rd ed. Falls Church, Va.: American Congress on Surveying and Mapping.

Trimble, G. R., Jr., and Smith, H. C. 1976. Stand structure and stocking control in Appalachian mixed hardwoods. *U.S.F.S. Northeast For. Exp. Sta. Res. Paper* NE-340.

Trorey, L. G. 1932. A mathematical expression for the construction of diameter height curves based on site. *Forestry Chron.* 18:3–14.

Turnbull, K. J. 1963. Population dynamics in mixed forest stands. A system of mathematical models of mixed stand growth and structure. Ph.D. thesis. Univ. of Washington.

Turnbull, K. J., and Hoyer, G. E. 1965. *Construction and analysis of comprehensive tree-*

volume tarif tables. Olympia, Wash.: Dept. of Nat. Resources Resource Management Report no. 8.

Turnbull, K. J., Little, G. R., and Hoyer, G. E. 1963. *Comprehensive tree-volume tarif tables*. Olymbia, Wash.: Dept. of Nat. Resources.

U.S. Forest Service. 1977. *National forest log scaling handbook*. Washington, D.C.: U.S. Forest Service.

Vaughan, C. L., Wollin, A. C., McDonald, K. A., and Bulgrin, E. H. 1966. Hardwood log grades for standard lumber. *U.S.F.S. For. Prod. Lab. Res. Paper* FPL-63.

Vuokila, Y. 1955. The Finnish curve calipers. *Jour. Forestry* 53:366–67.

Wager, J. A. 1974. Recreational carrying capacity reconsidered. *Jour. Forestry* 72:274–78.

Wahlenberg, W. G. 1941. Methods of forecasting timber growth in irregular stands. *U.S. Dept. Agric. Tech. Bull.* no. 796.

Walters, C. S., and Herrick, A. M. 1956. A comparison of two log-grading systems. *Univ. of Ill. Agric. Exp. Sta. Bull.* no. 603.

Ware, K. D. 1964. Some problems in the quantification of tree quality. *Proc. Soc. Amer. Foresters*: 211–17.

Ware, K. D. 1967. *Sampling properties of three-pee estimates*. Paper presented at 1967 Soc. Amer. Foresters–Canadian Inst. Forestry Meeting, Ottawa, Canada.

Ware, K. D., and Cunia, T. 1962. Continuous forest inventory with partial replacement of samples. *Forest Sci.* Monograph 3.

Wargo, P. M. 1978. Correlations of leaf area with length and width measurements of leaves of black oak, white oak and sugar maple. *U.S.F.S. Northeast. For. Exp. Sta. Res. Note* NE-256.

Wartluft, J. L. 1977. Weights for small Appalachian hardwood trees and components. *U.S.F.S. Northeast For. Exp. Sta. Res. Paper* NE-366.

Watt, R. F. 1950. Growth in understocked and overstocked western white pine stands. *U.S.F.S. Northern Rocky Mt. For. and Range Exp. Sta. Note* no. 78.

Wesley, R. 1956. *Measuring the height of trees*. London: Park Administration 21:80–84.

Westveld, M. 1954. Use of plant indicators as an index to site quality. *U.S.F.S. Northeast. For. Exp. Sta. Paper* no. 69.

Wheeler, P. R. 1962. Penta prism caliper for upper-stem diameter measurements. *Jour. Forestry* 60:877–78.

White, C. M. 1968. Productivity and dynamics of the white-tailed deer on the Crane Naval Ammunition Depot in Martin County, Indiana. Ph.D. Thesis, W. Lafayette, Ind.: Purdue University.

White, Z. W. 1971. SAF looks at metrication. *Jour. Forestry* 69:218–23.

Wiljamaa, L. E. 1942. A simple but practical Finnish instrument for measuring tree diameters. *Jour. Forestry* 40:506–7.

Williston, H. L. 1975. Selected bibliography on the growth and yield of the four major southern pines. Southeastern Area, State and Private Forestry, USDA Forest Service.

Winer, H. I. 1961. Notes on the analysis of pulpwood logging in the Southeast. Southeast. Tech. Comm. Amer. Pulpwood Assn. Logging School. Georgetown, S.C.

Wolf, P. R. 1974. *Elements of photogrammetry*. New York: McGraw-Hill Book Co.

Wright, W. J. 1962. Motion and time study techniques and their application to forest operations in northern Ireland. Master's thesis. Univ. of California.

Yerkes, V. P. 1966. Weight and cubic-foot relationships for Black Hills ponderosa pine sawlogs. *U.S.F.S. Rocky Mt. For. and Range Exp. Sta. Res. Note* RM-78.

Yocom, H. A. 1970. Vernier scales for diameter tapes. *Jour. Forestry* 68:725.

Young, H. E. 1964. The complete tree concept—a challenge and an opportunity. *Proc. Soc. Amer. Foresters*: 231–33.

Young, H. E. 1978. Forest biomass inventory: the basis for complete-tree utilization. *For. Products Jour.* 28:38–41.

Young, H. E., Robbins, W. C., and Wilson, S. 1967. Errors in volume determination of primary forest products. School of Forestry, Univ. of Maine (mimeographed).

Young, H. E., Strand, L., and Altenberger, R. 1964. Preliminary frest and dry weight tables for seven tree species in Maine. *Univ. Maine Agric. Exp. Sta. Tech. Bull.* no. 12.

Zeide, B. 1980. Plot size optimization. *Forest Sci.* 26:251–57.

Zeide, B., Troxell, J. K., and Haag, D. 1979. *Field instructions on point sampling*. Fort Collins, Colo.: Proc. Forest Resource Inventories Workshop, pp. 917–22.

Zenger, A. 1964. Systematic sampling in forestry. *Biometrics* 22:553–65.

Zobel, B., Ralston, J., and Roberds, J. H. 1965. Wood yields from loblolly pine stands of different age, site and stand density. *School of Forestry, North Carolina State Univ. Tech. Rep.* no. 26.

Zobel, B., Roberds, J. H., and Ralston, J. 1969. Dry wood weight yields of loblolly pine. *Jour. Forestry* 67:822–24.

Index

Abney level, 35
Accretion, 293
Accuracy, 11-12
  in forest inventory, 157
Aerial photographs:
  acquisition of, 202-203
  area determination from, 206-208
  classification of, 193
  forest classification and mapping from, 208-209
  interpretation of, 193-194, 203-206
  sampling error of area measurement from, 206-208
  scale formula for, 194-196
  scale variations on, 197-199
  volume estimates from, 209
Aerial photography, scale, 203
Aerial photo mensuration, 203-206
Age, 55-58
  of stand, 57-58
  of tree, 55-57
Angle gauges:
  for horizontal sampling, 241-244
    modification of, 247-248
    Panama, 241
    prism, 243-244
    relaskop, 241-242
    stick-type, 241-242
  for vertical sampling, 248-249
Annual ring, 56
Approach to normality, 304-305
Area:
  definition, 80
  of land, 86
  of leaves, 89
  measurement and determination, 80-86
    from aerial photographs, 206-208
    by coordinates, 80-81
    by dot grids, line transects, and weighing, 84
    by integration, 81-83
    by planimeter, 84-85
    sampling error of, 206-208
  of tree bole surface, 88

of tree cross section, 86

Bark:
  factor, 104-108
    bark volume determination, 107-108
    diameter increment determination, 286-289
  gauge, 19
  thickness, measurement of, 19, 105
  volume, determination of, 107-108
Barr and Stroud dendrometer, 25
Basal area:
  estimation by horizontal point sampling, 220
  factor, 231, 234-237
  of stand, 86
  table of, 373
  of tree, 86
Baseline, 50
Biareal plot sampling, 223
Bias, 11-12
  in forest inventory, 157
Biltmore stick, 21-22
Blodgett foot, 93
Blume-Leiss altimeter, 35
Board foot, 93-94
  log scale, 114
  mill tally, 114
Board foot-cord conversion, 95-96
Board foot-cubic foot conversion, 95
Bole-surface area:
  factor, 231, 234-237
  of tree, 88
Breithaupt-Todis dendrometer, 26

Calipers, 20
Census:
  spatial, 352
  temporal, 352
Chain, surveyor's, 42
Change-of-composition method, 356-358
Christen hypsometer, 30-31
Chronograph, 54-55
Circumference factor, 231, 234-237

397

Clinometers, 37
Clocks, 54
Coefficient of condition, 359
Color aerial photography, 203
Compass surveying, 46-49
  common errors, 49
  declination in, 47-49
Composite volume table, 134
Confidence limits, 157-159
Continuous forest inventory, 306, 311-313
  horizontal point sampling in, 315-317
  planning for, 313-314
Continuous variable, 10
Conversion:
  board foot-cord, 95-96
  board foot-cubic foot, 95
  cubic foot-cord, 95
  factor, per-acre, 227
  volume unit, 95-96
Cord, 94-95
  determining solid cubic contents of, 127-129
  scaling, 126-127
Cross-sectional area of tree, 86-87
Crown:
  competition factor, 332-333
  coverage, 89
  definition of, 97
  diameter, 28-29
    measurement on aerial photographs, 204
Cruise, 150
Cubic foot, 93
Cubic foot-cord conversion, 95
Cubic meter, 93
Cubic volume, of trees, *see* Volume, determination
Cull estimation:
  in logs, 121-124
  in trees, 137-141
Culmination of mean annual increment, 277
Current annual increment, 276, 277
Curtis-Bruce method of height measurement, 37

Declination:
  magnetic, 47
  variation of magnetic, 47-49
Deduction for defect:
  in logs, 121-124
  in trees, 137-141
Dendrograph, 28
Dendrometer, 23
  rangefinder, 25-26
Density, 65
  of stands, 328-334
  of wood, 65-67

Diameter:
  change, precise measurement of, 28
  distributions, 321-328
    of even-aged stands, 320-321
    of uneven-aged stands, 321-328
  factor, 231, 234-237
  measurement, 17-28
    effect of irregularity of form on, 19
    of standing tree, 19-23
    of upper stem, 23-28
  obviation, 241
  tape, 21
Diameter breast high (d.b.h.), 18
Discrete variable, 10
Displacement, volume determination by, 91
Distance measurement:
  by electronic devices, 41-42
  by optical devices, 41-42
  by pacing, 40-41
  by steel tapes, 41
  by topographic trailer tape, 42-46
Distance multiplier, 235, 237
Diversity index, 359-360
Dot grid, 84, 206
Double effective, 309
Double rising, 309
Doyle log rule, 117
Doyle rule of thumb, 117
Doyle-Scribner log rule, 119-120

Empirical yield table, 343, 346
English units of measurement, 8-9
Enumeration, 150
Even-aged stand:
  age of, 57-58
  diameter distribution of, 321
Even-aged yield table, 304-305, 343-346
Expansion constant, 231, 235, 237

Factor:
  concept, 227-229
  terminology, 227-229
Factors, 231, 234-237
  basal area, 231, 234-237
  bole surface area, 231, 234-237
  circumference, 231, 234-237
  diameter, 231, 234-237
  height, 231, 234-237
  quadratic height, 231, 234-237
  tree, 231, 234-237
  volume, 231, 234-237
Fecal-count method, 354-355
Fixed-area sampling units, 161-162

Forest:
  inventory:
    confidence limits for, 157-159
    definition of, 150
    design of, 153-154
    errors in, 157
    information obtained from, 151-153
    planning for, 151-153
    plot size and shape, 161-162
    sampling design, 159-160
    sampling terminology in, 156-157
  mapping, 208-210
  stocking, 328
Form:
  factors, 108-109
  point, 111-112
  quotients, 109-111
  of tree, 108
Forms, design of, 155, 275

Gauge constant:
  horizontal, 221, 234, 236
  vertical, 234, 236, 249
Gauges, see Angle gauges
Geometric mean diameter, 19
Girard form class, 110-111
Grading logs and trees, 142-149
  problems in, 142-143
  systems of, 143-144
Graphical methods, volume determination by, 91-92
Gross growth:
  including ingrowth, 291-292
  of initial volume, 291-292
Gross scale, 121
Growing season, 56
Growth:
  and age, 55-56
  curves, 277-279
    diameter growth, 279-280
    height growth, 279-280
  percentage, 285-286
  of tree, 276
    adjustment for environmental influences, 284-285
    basal area, 283
    bole surface area, 283
    diameter, 280
    environmental influences on, 276
    height, 280
    from increment cores, 286-290
    volume, 283-284
    weight, 284

and yield functions, 305-306, 346-351

Haga altimeter, 35
Harmonized-curve method, 134
Height:
  accumulation, 103-104
  factor, 231, 234-237
  measurement, 29-38
    methods of, individual tree, 30-38
    types of, individual tree, 29-30
Height-diameter curves, 281
Hoppus foot, 93
Huber's formula, 99-101
Hypsometers, 30-37

Increment, see Growth, of tree
Increment borer, 56-57
Increment cores:
  determination of growth from, 286-290
  taking and handling of, 56-57
Ingrowth, 291, 299
Initial point, 50
Intensity of sampling:
  with dot grid, 207-208
  in forest inventory, 165, 175, 185, 186, 249-253
  in log scaling, 125
International log rule, 117, 119
Interval scale, 4
Inventory, see Forest, inventory

Land area measurement, 86
Land surveys, 49-53
  baseline, 50
  initial point, 50
  principal meridian, 50
  standard parallels, 50
  subdivision of sections, 52-53
  subdivision of townships, 51-52
Leaf area of tree, 89
Linear measurement, 17
Line:
  plot sampling, 186
  sampling:
    horizontal, 224-231
    vertical, 224-231
List sampling, 214-216
Log:
  diameter, 121
  grading systems, 144-147
    comparison of, 145-147
    hardwood, 144-145
    softwood, 147
  length, 121

rules, 114-120
  construction of, 115-119
  scale, 114
  scaling, *see* Scaling of logs
  and tree grades, application of, 148-149
Lorey's mean height, 39, 265

Mark-recapture method, 356-358
Mass, 6, 63
Massachusetts log rule, 115
Mean annual increment, 277
Mean solar day, 54
Measurement:
  definition of, 2
  English system of, 8-9
  metric system of, 5-8
  on interval scale, 4
  on nominal scale, 2, 10-11
  on ordinal scale, 4, 11
  on ratio scale, 4
  units of, 4-9
Mensuration, forest
  definition of, 1
  importance of, 1-2
  scope of, 1
Meridian:
  magnetic, 47
  true, 47
Merritt hypsometer, 31-33
Method of control, 307-311
Metric units of measurement, 5-8
Mill tally, 114
Moisture content of wood, 67-68
Monareal plot sampling, 223, 224
Monitoring campsites, 362-363
Mortality, 291, 299
Movement ratio, 296-297
Multiphase sampling, 179
Multistage sampling, 176-179

Net growth:
  including ingrowth, 292
  of initial volume, 292
Net increase, 292
Net scale, 121-124
Newton's formula, 99-102
Nominal scale, 2, 10-11
Normal stand, 304-344
Normal yield table, 304-305, 343-346
Number of samples, *see* Intensity of sampling

Optical caliper, 25
Optical fork, 23-25

Ordinal scale, 4, 11
Overrun, 114-115
  ratio, 115

Pacing, 40-41
  accuracy, 41
Panama angle gauge, 241
Parts, of tree, 97
Pen, 95
Periodic annual increment, 276, 277
Periodic increment, 276-277
Planimeter:
  compared with dot grid, 85
  polar, 84-85
Point:
  density, 333-334
  sampling:
    horizontal, 220-231
    vertical, 224-231
Point-center extension, 302-303
Polyareal plot sampling, 214, 224-231
Population estimation, 352-353
  change-of-composition method, 356-358
  fecal-count method, 354-355
  mark-recapture method, 356-358
  strip-census method, 353-354
P. P. S. sampling, 213, 220-231
Precision, 11-12
  in forest inventory, 157
Pressler's growth per cent, 286
Principal meridian, 50
Prism angle gauge, 243-244
  calibration of, 244-247
  modification of, 247-248

Quadratic height factor, 231, 234-237
Quadratic mean diameter, 23, 259
Quality:
  index, 143
  quantification of log and tree, 142-143
Quarter girth, 93

Radioactive dating, 55
Random sampling:
  simple, 162-166
  stratified, 171-174
Ratio:
  estimation, 166-170
  scale, 4
Recreational use estimation, 360-362
Regression estimation, 166-171
Relaskop, 24, 241
Relative density, 328-329
Reliable minimum estimate, 159

Repeated measurements:
  in forest inventory, 188-192
  growth from, 280
Rounding off, 14

Sampling, 156
  confidence limits, 157-159
  design in forest inventory, 159-160
  error, 157, 182-183
  intensity of, 174-176, 178-179
  list, 214-216
  multiphase, 179
  multistage, 176-179
  with partial replacement, 191-192
  with replacement, 191
  without replacement, 189
  selective, 181-182
  simple random, 162-166
  stratified random, 171-174
  systematic, 182-188
  three-P, 216-220
  unit size and shape, 161-162
  with varying probabilities, 213
Scale stick, 121
Scaling of logs, 121-127
  in board feet, 121-124
  in cords, 126-127
  in cubic units, 124
  sample, 125
Scribner decimal rule, 116
Scribner log rule, 116
Sections, subdivision of, 52-53
Sector fork, 22
Selective sampling, 181-182
Significant digits, 12-14
  in arithmetic operations, 14-16
Site:
  factors, 276, 336-337
  index, 338
    curves, preparation of, 340-343
  quality, 276, 336
    measures of, 336-340
    plant indicators of, 340
Smalian's formula, 99-100
Sound volume, 294-295
Spacing, of plots and lines, *see* Intensity of
    sampling
Spaulding log rule, 116
Specific gravity, 65-66
Spiegel relaskop, *see* Relaskop
Stand:
  aerial volume tables, 209
  density:

    crown competition factor, 332-333
    definition of, 328
    point density, 333-334
    relative density, 328-329
    stand density index, 329-330
    tree-area ratio, 330-332
  density index, 329-330
  height, expressions of, 38-39
  structure, definition of, 320
    even-aged, 320-321
    uneven-aged, 321-328
  table, 153
  table projection, 295-300
Standard parallels, 50
Status:
  changer, 302
  of tree, 295
Stem analysis, 281-283
Stere, 95
Stereograms, preparation of, 212
Stereoscopic vision, 199-202
  alignment of photos for, 200-202
Stick-type angle gauge, 241-242
Stocking:
  forest, 328-329, 334
  normal, 344
Stock table, 153
Strata, 171
Stratification, 171-174
Strip-census method, 353-354
Structure, *see* Stand, structure
Surface area, bole, 88
Survivor growth, 293
Systematic sampling, 183-188
  with plots, 185-188
  with strips, 183-185

Taper:
  curves, 91, 112-113, 283
  formulas, 112-113
  tables, 112-113
Tarif tables, 137
Terrestrial stereoscopic photography, 210-212
Three-P sampling, 216-220
Time, measurement of, 54-55, 59-60
Time and motion study, 59-62
Topographic displacement, 197-199
Topographic trailer tape, 42-44
  accuracy, 46
  field operations, 45-46
Total stand projection, 301-304
Total volume, 294-295
Townships, subdivision of, 51-52

Tree:
  factor, 231, 234-237
  grading:
    application of, 148-149
    use of log grades in, 143-144
  measurement:
    bole-surface area, 88
    cross-sectional area, 86-87
    crown coverage, 89
    crown diameter, 28-29
    diameter, 17-28
    height, 29-38
    leaf area, 89
  volume equations, 134-135
Tree-area ratio, 330-332
Triareal plot sampling, 223
Two-way method, 301-302

Underrun, 115
Uneven-aged forest, balanced, 323
Uneven-aged stand, 321
Uneven-aged yield table, 343
Unmerchantable logs, 124
Upper-stem diameters, 23-28
U.S. Public Land Surveys, *see* Land surveys

Value measurement, 363-364
Variable:
  definition of, 10
  density yield table, 346
Variation, in magnetic declination, 47-49
Vertical exaggeration, 202
Volume:
  determination, 90-93
    of bark, 104-108
    by displacement, 91
    by formulas, 90, 98-103
    by graphical methods, 91-92
    by height accumulation, 103-104
    by integration, 90-91
  distribution of individual trees, 137-139
  factor, 231, 234-237
  tables, 129-137
    checking applicability of, 132-133
    composite, 134
    construction of, 133-136
    form class, 129
    information to accompany, 129-132
    local, 129, 136

    standard, 129
    tarif tables, 137
    use of in inventory, 180-181
  unit conversion, 95-96
    board foot-cord, 95-96
    board foot-cubic foot, 95
    cubic foot-cord, 95
Volume units:
  to measure stacked wood, 94-95
  to measure trees and logs, 93-94
  to measure wood, 93-95
    Blodgett foot, 93
    board foot, 93-94, 114-120
    cord, 94-95, 126-127
    cubic units, 93
    cunit, 93
    Hoppus foot, 93
    pen, 95
    stere, 9

Weighing:
  machines, 63-64
  methods of, 63-64
Weight:
  definition of, 63
  estimation or measurement, 68-79
    factors influencing, 65-68
    of forest products, 73-74
    of pulp, 72-73
    of pulpwood, 68-71
    of sawlogs, 71-72
    of stands, 74-78
    of trees, 74-78
    of yield and growth, 78-79
  use of, 64-65
  of wood:
    factors influencing, 65-68
    oven-dried, 67
Work sampling, 60-62

Yield:
  functions, 305-306, 346-351
  table:
    construction of, 343-346
    definition of, 343
    empirical, 346
    normal even-aged, 304-305, 343-346
    uneven-aged, 343
    variable density, 346